THE THEORY OF LIE DERIVATIVES AND ITS APPLICATIONS

THE THEORY OF LIE DERIVATIVES AND ITS APPLICATIONS

Kentaro Yano

Dover Publications, Inc.
Mineola, New York

This book is dedicated to
Prof. Dr. J. A. SCHOUTEN
who has been a pioneer in the field of modern differential geometry.

Bibliographical Note

This Dover edition, first published in 2020, is an unabridged republication of the work originally published in 1957 by the North-Holland Publishing Company, Amsterdam, as Volume III in "BiblioMathematica: A Series of Monographs on Pure and Applied Mathematics."

Library of Congress Cataloging-in-Publication Data

Names: Yano, Kentaro, 1912-1993, author.
Title: The theory of Lie derivatives and its applications / Kentaro Yano.
Description: Dover edition | Mineola, New York : Dover Publications, Inc., 2020. | Originally published: Amsterdam : North-Holland Publishing Company, 1957. | Includes bibliographical references and index. | Summary: "Differential geometry has become one of the most active areas of math publishing, yet a small list of older, unofficial classics continues to interest the contemporary generation of mathematicians and students. This advanced treatment of topics in differential geometry, first published in 1957, was praised as 'well written' by The American Mathematical Monthly and hailed as 'undoubtedly a valuable addition to the literature'."—Provided by publisher.
Identifiers: LCCN 2019049267 | ISBN 9780486842097 (trade paperback)
Subjects: LCSH: Projective differential geometry. | Geometry, Differential.
Classification: LCC QA660 .Y33 2020 | DDC 516.3/6—dc23
LC record available at https://lccn.loc.gov/2019049267

Manufactured in the United States by LSC Communications
84209601
www.doverpublications.com
2 4 6 8 10 9 7 5 3 1
2020

PREFACE

Since the theory of continuous groups of transformations was inaugurated by S. Lie and F. Engel, the groups of motions in Riemannian spaces were studied by L. Bianchi, G. Fubini, W. Killing, G. Ricci and others.

On the other hand, the idea of spaces with a linear connexion was introduced by E. Cartan, J. A. Schouten and H. Weyl and the affine and projective motions in these spaces were first considered by L. P. Eisenhart and M. S. Knebelman.

In 1931, W. Slebodzinski introduced a new differential operator, later called by D. van Dantzig that of Lie derivation, which can be applied to scalars, vectors, tensors and affine connexions and which proved to be a powerful instrument in the study of groups of automorphisms. Using this operator, D. van Dantzig showed that his n-dimensional projective space described by $n + 1$ homogeneous curvilinear coordinates can be regarded as an $(n + 1)$-dimensional space with a linear connexion which admits a one-parameter group of affine motions. He applied also the idea of Lie derivation to physics.

Since then the deformations of curves, subspaces and spaces themselves as well as groups of motions, affine motions, projective motions and conformal motions were extensively studied by L. Berwald, E. Cartan, N. Coburn, E. T. Davies, P. Dienes, A. Duschek, L. P. Eisenhart, F. A. Ficken, H. A. Hayden, V. Hlavatý, E. R. van Kampen, M. S. Knebelman, T. Levi-Civita, J. Levine, W. Mayer, A. J. McConnel, A. D. Michal, H. P. Robertson, S. Sasaki, J. A. Schouten, J. L. Synge, A. H. Taub, H. C. Wang, the present author and others.

The Lie derivatives of general geometric objects were studied by A. Nijenhuis, Y. Tashiro and the present author.

It is now a well-known fact that, if an n-dimensional space admits a group of motions, affine motions, projective motions or conformal motions of the maximum order $\frac{1}{2}n(n+1)$, n^2+n, n^2+2n or $\frac{1}{2}(n+1)(n+2)$ respectively, the space is of constant curvature, affinely flat, projectively Euclidean or conformally Euclidean.

In 1947, I. P. Egorov began the study of spaces which have a non-vanishing curvature tensor and which admit a group of automorphisms of the maximum order. Investigations in this direction were carried out by Y. Mutō, G. Vranceanu, H. C. Wang and the present author.

Chapters I—VII of the present book are devoted to the abovementioned publications.

The automorphisms in Finsler spaces, Cartan spaces, general affine and projective spaces of geodesics and general affine and projective spaces of k-spreads were studied also very extensively by the use of Lie derivatives by R. S. Clark, E. T. Davies, H. Hiramatu, Y. Katsurada, M. S. Knebelman, D. D. Kosambi, B. Laptev, Gy. Soós, B. Su, K. Takano, H. C. Wang, the present author and others. Chapter VIII contains the theory of Lie derivatives and its applications in these spaces.

Chapter IX is devoted to the study of global properties of the groups of motions in a compact orientable Riemannian space. The method used in this Chapter is due to S. Bochner and A. Lichnerowicz.

The last Chapter is devoted to a brief exposition on the almost complex spaces and to some problems which can be dealt with by the use of Lie derivatives.

There is a tendency of developing the theory of Lie derivatives from the point of view of the theory of fibre bundles. But such an investigation has just been started and it seems to the author that it is still premature to give an exposition of the results already obtained. We only refer to the recent papers by R. S. Palais, N. H. Kuiper and the present author.

The bibliography at the end of the book contains only the papers and books quoted in the text and those of which the author may suppose that they are of interest for the readers.

The author wishes to express here his hearty thanks to Prof. J. A. Schouten who read the manuscript and gave many valuable suggestions. The author wishes to thank also the editors of Bibliotheca Mathematica, Prof. D. van Dantzig, Prof. J. de Groot and Prof. N. G. de Bruijn for their most agreeable collaboration.

The author appreciates very much the kind help from his Dutch friends at the Mathematical Centre and the University of Amsterdam. Miss P. Brouwer looked through the manuscript and improved the English of the text. The author's sincere thanks go to all of them.

Amsterdam, April 14, 1955 KENTARO YANO

CONTENTS

CHAPTER I

INTRODUCTION

§ 1. Motions in a Riemannian space.

Consider an n-dimensional Riemannian space V_n [1] of class C^ω [2] covered by a set of neighbourhoods with coordinates $\xi^\varkappa$ and endowed with the fundamental quadratic differential form

$$(1.1) \qquad ds^2 = g_{\lambda\varkappa}(\xi)d\xi^\lambda d\xi^\varkappa, \text{[3] [4]}$$

where the Greek indices $\varkappa, \lambda, \mu, \nu, \ldots$ run over the range $1, 2, \ldots, n$. We write $(\varkappa)$ to denote the system of coordinates $\xi^\varkappa$.

In the V_n referred to $(\varkappa)$, we consider a point transformation

$$(1.2) \qquad T\colon\ '\xi^\varkappa = f^\varkappa(\xi^\nu); \qquad \operatorname{Det}(\partial_\lambda f^\varkappa) \neq 0$$

of class C^ω [5] which establishes a one-to-one correspondence between the points of a region R and those of some other region $'R$, where ∂_λ stands for the partial derivation $\partial/\partial\xi^\lambda$.

During this point transformation, a point $\xi^\varkappa$ in R is carried to a point $'\xi^\varkappa$ in $'R$ and a point $\xi^\varkappa + d\xi^\varkappa$ in R to a point $'\xi^\varkappa + d'\xi^\varkappa$ in $'R$.

---

[1] In principle, we follow, throughout the book, the standard notations which appear in the recent book by SCHOUTEN [8]. The number in parentheses refers to the Bibliography at the end of the book.

[2] A function is said to be of *class* C^r in some region if it is continuous and has continuous derivatives with respect to the coordinates up to the order r at each point of the region, and it is said to be of *class* C^ω if it is analytic. A space is said to be of class C^r (C^ω) if it can be covered by a set of coordinate neighbourhoods in such a way that the transformation of coordinates in an overlapping domain is represented by functions of class C^r (C^ω) in that domain.

[3] We adopt the *summation convention*: If an index appears twice in a term once as a subscript and once as superscript, summation has to be effected on the range of the index.

[4] The $g_{\lambda\varkappa}(\xi)$ means the value of $g_{\lambda\varkappa}$ at the point ξ whose coordinates with respect to $(\varkappa)$ are $\xi^\varkappa$. The $f^\varkappa(\xi^\nu)$ in (1.2) denotes n functions of coordinates ξ^ν.

[5] A point transformation is said to be of class C^r (C^ω) if the functions defining it are of class C^r (C^ω).

If the distance $d's$ between two displaced points $'\xi^\varkappa$ and $'\xi^\varkappa + d'\xi^\varkappa$ is always equal to the distance between the two original points $\xi^\varkappa$ and $\xi^\varkappa + d\xi^\varkappa$, the point transformation (1.2) is called a *motion* [1] or an *isometry* in the V_n.

Now in order to formulate the condition for (1.2) to be a motion in a V_n, we proceed as follows:

The point transformation T carries a point $\xi^\varkappa$ in R to a point $'\xi^\varkappa$ in $'R$ and consequently the point transformation T^{-1} inverse to T carries the point $'\xi^\varkappa$ in $'R$ to the point $\xi^\varkappa$ in R. With this inverse point transformation $T^{-1} : {'\xi} \rightarrow \xi$, we can associate a coordinate transformation $(\varkappa) \rightarrow (\varkappa')$ such that the transform in R of a point in $'R$ by T^{-1} has the same coordinates with respect to $(\varkappa')$ as the original point in $'R$ had with respect to $(\varkappa)$. This coordinate transformation is given by the equation

$$\xi^{\varkappa'} = {'\xi^\varkappa} \tag{1.3}$$ [2]

that is

$$\xi^{\varkappa'} = f^\varkappa(\xi^\nu). \tag{1.4}$$

This process $(\varkappa) \rightarrow (\varkappa')$ is called the *dragging along* of the coordinate system $(\varkappa)$ by the point transformation T^{-1}: $'\xi \rightarrow \xi$ and $(\varkappa')$ is called the *coordinate system dragged along by* T^{-1}.

By this dragging along of $(\varkappa)$ the $d'\xi^\varkappa$ at $'\xi^\varkappa$ becomes $d\xi^{\varkappa'}$ at $\xi^{\varkappa'}$ and we have

$$d\xi^{\varkappa'} = d'\xi^\varkappa. \tag{1.5}$$

Now the distance $d's$ between $'\xi^\varkappa$ and $'\xi^\varkappa + d'\xi^\varkappa$ is given by

$$d's^2 = g_{\lambda\varkappa}('\xi)d'\xi^\lambda\, d'\xi^\varkappa \tag{1.6}$$

and the distance ds between $\xi^\varkappa$ and $\xi^\varkappa + d\xi^\varkappa$ is given by (1.1). But in the coordinate system $(\varkappa')$, (1.1) can be written as

$$ds^2 = g_{\lambda'\varkappa'}(\xi)d\xi^{\lambda'}\, d\xi^{\varkappa'} \tag{1.7}$$

where

$$g_{\lambda'\varkappa'}(\xi) = A^{\lambda\varkappa}_{\lambda'\varkappa'}\, g_{\lambda\varkappa}(\xi) \tag{1.8}$$ [3].

---

[1] Following this definition, the reflexion is a motion.

[2] Cf. SCHOUTEN [8], p. 102. This is written more elaborately $\xi^{\varkappa'} = \delta^{\varkappa'}_{\varkappa}{'\xi^\varkappa}$ where $\delta^{\varkappa'}_{\varkappa}$ is the general Kronecker delta. In all cases where no ambiguity can arise, we drop the symbol $\delta^{\varkappa'}_{\varkappa}$ for the sake of shortness.

[3] $A^{\lambda\varkappa}_{\lambda'\varkappa'} \overset{\text{def}}{=} A^{\lambda}_{\lambda'}A^{\varkappa}_{\varkappa'}$ and $A^{\varkappa}_{\varkappa'} \overset{\text{def}}{=} \partial_{\varkappa'}\xi^\varkappa$, $A^{\varkappa'}_{\varkappa} \overset{\text{def}}{=} \partial_{\varkappa}\xi^{\varkappa'}$.

Thus comparing (1.6) with (1.7) and taking account of (1.5), we have

$$g_{\lambda\kappa}('\xi) = g_{\lambda'\kappa'}(\xi) \tag{1.9}$$

for a motion in the V_n.

Now the field $g_{\lambda\kappa}(\xi)$ is given at each point ξ of the space and consequently we have the field $g_{\lambda\kappa}('\xi)$ at $'\xi$ in $'R$. Starting from this field $g_{\lambda\kappa}('\xi)$ at $'\xi$, we form a new field $'g_{\lambda\kappa}(\xi)$ at ξ in R in the following way:

We define a new field $'g_{\lambda\kappa}(\xi)$ at ξ in R as a field whose components $'g_{\lambda'\kappa'}(\xi)$ with respect to (κ') at each point ξ in R are equal to the $g_{\lambda\kappa}('\xi)$ at the corresponding point $'\xi$ in $'R$, that is,

$$'g_{\lambda'\kappa'}(\xi) \stackrel{\text{def}}{=} g_{\lambda\kappa}('\xi) \tag{1.10}$$

Since

$$'g_{\lambda\kappa}(\xi) = A_{\lambda\ \kappa}^{\lambda'\kappa'}\,'g_{\lambda'\kappa'}(\xi),$$

we have from (1.4) and (1.10),

$$'g_{\lambda\kappa}(\xi) = (\partial_\lambda f^\sigma)(\partial_\kappa f^\rho) g_{\sigma\rho}('\xi). \tag{1.11}$$

This process $g_{\lambda\kappa} \to 'g_{\lambda\kappa}$ is called the *dragging along* of the field $g_{\lambda\kappa}$ by the point transformation T^{-1} and the field $'g_{\lambda\kappa}$ is called the field *dragged along*. We say also that the point transformation T^{-1} has *deformed* the tensor $g_{\lambda\kappa}$ into $'g_{\lambda\kappa}$ and we call $'g_{\lambda\kappa}$ the *deformed tensor* of $g_{\lambda\kappa}$ by T^{-1}.

Now comparing (1.9) with (1.10) we have

$$'g_{\lambda'\kappa'}(\xi) = g_{\lambda'\kappa'}(\xi) \tag{1.12}$$

with respect to (κ') and

$$'g_{\lambda\kappa}(\xi) = g_{\lambda\kappa}(\xi) \tag{1.13}$$

with respect to (κ) for a motion in V_n. Hence we have

THEOREM 1.1. *In order that* (1.2) *be a motion in a* V_n, *it is necessary and sufficient that the transformation* $'\xi \to \xi$ *do not deform the fundamental tensor of the* V_n.

We call $'g_{\lambda\kappa} - g_{\lambda\kappa}$ the *Lie difference* of $g_{\lambda\kappa}$ with respect to (1.2). The Lie difference of $g_{\lambda\kappa}$ is a tensor of the same type as $g_{\lambda\kappa}$, because it is the difference of two tensors of this type. In order that (1.2) be a motion in a V_n, it is necessary and sufficient that the Lie difference of the fundamental tensor of V_n with respect to (1.2) vanish.

We now consider the case in which the point transformation (1.2) is an infinitesimal one

$$'\xi^\kappa = \xi^\kappa + v^\kappa dt, \tag{1.14}$$

where $v^\varkappa$ is a contravariant vector field and dt is an infinitesimal. For the coordinate transformation (1.4) we have

$$\xi^{\varkappa'} = f^\varkappa(\xi^\nu) = \xi^\varkappa + v^\varkappa dt, \tag{1.15}$$

from which

$$\partial_\lambda f^\varkappa = \delta_\lambda^\varkappa + \partial_\lambda v^\varkappa dt \tag{1.16}$$

up to infinitesimals of the first order with respect to dt. In the following we shall always neglect quantities of an order higher than the first with respect to dt. Of course the equalities (1.14) and (1.16) should be written with the use of the sign * [1] because they are only valid for special coordinate systems. But we may accept as a general rule that * will be dropped in cases where no ambiguity can arise.

Substituting (1.16) in (1.11), we find

$$'g_{\lambda\varkappa} = (\delta_\lambda^\sigma + \partial_\lambda v^\sigma dt)(\delta_\varkappa^\rho + \partial_\varkappa v^\rho dt)(g_{\sigma\rho} + v^\mu \partial_\mu g_{\sigma\rho} dt),$$

from which

$$'g_{\lambda\varkappa} = g_{\lambda\varkappa} + (v^\mu \partial_\mu g_{\lambda\varkappa} + g_{\rho\varkappa}\partial_\lambda v^\rho + g_{\lambda\rho}\partial_\varkappa v^\rho)dt. \tag{1.17}$$

Thus we have

THEOREM 1.2. *In order that* (1.14) *be a motion in a* V_n, *it is necessary and sufficient that*

$$v^\mu \partial_\mu g_{\lambda\varkappa} + g_{\rho\varkappa}\partial_\lambda v^\rho + g_{\lambda\rho}\partial_\varkappa v^\rho = 0. \tag{1.18}$$

We call

$$\underset{v}{\pounds} g_{\lambda\varkappa} dt \overset{\text{def}}{=} {'g_{\lambda\varkappa}} - g_{\lambda\varkappa} \;[2] \tag{1.19}$$
$$= (v^\mu \partial_\mu g_{\lambda\varkappa} + g_{\rho\varkappa}\partial_\lambda v^\rho + g_{\lambda\rho}\partial_\varkappa v^\rho)dt$$

---

[1] The sign * is used to emphasize the fact that an equation is only valid or that its validity is only asserted for the coordinate system or coordinate systems occuring explicitly in the formula itself. Cf. SCHOUTEN [8], p. 2.

[2] In the coordinate system $(\varkappa')$ which only differs infinitesimally from $(\varkappa)$, this equation can be written as

$$\underset{v}{\pounds} g_{\lambda'\varkappa'} dt = {'g_{\lambda'\varkappa'}}(\xi) - g_{\lambda'\varkappa'}(\xi) = g_{\lambda\varkappa}('\xi) - g_{\lambda'\varkappa'}(\xi).$$

But as is stated below, $\underset{v}{\pounds} g_{\lambda\varkappa}$ is a tensor and consequently

$$\underset{v}{\pounds} g_{\lambda'\varkappa'} dt = A_{\lambda'\varkappa'}^{\lambda\varkappa} \underset{v}{\pounds} g_{\lambda\varkappa} dt = \underset{v}{\pounds} g_{\lambda\varkappa} dt + (\text{term of higher order}),$$

from which

$$\underset{v}{\pounds} g_{\lambda\varkappa} dt = g_{\lambda\varkappa}('\xi) - g_{\lambda'\varkappa'}(\xi).$$

This is the usual definition of the Lie derivative. See YANO [13].

the *Lie differential* of $g_{\lambda\varkappa}$ with respect to (1.14) or with respect to the vector field $v^{\varkappa}$ and $\underset{v}{\pounds}\, g_{\lambda\varkappa}$ the *Lie derivative* [1] of $g_{\lambda\varkappa}$.

The Lie differential of $g_{\lambda\varkappa}$ is a tensor of the same type as $g_{\lambda\varkappa}$. Thus the Lie derivative of $g_{\lambda\varkappa}$ is also a tensor of the same type.

In fact, using the relations

$$\nabla_\mu g_{\lambda\varkappa} \overset{\text{def}}{=} \partial_\mu g_{\lambda\varkappa} - g_{\rho\varkappa}\{^{\rho}_{\mu\lambda}\} - g_{\lambda\rho}\{^{\rho}_{\mu\varkappa}\} = 0, \text{ [2] [3]}$$

$$\nabla_\mu v^{\varkappa} \overset{\text{def}}{=} \partial_\mu v^{\varkappa} + \{^{\varkappa}_{\mu\lambda}\} v^{\lambda},$$

we can write the Lie derivative of $g_{\lambda\varkappa}$ in the form

$$(1.20) \qquad \boxed{\underset{v}{\pounds}\, g_{\lambda\varkappa} = 2\nabla_{(\lambda} v_{\varkappa)} \text{ [4]}} \; ; \quad v_{\varkappa} \overset{\text{def}}{=} v^{\lambda} g_{\lambda\varkappa}, \text{ [5]}$$

which shows explicitly the tensor character of $\underset{v}{\pounds}\, g_{\lambda\varkappa}$.

Thus we have

THEOREM 1.3. *In order that* (1.14) *be a motion in a* V_n *it is necessary and sufficient that the Lie derivative of* $g_{\lambda\varkappa}$ *with respect to* (1.14) *vanish*:

$$(1.22) \qquad \underset{v}{\pounds}\, g_{\lambda\varkappa} = 2\nabla_{(\lambda} v_{\varkappa)} = 0.$$

The equation (1.22) is called after Killing [6] and a vector field satisfying a Killing equation is called a *Killing vector*.

Myers and Steenrod [7] proved

THEOREM 1.4. *Any closed group of motions in a* V_n *of class* C^r $(r \geq 2)$ *is a Lie group of motions.*

---

[1] The name "Lie derivative" was introduced by VAN DANTZIG [2, 3].

[2] We use the notations $\delta\Phi$ and $\nabla_\mu \Phi$ to denote the covariant differential and the covariant derivative of Φ respectively. Cf. SCHOUTEN [8], p. 124.

[3] The $\{^{\varkappa}_{\mu\lambda}\}$ denotes the Christoffel symbol: $\{^{\varkappa}_{\mu\lambda}\} \overset{\text{def}}{=} \frac{1}{2} g^{\varkappa\rho}(\partial_\mu g_{\lambda\rho} + \partial_\lambda g_{\mu\rho} - \partial_\rho g_{\mu\lambda})$. Cf. SCHOUTEN [8], p. 132.

[4] The round brackets denote the symmetric part, e.g. $2\nabla_{(\lambda} v_{\varkappa)} = \nabla_\lambda v_\varkappa + \nabla_\varkappa v_\lambda$, while the square brackets denote the alternating part, e.g. $\Gamma^{\varkappa}_{[\mu\lambda]} = \frac{1}{2}(\Gamma^{\varkappa}_{\mu\lambda} - \Gamma^{\varkappa}_{\lambda\mu})$. Cf. SCHOUTEN [8], p. 14.

[5] In the following we distinguish the contravariant, covariant and mixed components of a tensor by the position of the indices, the same kernel being used in all cases. Cf. SCHOUTEN [8], p. 44.

[6] KILLING [1].

[7] MYERS and STEENROD [1].

§ 2. Affine motions in a space with a linear connexion.

We consider in this section n-dimensional space L_n [1] provided with a linear connexion $\Gamma^{\varkappa}_{\mu\lambda}(\xi)$. In an L_n the parallelism between a vector $u^{\varkappa}$ at a point $\xi^{\varkappa}$ and a vector $u^{\varkappa} + du^{\varkappa}$ at a point $\xi^{\varkappa} + d\xi^{\varkappa}$ is defined by

$$\delta u^{\varkappa} \stackrel{\text{def}}{=} du^{\varkappa} + \Gamma^{\varkappa}_{\mu\lambda} u^{\lambda} d\xi^{\mu} = 0. \tag{2.1}$$

When we effect a point transformation (1.2), the differentials $d\xi^{\varkappa}$ at $\xi^{\varkappa}$ are transformed into the differentials

$$d'\xi^{\varkappa} = \frac{\partial f^{\varkappa}}{\partial \xi^{\nu}} d\xi^{\nu} \tag{2.2}$$

at $'\xi^{\varkappa}$. Now if we make the condition that the vector $u^{\varkappa}$ at $\xi^{\varkappa}$ is transformed from $\xi^{\varkappa}$ to $'\xi^{\varkappa}$ in the same way as the linear elements $d\xi^{\varkappa}$ at $\xi^{\varkappa}$, then the corresponding vector at $'\xi$ is

$$\overset{m}{u}{}^{\varkappa}('\xi) = \frac{\partial f^{\varkappa}}{\partial \xi^{\nu}} u^{\nu}(\xi). \tag{2.3}$$

When a point transformation (1.2) transforms any pair of parallel vectors into a pair of parallel vectors, (1.2) is called an *affine motion* [2] in an L_n.

For an affine motion, we must have

$$\delta \overset{m}{u}{}^{\varkappa}('\xi) \stackrel{\text{def}}{=} d\overset{m}{u}{}^{\varkappa}('\xi) + \Gamma^{\varkappa}_{\mu\lambda}('\xi)\overset{m}{u}{}^{\lambda}('\xi)d'\xi^{\mu} = 0. \tag{2.4}$$

Now we introduce the coordinate transformation $\xi^{\varkappa'} = {}'\xi^{\varkappa}$. Then with respect to $(\varkappa')$ dragged along by $T^{-1} : {}'\xi \to \xi$, the equation (2.1) can be written as

$$\delta u^{\varkappa'} \stackrel{\text{def}}{=} du^{\varkappa'}(\xi) + \Gamma^{\varkappa'}_{\mu'\lambda'}(\xi)u^{\lambda'}(\xi)d\xi^{\mu'} = 0, \tag{2.5}$$

where

$$u^{\varkappa'}(\xi) = A^{\varkappa'}_{\varkappa} u^{\varkappa}(\xi) \tag{2.6}$$

and

$$\Gamma^{\varkappa'}_{\mu'\lambda'}(\xi) = (A^{\mu\lambda}_{\mu'\lambda'}\Gamma^{\varkappa}_{\mu\lambda}(\xi) + \partial_{\mu'} A^{\varkappa}_{\lambda'})A^{\varkappa'}_{\varkappa}, \tag{2.7}$$

and (2.3) can now be written as

$$\overset{m}{u}{}^{\varkappa}('\xi) = u^{\varkappa'}(\xi). \tag{2.8}$$

---

[1] An n-dimensional space with a linear connexion is called an L_n. Cf. SCHOUTEN [8], p. 125.

[2] An affine motion was first defined by SLEBODZINSKI [2].

From this we see that $\overset{m}{u}{}^{\varkappa}('\xi)$ is exactly the field value at $'\xi^{\varkappa}$ of the field $u^{\varkappa}$ dragged along by $\xi \to \xi'$.

Hence, from (2.4) and (2.5), we have

$$\Gamma^{\varkappa}_{\mu\lambda}('\xi) = \Gamma^{\varkappa'}_{\mu'\lambda'}(\xi) \tag{2.9}$$

as the necessary and sufficient condition for an affine motion in an L_n.

We now define a new linear connexion $'\Gamma^{\varkappa}_{\mu\lambda}(\xi)$ in R as a linear connexion whose components $'\Gamma^{\varkappa'}_{\mu'\lambda'}(\xi)$ with respect to $(\varkappa')$ are equal to the $\Gamma^{\varkappa}_{\mu\lambda}('\xi)$ at the corresponding point $'\xi$ in $'R$, that is

$$'\Gamma^{\varkappa'}_{\mu'\lambda'}(\xi) \overset{\text{def}}{=} \Gamma^{\varkappa}_{\mu\lambda}('\xi) \tag{2.10}$$

with respect to $(\varkappa')$.

Since

$$A^{\varkappa'}_{\varkappa}\, '\Gamma^{\varkappa}_{\mu\lambda}(\xi) = A^{\mu'\lambda'}_{\mu\ \lambda}\, '\Gamma^{\varkappa'}_{\mu'\lambda'}(\xi) + \partial_{\mu} A^{\varkappa'}_{\lambda},$$

we have, from $\xi^{\varkappa'} = f^{\varkappa}(\xi)$ and (2.10),

$$(\partial_{\varkappa} f^{\rho})'\Gamma^{\varkappa}_{\mu\lambda}(\xi) = (\partial_{\mu} f^{\tau})(\partial_{\lambda} f^{\sigma})\Gamma^{\rho}_{\tau\sigma}('\xi) + \partial_{\mu}\partial_{\lambda} f^{\rho}. \tag{2.11}$$

This process $\Gamma^{\varkappa}_{\mu\lambda} \to {}'\Gamma^{\varkappa}_{\mu\lambda}$ is called the *dragging along* of the linear connexion $\Gamma^{\varkappa}_{\mu\lambda}$ by the point transformation $'\xi \to \xi$ and $'\Gamma^{\varkappa}_{\mu\lambda}$ is called the linear connexion *dragged along*. We say also that the point transformation has *deformed* the linear connexion $\Gamma^{\varkappa}_{\mu\lambda}$ into $'\Gamma^{\varkappa}_{\mu\lambda}$ and we call $'\Gamma^{\varkappa}_{\mu\lambda}$ the *deformed linear connexion* of $\Gamma^{\varkappa}_{\mu\lambda}$.

Now comparing (2.9) with (2.10), we find

$$'\Gamma^{\varkappa'}_{\mu'\lambda'}(\xi) = \Gamma^{\varkappa'}_{\mu'\lambda'}(\xi) \tag{2.12}$$

with respect to $(\varkappa')$ and

$$'\Gamma^{\varkappa}_{\mu\lambda}(\xi) = \Gamma^{\varkappa}_{\mu\lambda}(\xi) \tag{2.13}$$

with respect to $(\varkappa)$ for an affine motion in an L_n. Hence we have

THEOREM 2.1. *In order that* (1.2) *be an affine motion in an* L_n *it is necessary and sufficient that the transformation* $'\xi \to \xi$ *do not deform the linear connexion of* L_n.

We call $'\Gamma^{\varkappa}_{\mu\lambda} - \Gamma^{\varkappa}_{\mu\lambda}$ the *Lie difference* of $\Gamma^{\varkappa}_{\mu\lambda}$ with respect to (1.2). The Lie difference of $\Gamma^{\varkappa}_{\mu\lambda}$ is the difference of two linear connexions and consequently it is a mixed tensor of contravariant valence 1 and covariant valence 2. In order that (1.2) be an affine motion in an L_n it is necessary and sufficient that the Lie difference of the linear connexion with respect to (1.2) vanish.

We next consider the case in which the point transformation (1.2)

becomes an infinitesimal one (1.14). Substituting (1.16) in (2.11), we find

$$(\delta^\rho_\varkappa + \partial_\varkappa v^\rho dt)'\Gamma^\varkappa_{\mu\lambda} = (\delta^\tau_\mu + \partial_\mu v^\tau dt)(\delta^\sigma_\lambda + \partial_\lambda v^\sigma dt)(\Gamma^\rho_{\tau\sigma} + v^\nu \partial_\nu \Gamma^\rho_{\tau\sigma} dt) + \partial_\mu \partial_\lambda v^\rho dt,$$

from which

$$(2.14)\quad '\Gamma^\varkappa_{\mu\lambda} = \Gamma^\varkappa_{\mu\lambda} + [\partial_\mu \partial_\lambda v^\varkappa + v^\nu \partial_\nu \Gamma^\varkappa_{\mu\lambda} - \Gamma^\rho_{\mu\lambda} \partial_\rho v^\varkappa + \Gamma^\varkappa_{\rho\lambda} \partial_\mu v^\rho + \Gamma^\varkappa_{\mu\rho} \partial_\lambda v^\rho]dt.$$

Thus we have

THEOREM 2.2. *In order that* (1.14) *be an affine motion in an* L_n, *it is necessary and sufficient that*

$$(2.15)\quad \partial_\mu \partial_\lambda v^\varkappa + v^\nu \partial_\nu \Gamma^\varkappa_{\mu\lambda} - \Gamma^\rho_{\mu\lambda} \partial_\rho v^\varkappa + \Gamma^\varkappa_{\rho\lambda} \partial_\mu v^\rho + \Gamma^\varkappa_{\mu\rho} \partial_\lambda v^\rho = 0.$$

We call

$$(2.16)\quad \underset{v}{\pounds} \Gamma^\varkappa_{\mu\lambda} dt \overset{\text{def}}{=} {}'\Gamma^\varkappa_{\mu\lambda} - \Gamma^\varkappa_{\mu\lambda}\,{}^{1}$$
$$= (\partial_\mu \partial_\lambda v^\varkappa + v^\nu \partial_\nu \Gamma^\varkappa_{\mu\lambda} - \Gamma^\rho_{\mu\lambda} \partial_\rho v^\varkappa + \Gamma^\varkappa_{\rho\lambda} \partial_\mu v^\rho + \Gamma^\varkappa_{\mu\rho} \partial_\lambda v^\rho)dt$$

the *Lie differential* of $\Gamma^\varkappa_{\mu\lambda}$ with respect to (1.14) or with respect to the vector $v^\varkappa$ and $\underset{v}{\pounds} \Gamma^\varkappa_{\mu\lambda}$ the *Lie derivative* of $\Gamma^\varkappa_{\mu\lambda}$.

The Lie differential and the Lie derivative of $\Gamma^\varkappa_{\mu\lambda}$ are mixed tensors of contravariant valence 1 and of covariant valence 2.

In fact putting

$$(2.17)\quad v_\lambda^{\cdot\varkappa} \overset{\text{def}}{=} \nabla_\lambda v^\varkappa + 2S_{\rho\lambda}^{\cdot\cdot\varkappa} v^\rho = \partial_\lambda v^\varkappa + \Gamma^\varkappa_{\rho\lambda} v^\rho,$$

we can write the Lie derivative of $\Gamma^\varkappa_{\mu\lambda}$ in the form

$$(2.18)\quad \boxed{\underset{v}{\pounds} \Gamma^\varkappa_{\mu\lambda} = \nabla_\mu v_\lambda^{\cdot\varkappa} + R_{\nu\mu\lambda}^{\cdot\cdot\cdot\varkappa} v^\nu}$$

which shows explicitly its tensor character. In these formulae $S_{\mu\lambda}^{\cdot\cdot\varkappa}$ and $R_{\nu\mu\lambda}^{\cdot\cdot\cdot\varkappa}$ are respectively the torsion tensor and the curvature tensor of the space:

$$(2.19)\quad S_{\mu\lambda}^{\cdot\cdot\varkappa} \overset{\text{def}}{=} \Gamma^\varkappa_{[\mu\lambda]},$$

$$(2.20)\quad R_{\nu\mu\lambda}^{\cdot\cdot\cdot\varkappa} \overset{\text{def}}{=} 2\partial_{[\nu} \Gamma^\varkappa_{\mu]\lambda} + 2\Gamma^\varkappa_{[\nu|\rho|} \Gamma^\rho_{\mu]\lambda}.$$

Thus we have

THEOREM 2.3. *In order that* (1.14) *be an affine motion in an* L_n, *it is necessary and sufficient that the Lie derivative of* $\Gamma^\varkappa_{\mu\lambda}$ *with respect to* (1.14) *vanish*:

$$(2.21)\quad \underset{v}{\pounds} \Gamma^\varkappa_{\mu\lambda} = \nabla_\mu v_\lambda^{\cdot\varkappa} + R_{\nu\mu\lambda}^{\cdot\cdot\cdot\varkappa} v^\nu = 0.$$

---

[1] The remark made in the footnote 2 of p. 4 on $\underset{v}{\pounds}\, g_{\lambda\varkappa}$ is also valid for $\underset{v}{\pounds} \Gamma^\varkappa_{\mu\lambda}$.

When the linear connexion $\Gamma^{\varkappa}_{\mu\lambda}$ is symmetric the space L_n is called an A_n.[1] In an A_n the linear connexion determines geodesics by means of the equation

$$\frac{d^2\xi^{\varkappa}}{ds^2} + \Gamma^{\varkappa}_{\mu\lambda}\frac{d\xi^{\mu}}{ds}\frac{d\xi^{\lambda}}{ds} = 0, \tag{2.22}$$

and this equation also determines on each geodesic an *affine parameter* s but for an affine transformation with constant coefficients.

Conversely when a system of geodesics and affine parameters on them are given by (2.22), a linear connexion is uniquely determined by the coefficients $\Gamma^{\varkappa}_{\mu\lambda}$.

Thus it is evident that an affine motion in an A_n carries a geodesic into a geodesic and does not change the affine parameter on it but for an affine transfoımation with constant coefficients. Conversely a point transformation which carries every geodesic into a geodesic and leaves invariant the affine parameter on it but for an affine transformation with constant coefficients is an affine motion in the A_n. Thus we have

THEOREM 2.4. *In order that a point transformation* (1.2) *in an* A_n *change every geodesic into a geodesic and every affine parameter into an affine parameter, it is necessary and sufficient that* (1.2) *be an affine motion in the* A_n.[2]

In an A_n, the Lie derivative $\underset{v}{\pounds}\Gamma^{\varkappa}_{\mu\lambda}$ of $\Gamma^{\varkappa}_{\mu\lambda}$ can be written as

$$\boxed{\underset{v}{\pounds}\Gamma^{\varkappa}_{\mu\lambda} = \nabla_{\mu}\nabla_{\lambda}v^{\varkappa} + R_{\nu\mu\lambda}{}^{\varkappa}v^{\nu}.} \tag{2.23}$$

Nomizu[3] proved

THEOREM 2.5. *The group of affine motions in a complete* L_n *of class* C^{∞} *is a Lie group.*

§ 3. Lie derivatives of scalars, vectors and tensors.

In the preceding sections, we have seen some examples of Lie derivatives. In the present section, we define systematically Lie derivatives of scalars, vectors and tensors.

---

[1] Cf. SCHOUTEN [8], p. 126.

[2] Some authors call an affine motion in an A_n an *affine collineation.*

[3] NOMIZU [1]. An L_n is said to be *complete* if every geodesic can be extended for any large value of the affine parameter on it.

Take a scalar field $p(\xi)$ in an n-dimensional space X_n and consider an infinitesimal point transformation

$$T\colon\ '\xi^\varkappa = \xi^\varkappa + v^\varkappa dt. \tag{3.1}$$

The dragging along $(\varkappa) \to (\varkappa')$ of the coordinate system $(\varkappa)$ by the infinitesimal point transformation T^{-1} : $'\xi \to \xi$ inverse to T is given by

$$\xi^{\varkappa'} = \xi^\varkappa + v^\varkappa dt. \tag{3.2}$$

We define a new scalar field $'p(\xi)$ at ξ as a scalar field whose components with respect to $(\varkappa')$ at each point ξ is equal to $p('\xi)$ at the corresponding point $'\xi$, that is,

$$'p(\xi) \overset{\text{def}}{=} p('\xi) \tag{3.3}$$

with respect to $(\varkappa')$. But since $'p$ and p are both scalar fields, the equation (3.3) is valid also with respect to $(\varkappa)$.

The process $p(\xi) \to {'p}(\xi)$ is the *dragging along* of the scalar field by T^{-1}: $'\xi \to \xi$ and $'p(\xi)$ is the scalar field *dragged along*.

From (3.3) we have

$$'p(\xi) = p(\xi) + v^\mu \partial_\mu p dt. \tag{3.4}$$

We call

$$\begin{aligned} \underset{v}{\pounds} p\, dt &\overset{\text{def}}{=} {'p}(\xi) - p(\xi) \\ &= v^\mu \partial_\mu p dt \end{aligned} \tag{3.5}$$

the *Lie differential* of the scalar field p with respect to (3.1), and

$$\boxed{\underset{v}{\pounds} p = v^\mu \partial_\mu p} \tag{3.6}$$

the *Lie derivative* of p. We call $'p = p + \underset{v}{\pounds} p dt$ the *deformed scalar* of p.

Take next a contravariant vector field $u^\varkappa(\xi)$ in X_n.

We define a new contravariant vector field $'u^\varkappa(\xi)$ at ξ as a field whose components $'u^{\varkappa'}(\xi)$ with respect to $(\varkappa')$ at ξ are equal to the $u^\varkappa('\zeta)$ at the corresponding point $'\xi$, that is,

$$'u^{\varkappa'}(\xi) \overset{\text{def}}{=} u^\varkappa('\xi) \tag{3.7}$$

with respect to $(\varkappa')$. Since

$$'u^\varkappa(\xi) = A^\varkappa_{\varkappa'}\, {'u}^{\varkappa'}(\xi),$$

we have from (3.2) and (3.7)

$$'u^\varkappa(\xi) = (\delta^\varkappa_\rho - \partial_\rho v^\varkappa dt)(u^\rho(\xi) + v^\mu \partial_\mu u^\rho dt),$$

from which

$$'u^\varkappa(\xi) = u^\varkappa(\xi) + (v^\mu \partial_\mu u^\varkappa - u^\mu \partial_\mu v^\varkappa)dt. \tag{3.8}$$

We call

$$\begin{aligned} \underset{v}{\pounds} u^\varkappa \, dt &\overset{\text{def}}{=} {}'u^\varkappa(\xi) - u^\varkappa(\xi) \\ &= (v^\mu \partial_\mu u^\varkappa - u^\mu \partial_\mu v^\varkappa)dt \end{aligned} \tag{3.9}$$

the *Lie differential* of the contravariant vector field $u^\varkappa$ with respect to (3.1) and

$$\boxed{\underset{v}{\pounds} u^\varkappa = v^\mu \partial_\mu u^\varkappa - u^\mu \partial_\mu v^\varkappa} \tag{3.10}$$

the *Lie derivative* of $u^\varkappa$. We call $'u^\varkappa = u^\varkappa + \underset{v}{\pounds} u^\varkappa dt$ the *deformed contravariant vector* of $u^\varkappa$.

The Lie differential $\underset{v}{\pounds} u^\varkappa dt$ is a contravariant vector because it is the difference between two contravariant vectors. Thus the Lie derivative $\underset{v}{\pounds} u^\varkappa$ is also a contravariant vector.

In fact, when X_n is provided with a linear connexion (3.10) can be written also as (cf. 2.17)

$$\boxed{\underset{v}{\pounds} u^\varkappa = v^\mu \nabla_\mu u^\varkappa - u^\lambda v_{;\lambda}^{\cdot\varkappa},} \tag{3.11}$$

which shows explicitly the vector character of $\underset{v}{\pounds} u^\varkappa$.

The Lie derivative of a contravariant vector with respect to an infinitesimal transformation can be defined whenever the field value of the contravariant vector at the transformed point is defined. We give an important example.

When a curve is given by its parametric expression $\xi^\varkappa(z)$, the tangent $\dfrac{d\xi^\varkappa}{dz}$ is defined at each point $\xi^\varkappa(z)$ on the curve. By an infinitesimal point transformation $\xi^\varkappa \to {}'\xi^\varkappa = \xi^\varkappa + v^\varkappa(\xi)dt$, the curve $\xi^\varkappa(z)$ is transformed into the curve $'\xi^\varkappa(z)$ and the tangent $\dfrac{d\xi^\varkappa}{dz}$ at $\xi^\varkappa(z)$ into the tangent $\dfrac{d'\xi^\varkappa}{dz}$ at $'\xi^\varkappa(z)$ provided that the parameter z is not changed by the transformation.

To find the Lie derivative of $\dfrac{d\xi^\varkappa}{dz}$ we proceed as follows. We define a new contravariant vector $'\left(\frac{d\xi^{\varkappa'}}{dz}\right)$ at ξ as a vector whose components $'\left(\frac{d\xi^{\varkappa'}}{dz}\right)$ with respect to $(\varkappa')$ at ξ are equal to the $\frac{d'\xi^{\varkappa}}{dz}$ at the corresponding point $'\xi$, that is,

(3.12) $$'\left(\frac{d\xi^{\varkappa'}}{dz}\right) \overset{\text{def}}{=} \frac{d'\xi^{\varkappa}}{dz} = \frac{d\xi^{\varkappa'}}{dz}$$

with respect to $(\varkappa')$. Since

$$'\left(\frac{d\xi^{\varkappa}}{dz}\right) = A^{\varkappa}_{\varkappa'}\,'\left(\frac{d\xi^{\varkappa'}}{dz}\right)$$

we find

$$'\left(\frac{d\xi^{\varkappa}}{dz}\right) = A^{\varkappa}_{\varkappa'}\frac{d\xi^{\varkappa'}}{dz} = \frac{d\xi^{\varkappa}}{dz},$$

from which

$$\underset{v}{\pounds}\frac{d\xi^{\varkappa}}{dz} = 0$$

or

(3.13) $$\boxed{\underset{v}{\pounds}\, d\xi^{\varkappa} = 0,}$$

because z is supposed to be invariant during the point transformation.

Take next a covariant vector field $w_\lambda(\xi)$ in X_n. We define a new covariant vector field $'w_\lambda(\xi)$ at ξ as a field whose components $'w_{\lambda'}(\xi)$ with respect to $(\varkappa')$ at ξ are equal to $w_\lambda('\xi)$ at the corresponding point $'\xi$, that is,

(3.14) $$'w_{\lambda'}(\xi) \overset{\text{def}}{=} w_\lambda('\xi)$$

with respect to $(\varkappa')$. We call

(3.15) $$\begin{aligned}\underset{v}{\pounds}\, w_\lambda dt &\overset{\text{def}}{=} 'w_\lambda(\xi) - w_\lambda(\xi)\\ &= (v^\mu \partial_\mu w_\lambda + w_\mu \partial_\lambda v^\mu)dt\end{aligned}$$

the *Lie differential* of the covariant vector field w_λ with respect to (3.1) and call

(3.16) $$\boxed{\underset{v}{\pounds}\, w_\lambda = v^\mu \partial_\mu w_\lambda + w_\mu \partial_\lambda v^\mu}$$

the *Lie derivative* of w_λ. $\underset{v}{\pounds}\, w_\lambda$ is a covariant vector and $'w_\lambda = w_\lambda + \underset{v}{\pounds}\, w_\lambda dt$ is called the *deformed covariant vector* of w_λ.

Van Dantzig [1] showed that the equations of motion of a dynamical system, found by variation of $\int d\Lambda = \int p_\lambda d\xi^\lambda$ can be written in a very simple form $\underset{v}{\pounds}\, p_\lambda = 0$. The equations of motions are

$$(3.17) \qquad dp_\lambda - \partial d\Lambda/\partial\xi^\lambda = 0.$$

But, following Euler's condition, the p_λ are homogeneous of degree zero in the $d\xi^\varkappa$. Hence

$$\partial d\Lambda/\partial\xi^\lambda = (\partial_\lambda p_\mu) d\xi^\mu,$$

and consequently, the equation of motion (3.17) is equivalent with

$$(3.18) \qquad 2d\xi^\mu \partial_{[\mu} p_{\lambda]} = 0.$$

Now, if we put

$$v^\lambda = d\xi^\lambda/d\Lambda,$$

we have

$$(3.19) \qquad p_\lambda v^\lambda = 1,$$

from which

$$(3.20) \qquad (\partial_\mu p_\lambda) v^\lambda + p_\lambda \partial_\mu v^\lambda = 0.$$

Thus (3.18) can be written as

$$(3.21) \qquad \underset{v}{\pounds}\, p_\lambda = v^\mu \partial_\mu p_\lambda + p_\mu \partial_\lambda v^\mu = 0. \text{ [2]}$$

If the X_n is provided with a linear connexion (3.16) can be written as

$$(3.22) \qquad \boxed{\underset{v}{\pounds}\, w_\lambda = v^\mu \nabla_\mu w_\lambda + w_\mu v_\lambda^{;\mu},}$$

which also shows the vector character of $\underset{v}{\pounds}\, w_\lambda$.

Quite similarly the *Lie differential* and the *Lie derivative* of a general tensor, for instance, $P^{\lambda\varkappa}_{\cdot\cdot\mu}$ are defined by

$$(3.23) \qquad \underset{v}{\pounds}\, P^{\varkappa\lambda}_{\cdot\cdot\mu}\, dt = {}'P^{\varkappa\lambda}_{\cdot\cdot\mu} - P^{\varkappa\lambda}_{\cdot\cdot\mu},$$

where

$$(3.24) \qquad {}'P^{\varkappa'\lambda'}_{\cdot\cdot\mu'}(\xi) \overset{\text{def}}{=} P^{\varkappa\lambda}_{\cdot\cdot\mu}({}'\xi)$$

---

[1] VAN DANTZIG [4].

[2] Van Dantzig showed also a beautiful application of the Lie derivatives in thermo-hydrodynamics of perfectly perfect fluids. See VAN DANTZIG [5].

and consequently

$$'P^{\varkappa\lambda}_{\cdot\cdot\mu}(\xi) = A^{\varkappa}_{\varkappa'} A^{\lambda}_{\lambda'} A^{\mu'}_{\mu}\, 'P^{\varkappa'\lambda'}_{\cdot\cdot\mu'}(\xi)$$
$$= (\delta^{\varkappa}_{\rho} - \partial_{\rho} v^{\varkappa} dt)(\delta^{\lambda}_{\sigma} - \partial_{\sigma} v^{\lambda} dt) \times$$
$$\times (\delta^{\tau}_{\mu} + \partial_{\mu} v^{\tau} dt)(P^{\rho\sigma}_{\cdot\cdot\tau}(\xi) + v^{\nu}\partial_{\nu} P^{\rho\sigma}_{\cdot\cdot\tau} dt),$$

that is,

(3.25) $$'P^{\varkappa\lambda}_{\cdot\cdot\mu}(\xi) = P^{\varkappa\lambda}_{\cdot\cdot\mu}(\xi) + (v^{\nu}\partial_{\nu} P^{\varkappa\lambda}_{\cdot\cdot\mu} - P^{\rho\lambda}_{\cdot\cdot\mu}\partial_{\rho} v^{\varkappa} - P^{\varkappa\rho}_{\cdot\cdot\mu}\partial_{\rho} v^{\lambda} + P^{\varkappa\lambda}_{\cdot\cdot\rho}\partial_{\mu} v^{\rho})dt.$$

Thus the Lie derivative of a tensor $P^{\varkappa\lambda}_{\cdot\cdot\mu}$ with respect to (3.1) is given by

(3.26) $$\underset{v}{\pounds} P^{\varkappa\lambda}_{\cdot\cdot\mu} = v^{\nu}\partial_{\nu} P^{\varkappa\lambda}_{\cdot\cdot\mu} - P^{\rho\lambda}_{\cdot\cdot\mu}\partial_{\rho} v^{\varkappa} - P^{\varkappa\rho}_{\cdot\cdot\mu}\partial_{\rho} v^{\lambda} + P^{\varkappa\lambda}_{\cdot\cdot\rho}\partial_{\mu} v^{\rho},$$

or

(3.27) $$\underset{v}{\pounds} P^{\varkappa\lambda}_{\cdot\cdot\mu} = v^{\nu}\nabla_{\nu} P^{\varkappa\lambda}_{\cdot\cdot\mu} - P^{\rho\lambda}_{\cdot\cdot\mu} v^{\cdot\varkappa}_{\rho} - P^{\varkappa\rho}_{\cdot\cdot\mu} v^{\cdot\lambda}_{\rho} + P^{\varkappa\lambda}_{\cdot\cdot\rho} v^{\cdot\rho}_{\mu}$$

which shows the tensor character of the Lie derivative.

Finally the Lie derivative of a general tensor density, for example, $\mathfrak{P}^{\varkappa\lambda}_{\cdot\cdot\mu}$ of weight w can be found to be

(3.28) $$\underset{v}{\pounds} \mathfrak{P}^{\varkappa\lambda}_{\cdot\cdot\mu} = v^{\nu}\partial_{\nu} \mathfrak{P}^{\varkappa\lambda}_{\cdot\cdot\mu} - \mathfrak{P}^{\rho\lambda}_{\cdot\cdot\mu}\partial_{\rho} v^{\varkappa} - \mathfrak{P}^{\varkappa\rho}_{\cdot\cdot\mu}\partial_{\rho} v^{\lambda} + \mathfrak{P}^{\varkappa\lambda}_{\cdot\cdot\rho}\partial_{\mu} v^{\rho} + w\mathfrak{P}^{\varkappa\lambda}_{\cdot\cdot\mu}\partial_{\rho} v^{\rho},$$

or

(3.29) $$\underset{v}{\pounds} \mathfrak{P}^{\varkappa\lambda}_{\cdot\cdot\mu} = v^{\nu}\nabla_{\nu} \mathfrak{P}^{\varkappa\lambda}_{\cdot\cdot\mu} - \mathfrak{P}^{\rho\lambda}_{\cdot\cdot\mu} v^{\cdot\varkappa}_{\rho} - \mathfrak{P}^{\varkappa\rho}_{\cdot\cdot\mu} v^{\cdot\lambda}_{\rho} + \mathfrak{P}^{\varkappa\lambda}_{\cdot\cdot\rho} v^{\cdot\rho}_{\mu} + w\mathfrak{P}^{\varkappa\lambda}_{\cdot\cdot\mu} v^{\cdot\rho}_{\rho},$$

which shows that $\underset{v}{\pounds} \mathfrak{P}^{\varkappa\lambda}_{\cdot\cdot\mu}$ is a tensor density of the same type as $\mathfrak{P}^{\varkappa\lambda}_{\cdot\cdot\mu}$.

From (3.29) it follows that the following rules, hold for the application of the Lie derivation to quantities: [1]

1. The Lie derivative of a sum of quantities of the same kind is equal to the sum of the Lie derivatives of the summands.
2. The Lie derivative of a contraction is equal to the contraction of the Lie derivative.
3. For a product or a transvection of two quantities Φ and Ψ, the rule of Leibniz

$$\underset{v}{\pounds}\Phi\Psi = (\underset{v}{\pounds}\Phi)\Psi + \Phi(\underset{v}{\pounds}\Psi)$$

holds.

---

[1] By *quantities* we mean here scalars, vectors, tensors and tensor densities. Cf. Schouten [8], p. 6.

§ 4. The Lie derivative of a linear connexion.

When we consider in an L_n an infinitesimal point transformation

$$'\xi^\varkappa = \xi^\varkappa + v^\varkappa dt, \tag{4.1}$$

the deform of a contravariant vector $u^\varkappa$ is defined by

$$'u^{\varkappa'}(\xi) \overset{\text{def}}{=} u^\varkappa('\xi) \tag{4.2}$$

and that of the linear connexion $\Gamma^\varkappa_{\mu\lambda}$ by

$$'\Gamma^{\varkappa'}_{\mu'\lambda'}(\xi) \overset{\text{def}}{=} \Gamma^\varkappa_{\mu\lambda}('\xi) \tag{4.3}$$

If we now denote by δ the covariant differential with respect to $\Gamma^\varkappa_{\mu\lambda}$ and by $'\delta$ the covariant differential with respect to $'\Gamma^\varkappa_{\mu\lambda}$, we have

$$\begin{aligned} '\delta' u^{\varkappa'}(\xi) &= d'u^{\varkappa'}(\xi) + '\Gamma^{\varkappa'}_{\mu'\lambda'}(\xi) u^{\lambda'}(\xi) d\xi^{\mu'} \\ &= du^\varkappa('\xi) + \Gamma^\varkappa_{\mu\lambda}('\xi) u^\lambda('\xi) d'\xi^\mu \\ &= \delta u^\varkappa('\xi). \end{aligned}$$

On the other hand, for the deform of $\delta u^\varkappa$, we have

$$'(\delta u^{\varkappa'}(\xi)) = \delta u^\varkappa('\xi).$$

From these two equations, we have

$$'\delta' u^{\varkappa'} = '(\delta u^{\varkappa'}) \tag{4.4}$$

holding with respect to every coordinate system and consequently

$$'\delta(u^\varkappa + \underset{v}{\pounds} u^\varkappa dt) = \delta u^\varkappa + \underset{v}{\pounds} \delta u^\varkappa dt \tag{4.5}$$

with respect to $(\varkappa)$. Thus we have

THEOREM 4.1. *The covariant differential of the deform of a contravariant vector with respect to the deformed linear connexion is equal to the deform of the covariant differential of the vector with respect to the original linear connexion.*

Since

$$\begin{aligned} '\delta(u^\varkappa + \underset{v}{\pounds} u^\varkappa dt) &= d(u^\varkappa + \underset{v}{\pounds} u^\varkappa dt) + (\Gamma^\varkappa_{\mu\lambda} + \underset{v}{\pounds} \Gamma^\varkappa_{\mu\lambda} dt)(u^\lambda + \underset{v}{\pounds} u^\lambda dt) d\xi^\mu \\ &= \delta u^\varkappa + \delta \underset{v}{\pounds} u^\varkappa dt + (\underset{v}{\pounds} \Gamma^\varkappa_{\mu\lambda}) u^\lambda d\xi^\mu dt, \end{aligned}$$

we have from (4.5)

$$\underset{v}{\pounds} \delta u^\varkappa - \delta \underset{v}{\pounds} u^\varkappa = (\underset{v}{\pounds} \Gamma^\varkappa_{\mu\lambda}) u^\lambda d\xi^\mu. \tag{4.6}$$

Taking account of (3.13), we have from (4.6)

$$\underset{v}{\pounds}\nabla_\mu u^\kappa - \nabla_\mu \underset{v}{\pounds} u^\kappa = (\underset{v}{\pounds}\Gamma^\kappa_{\mu\lambda})u^\lambda. \tag{4.7}$$ [1]

Formula (4.7) can be generalized for a covariant vector w_λ and for a general tensor $P^{\kappa\lambda}_{\cdot\cdot\mu}$ as follows:

$$\underset{v}{\pounds}\nabla_\mu w_\lambda - \nabla_\mu \underset{v}{\pounds} w_\lambda = -(\underset{v}{\pounds}\Gamma^\kappa_{\mu\lambda})w_\kappa, \tag{4.8}$$

$$\underset{v}{\pounds}\nabla_\nu P^{\kappa\lambda}_{\cdot\cdot\mu} - \nabla_\nu \underset{v}{\pounds} P^{\kappa\lambda}_{\cdot\cdot\mu} = (\underset{v}{\pounds}\Gamma^\kappa_{\nu\rho})P^{\rho\lambda}_{\cdot\cdot\mu} + (\underset{v}{\pounds}\Gamma^\lambda_{\nu\rho})P^{\kappa\rho}_{\cdot\cdot\mu} - (\underset{v}{\pounds}\Gamma^\rho_{\nu\mu})P^{\kappa\lambda}_{\cdot\cdot\rho}. \tag{4.9}$$

From these equations we have

THEOREM 4.2.[2] *In order that* (4.1) *be an affine motion in an* L_n, *it is necessary and sufficient that the covariant differentiation and the Lie derivation with respect to* (4.1) *be commutative.*

Now since the deformed linear connexion is given by

$$'\Gamma^\kappa_{\mu\lambda} = \Gamma^\kappa_{\mu\lambda} + \underset{v}{\pounds}\Gamma^\kappa_{\mu\lambda}dt, \tag{4.10}$$

it follows immediately that

$$'S_{\mu\lambda}^{\cdot\cdot\kappa} = S_{\mu\lambda}^{\cdot\cdot\kappa} + \underset{v}{\pounds}S_{\mu\lambda}^{\cdot\cdot\kappa}dt. \tag{4.11}$$ [3]

It is also evident that the deformed curvature tensor is given by

$$'R_{\nu\mu\lambda}^{\cdot\cdot\cdot\kappa} = R_{\nu\mu\lambda}^{\cdot\cdot\cdot\kappa} + \underset{v}{\pounds}R_{\nu\mu\lambda}^{\cdot\cdot\cdot\kappa}dt. \tag{4.12}$$

In fact substituting (4.10) into

$$'R_{\nu\mu\lambda}^{\cdot\cdot\cdot\kappa} = 2\partial_{[\nu}{}'\Gamma^\kappa_{\mu]\lambda} + 2'\Gamma^\kappa_{[\nu|\rho|}{}'\Gamma^\rho_{\mu]\lambda},$$

we find

$$'R_{\nu\mu\lambda}^{\cdot\cdot\cdot\kappa} = R_{\nu\mu\lambda}^{\cdot\cdot\cdot\kappa} + (\nabla_\nu \underset{v}{\pounds}\Gamma^\kappa_{\mu\lambda} - \nabla_\mu \underset{v}{\pounds}\Gamma^\kappa_{\nu\lambda} + 2S_{\nu\mu}^{\cdot\cdot\rho}\underset{v}{\pounds}\Gamma^\kappa_{\rho\lambda})dt. \tag{4.13}$$ [4]

---

[1] This equation can be deduced also by a direct calculation without using (3.13).

[2] W. SLEBODZINSKI [1, 2].

[3] E. T. DAVIES [1].

[4] PALATINI [1].

On the other hand, by virtue of the Ricci identity [1]:

$$2\nabla_{[\nu}\nabla_{\mu]} v_{\lambda}^{\cdot\varkappa} = R_{\nu\mu\rho}^{\cdot\cdot\cdot\varkappa} v_{\lambda}^{\cdot\rho} - R_{\nu\mu\lambda}^{\cdot\cdot\cdot\rho} v_{\rho}^{\cdot\varkappa} - 2S_{\nu\mu}^{\cdot\cdot\rho}\nabla_{\rho} v_{\lambda}^{\cdot\varkappa}$$

and of the second Bianchi identity [2]:

$$\nabla_{[\nu} R_{\rho\mu]\lambda}^{\cdot\cdot\cdot\varkappa} = 2S_{[\nu\rho}^{\cdot\cdot\sigma} R_{\mu]\sigma\lambda}^{\cdot\cdot\cdot\varkappa}$$

we find

$$\nabla_{\nu}\underset{v}{\pounds}\Gamma_{\mu\lambda}^{\varkappa} - \nabla_{\mu}\underset{v}{\pounds}\Gamma_{\nu\lambda}^{\varkappa} + 2S_{\nu\mu}^{\cdot\cdot\rho}\underset{v}{\pounds}\Gamma_{\rho\lambda}^{\varkappa}$$

$$= v^{\rho}\nabla_{\rho} R_{\nu\mu\lambda}^{\cdot\cdot\cdot\varkappa} - R_{\nu\mu\lambda}^{\cdot\cdot\cdot\rho} v_{\rho}^{\cdot\varkappa} + R_{\rho\mu\lambda}^{\cdot\cdot\cdot\varkappa} v_{\nu}^{\cdot\rho} + R_{\nu\rho\lambda}^{\cdot\cdot\cdot\varkappa} v_{\mu}^{\cdot\rho} + R_{\nu\mu\rho}^{\cdot\cdot\cdot\varkappa} v_{\lambda}^{\cdot\rho}$$

or

$$\boxed{\nabla_{\nu}\underset{v}{\pounds}\Gamma_{\mu\lambda}^{\varkappa} - \nabla_{\mu}\underset{v}{\pounds}\Gamma_{\nu\lambda}^{\varkappa} + 2S_{\nu\mu}^{\cdot\cdot\rho}\underset{v}{\pounds}\Gamma_{\rho\lambda}^{\varkappa} = \underset{v}{\pounds} R_{\nu\mu\lambda}^{\cdot\cdot\cdot\varkappa}.} \tag{4.14}$$

The equations (4.13) and (4.14) prove (4.12) [3].

---

[1] SCHOUTEN [8], p. 139.
[2] SCHOUTEN [8] p. 146.
[3] DAVIES [1].

CHAPTER II

LIE DERIVATIVES OF GENERAL GEOMETRIC OBJECTS

§ 1. Geometric objects.

Consider an n-dimensional space X_n of class C^u. An object which has the following properties is called a *geometric object* of class p $(\leq u)$.[1]

(i) In each coordinate system $(\varkappa)$, it has a well determined set of N components $\Omega^\Lambda(\xi)$, where capital Greek indices Λ, Σ, Π run over the range 1, 2, ..., N.

(ii) When we effect a coordinate transformation

$$\xi^{\varkappa'} = f^\varkappa(\xi^1, \xi^2, \ldots, \xi^n), \tag{1.1}$$

the new components $\Omega^{\Lambda'}(\xi)$ of the object with respect to the new coordinate system $(\varkappa')$ can be represented as well determined functions of class $u - p$ of the old components $\Omega^\Lambda(\xi)$, of the old coordinates $\xi^\varkappa$, of the functions $f^\varkappa$ and of their s-th partial derivatives $(1 \leq s \leq p \leq u)$, that is, the new components $\Omega^{\Lambda'}(\xi)$ of the object can be represented by equations of the form

$$\Omega^{\Lambda'} = F^\Lambda(\Omega^\Sigma, \xi^\varkappa, f^\varkappa, \partial_\lambda f^\varkappa, \ldots, \partial_{\lambda_p \ldots \lambda_1} f^\varkappa), \tag{1.2}$$

where

$$\partial_{\lambda_p \ldots \lambda_1} f^\varkappa \stackrel{\text{def}}{=} \partial_{\lambda_p} \ldots \partial_{\lambda_1} f^\varkappa.$$

For the sake of simplicity we sometimes denote the right-hand side of (1.2) by $F^\Lambda(\Omega, \xi^\nu, \xi^{\nu'})$.

(iii) The functions $F^\Lambda(\Omega, \xi^\nu, \xi^{\nu'})$ have the group properties, that is, they satisfy the following relations:

$$(1.3) \quad \begin{cases} \text{(a)} \quad F^\Lambda(F(\Omega, \xi^\nu, \xi^{\nu'}), \xi^{\nu'}, \xi^{\nu''}) = F^\Lambda(\Omega, \xi^\nu, \xi^{\nu''}), \\ \text{(b)} \quad F^\Lambda(F(\Omega, \xi^\nu, \xi^{\nu'}), \xi^{\nu'}, \xi^\nu) = \Omega^\Lambda. \\ \qquad \text{Combining these two we have} \\ \text{(c)} \quad F^\Lambda(\Omega, \xi^\nu, \xi^\nu) = \Omega^\Lambda. \end{cases}$$

---

[1] Cf. SCHOUTEN and HAANTJES [2]; GOLAB [1]; NIJENHUIS [2], Ch. I, § 6, p. 26; TASHIRO [1, 2], YANO and TASHIRO [1].

When the functions $F^{\Lambda}(\Omega, \xi^{\nu}, \xi^{\nu'})$ contain only Ω^{Σ} and the partial derivatives of the functions $f^{\varkappa}$ with respect to $\xi^{\varkappa}$ but not $\xi^{\varkappa}$ and $f^{\varkappa}$, the geometric object is said to be *differential.* [1]

When the functions $F^{\Lambda}(\Omega, \xi^{\nu}, \xi^{\nu'})$ are of the form

$$F^{\Lambda}(\Omega, \xi^{\nu}, \xi^{\nu'}) = F^{\Lambda}_{\Sigma}(\xi^{\nu}, \xi^{\nu'})\Omega^{\Sigma}, \tag{1.4}$$

the geometric object is said to be *linear homogeneous* and when the functions $F^{\Lambda}(\Omega, \xi^{\nu}, \xi^{\nu'})$ are of the form

$$F^{\Lambda}(\Omega, \xi^{\nu}, \xi^{\nu'}) = F^{\Lambda}_{\Sigma}(\xi^{\nu}, \xi^{\nu'})\Omega^{\Sigma} + G^{\Lambda}(\xi^{\nu}, \xi^{\nu'}), \tag{1.5}$$

the geometric object is said to be *linear*.

A tensor is a differential linear homogeneous object and a linear connexion is a differential linear object.

When the components $\Psi^{\cdot\Gamma}$ $(\Gamma = 1, 2, \ldots, M)$ of a geometric object are functions of another geometric object Ω^{Λ}:

$$\Psi^{\cdot\Gamma} = \Psi^{\cdot\Gamma}(\Omega) \tag{1.6}$$

and the functional forms of $\Psi^{\cdot\Gamma}(\Omega)$ do not depend on the choice of coordinate systems, $\Psi^{\cdot\Gamma}$ is said to be *a function* of the geometric object Ω^{Λ}.

§ 2. The Lie derivative of a geometric object.

Suppose that there is given a field of a geometric object $\Omega^{\Lambda}(\xi)$ with the transformation law $\Omega^{\Lambda'} = F^{\Lambda}(\Omega, \xi^{\nu}, \xi^{\nu'})$ in X_n and consider an infinitesimal point transformation

$$'\xi^{\varkappa} = \xi^{\varkappa} + v^{\varkappa}dt. \tag{2.1}$$

We define a new field of a geometric object $'\Omega^{\Lambda}$ of the same type as Ω^{Λ} as a field whose components $'\Omega^{\Lambda'}(\xi)$ at ξ with respect to the coordinate system

$$\xi^{\varkappa'} = {}'\xi^{\varkappa} = \xi^{\varkappa} + v^{\varkappa}dt \tag{2.2}$$

are equal to the $\Omega^{\Lambda}('\xi)$ at the corresponding point $'\xi$, that is

$$'\Omega^{\Lambda'}(\xi) \overset{\text{def}}{=} \Omega^{\Lambda}('\xi). \tag{2.3}$$

Because $'\Omega^{\Lambda}$ is an object of the same type as Ω^{Λ} we have for the relation between $'\Omega^{\Lambda'}(\xi)$ and $'\Omega^{\Lambda}(\xi)$

$$'\Omega^{\Lambda'}(\xi) = F^{\Lambda}('\Omega(\xi), \xi^{\nu}, \xi^{\nu'}). \tag{2.4}$$

Now the *Lie differential* $\underset{v}{\pounds}\,\Omega^{\Lambda}dt$ and the *Lie derivative* $\underset{v}{\pounds}\,\Omega^{\Lambda}$ of the

---

[1] GOLAB [1]; NIJENHUIS [2] p. 37.

geometric object Ω^Λ with respect to (2.1) are defined by

$$\underset{v}{\pounds}\,\Omega^\Lambda dt = {}'\Omega^\Lambda(\xi) - \Omega^\Lambda(\xi). \tag{2.5}$$

From (2.3) and (2.4), we have

$$\Omega^\Lambda('\xi) = F^\Lambda('\Omega(\xi), \xi^\nu, \xi^{\nu'}). \tag{2.6}$$

On the other hand, from

$$f^\varkappa(\xi) = \xi^\varkappa + v^\varkappa dt,$$

we find

$$\begin{aligned}
\partial_\lambda f^\varkappa &= \delta^\varkappa_\lambda + \partial_\lambda v^\varkappa dt,\\
\partial_{\lambda_2\lambda_1} f^\varkappa &= \partial_{\lambda_2\lambda_1} v^\varkappa dt,\\
&\cdots\cdots\cdots\cdots\\
\partial_{\lambda_p\ldots\lambda_1} f^\varkappa &= \partial_{\lambda_p\ldots\lambda_1} v^\varkappa dt,
\end{aligned}$$

and consequently, substituting these equations in (2.6) and neglecting all differentials of higher order in dt, we find

$$\Omega^\Lambda(\xi) + v^\rho\partial_\rho\Omega^\Lambda dt = {}'\Omega^\Lambda(\xi) + \Sigma^p_{s=0} F^{\lambda(s)\Lambda}_\varkappa(\Omega, \xi)\partial_{\lambda(s)} v^\varkappa dt, \tag{2.7}$$

where

$$\left\{\begin{array}{ll}
\partial_{\lambda(0)} v^\varkappa \overset{\text{def}}{=} v^\varkappa, & \partial_{\lambda(s)} v^\varkappa \overset{\text{def}}{=} \partial_{\lambda_s\ldots\lambda_1} v^\varkappa,\\
F^{\lambda(0)\Lambda}_\varkappa(\Omega, \xi) \overset{\text{def}}{=} \left[\dfrac{\partial F^\Lambda}{\partial f^\varkappa}\right]_\xi, & F^{\lambda(s)\Lambda}_\varkappa(\Omega, \xi) \overset{\text{def}}{=} \left[\dfrac{\partial F^\Lambda}{\partial(\partial_{\lambda_s\ldots\lambda_1} f^\varkappa)}\right]_\xi,
\end{array}\right. \tag{2.8}$$

$[\;]_\xi$ denoting the evaluation of the expression in parentheses at

$$f^\varkappa = \xi^\varkappa,\ \partial_\lambda f^\varkappa = \delta^\varkappa_\lambda,\ \partial_{\lambda_2\lambda_1} f^\varkappa = 0,\ \ldots,\ \partial_{\lambda_p\ldots\lambda_1} f^\varkappa = 0.$$

Thus from (2.5) and (2.7), we obtain

$$\boxed{\underset{v}{\pounds}\,\Omega^\Lambda = v^\rho\partial_\rho\Omega^\Lambda - \Sigma^p_{s=0} F^{\lambda(s)\Lambda}_\varkappa(\Omega, \xi)\partial_{\lambda(s)} v^\varkappa.} \tag{2.9}$$

It is to be noticed that the functions $F^{\lambda(s)\Lambda}_\varkappa(\Omega, \xi)$ depend only on the Ω^Λ and the $\xi^\varkappa$. If the object is differential, then $F^{\lambda(0)\Lambda}_\varkappa = 0$ and the $F^{\lambda(s)\Lambda}_\varkappa$ depend only on the Ω^Λ.

Now consider a general coordinate transformation $(\varkappa) \to (\varkappa')$, then the components $'\Omega^\Lambda$ are transformed into

$$'\Omega^{\Lambda'} = F^\Lambda('\Omega, \xi^\nu, \xi^{\nu'})$$

and the Ω^Λ into

$$\Omega^{\Lambda'} = F^\Lambda(\Omega, \xi^\nu, \xi^{\nu'}),$$

from which we have

$$\boxed{\underset{v}{\pounds}\,\Omega^{\Lambda'} = \frac{\partial F^{\Lambda}}{\partial \Omega^{\Pi}} \underset{v}{\pounds}\,\Omega^{\Pi}.} \tag{2.10}$$

This equation gives the transformation law of the Lie derivative $\underset{v}{\pounds}\,\Omega^{\Lambda}$ during a coordinate transformation $(\varkappa) \to (\varkappa')$. Since the partial derivatives $\partial F^{\Lambda}/\partial \Omega^{\Pi}$ contain in general Ω^{Λ}, the Lie derivative of a general geometric object is not necessarily a geometric object.

The Lie derivative of a geometric object is a geometric object if and only if the partial derivatives $\partial F^{\Lambda}/\partial \Omega^{\Pi}$ do not contain Ω^{Λ}, that is, if and only if the $F^{\Lambda}(\Omega, \xi^{\nu}, \xi^{\nu'})$ have the form

$$F^{\Lambda}(\Omega, \xi^{\nu}, \xi^{\nu'}) = F^{\Lambda}_{\Pi}(\xi^{\nu}, \xi^{\nu'})\Omega^{\Pi} + G^{\Lambda}(\xi^{\nu}, \xi^{\nu'}). \tag{2.11}$$

Thus we have

THEOREM 2.1. *In order that the Lie derivative of a geometric object be again a geometric object, it is necessary and sufficient that the geometric object be linear.*

If Ω^{Λ} is a linear geometric object whose transformation law is given by (2.11), then the Lie derivative of Ω^{Λ} is given by

$$\underset{v}{\pounds}\,\Omega^{\Lambda} = v^{\rho}\,\partial_{\rho}\,\Omega^{\Lambda} - \Sigma_{s=0}^{p}\,(F^{\lambda(s)\Lambda}_{\varkappa\ \ \Sigma}(\xi)\Omega^{\Sigma} + G^{\lambda(s)\Lambda}_{\varkappa}(\xi))\partial_{\lambda_{(s)}}\,v^{\varkappa}, \tag{2.12}$$

where

$$F^{\lambda(s)\Lambda}_{\varkappa\ \ \Sigma}(\xi) \overset{\text{def}}{=} \left[\frac{\partial F^{\Lambda}_{\Sigma}}{\partial(\partial_{\lambda_{(s)}} f^{\varkappa})}\right]_{\xi}, \qquad G^{\lambda(s)\Lambda}_{\varkappa}(\xi) = \left[\frac{\partial G^{\Lambda}}{\partial(\partial_{\lambda_{(s)}} f^{\varkappa})}\right]_{\xi}. \tag{2.13}$$

Similarly if Φ^{Λ} is a linear homogeneous geometric object whose transformation law is

$$\Phi^{\Lambda'} = F^{\Lambda}_{\Sigma}(\xi^{\nu}, \xi^{\nu'})\Phi^{\Sigma}, \tag{2.14}$$

then we have

$$\underset{v}{\pounds}\,\Phi^{\Lambda} = v^{\rho}\,\partial_{\rho}\Phi^{\Lambda} - \Sigma_{s=0}^{p}\,F^{\lambda(s)\Lambda}_{\varkappa\ \ \Sigma}(\xi)\Phi^{\Sigma}\,\partial_{\lambda_{(s)}}\,v^{\varkappa}. \tag{2.15}$$

From (2.12) and (2.15), we find

$$\underset{v}{\pounds}\,(\Omega^{\Lambda} \pm \Phi^{\Lambda}) = \underset{v}{\pounds}\,\Omega^{\Lambda} \pm \underset{v}{\pounds}\,\Phi^{\Lambda}. \tag{2.16}$$

This formula is valid also when the Ω^{Λ} and Φ^{Λ} are both linear homogeneous and have the same transformation law.

If the Ω^{Λ} and Ψ^{Λ} are both linear geometric objects having the same transformation law, we see from (2.12) that

$$\mathop{\pounds}_{v}(\Omega^\Lambda - \Psi^\Lambda) = \mathop{\pounds}_{v}\Omega^\Lambda - \mathop{\pounds}_{v}\Psi^\Lambda. \tag{2.17}$$

If the product of two geometric objects Ω and Φ is again a geometric object (this happens, for example, if two geometric objects are both linear homogeneous), then

$$\begin{aligned}'(\Omega\Phi) = {}'\Omega\, {}'\Phi &= (\Omega + \mathop{\pounds}_{v}\Omega dt)(.\Phi + \mathop{\pounds}_{v}\Phi dt) \\ &= \Omega\Phi + (\mathop{\pounds}_{v}\Omega)\Phi dt + \Omega(\mathop{\pounds}_{v}\Phi)dt,\end{aligned}$$

from which

$$\mathop{\pounds}_{v}(\Omega \cdot \Phi) = (\mathop{\pounds}_{v}\Omega)\Phi + \Omega(\mathop{\pounds}_{v}\Phi). \tag{2.18}$$

§ 3. Miscellaneous examples of Lie derivatives.

In Ch. I we saw already some examples of Lie derivatives of geometric objects. In this section we shall give some other examples.

We take an arbitrary field $\Omega^\Lambda(\xi)$ of a geometric object whose transformation law under a coordinate transformation $(\varkappa) \to (\varkappa')$ is

$$\Omega^\Lambda = F^\Lambda(\Omega^{\Sigma'}, \xi^{\nu'}, \xi^\nu). \tag{3.1}$$

Then the law of transformation of $\partial_\mu\Omega^\Lambda$ is given by

$$\partial_\mu\Omega^\Lambda = F^\Lambda_\mu(\Omega^{\Sigma'}, \partial_{\mu'}\Omega^{\Sigma'}, \xi^{\nu'}, \xi^\nu), \tag{3.2}$$

where the functions F^Λ_μ are obtained from F^Λ by partial differentiation with respect to ξ^μ. Thus $\partial_\mu\Omega^\Lambda$ are not components of a geometric object but $(\Omega^\Sigma, \partial_\mu\Omega^\Sigma)$ are components of a geometric object whose transformation law is given by (3.1) and (3.2). Since we have

$$\begin{cases} '\Omega^\Lambda = F^\Lambda(\Omega^\Sigma('\xi), \xi^{\nu'}, \xi^\nu), \\ \partial_\mu{}'\Omega^\Lambda = F^\Lambda_\mu(\Omega^\Sigma('\xi), \partial_\mu\Omega^\Sigma('\xi), \xi^{\nu'}, \xi^\nu) \end{cases} \tag{3.3}$$

we can see that

$$\mathop{\pounds}_{v}(\Omega^\Lambda, \partial_\mu\Omega^\Lambda) = (\mathop{\pounds}_{v}\Omega^\Lambda, \partial_\mu\mathop{\pounds}_{v}\Omega^\Lambda). \tag{3.4}$$

On the other hand, (3.3) can be written as

$$\begin{cases} '\Omega^\Lambda = F^\Lambda(\Omega^\Sigma('\xi), \xi^{\nu'}, \xi^\nu), \\ '(\partial_\mu\Omega^\Lambda) = F^\Lambda_\mu(\Omega^\Sigma('\xi), \partial_\mu\Omega^\Sigma('\xi), \xi^{\nu'}, \xi^\nu) \end{cases} \tag{3.5}$$

and consequently we have

$$\mathop{\pounds}_{v}(\Omega^\Lambda, \partial_\mu\Omega^\Lambda) = (\mathop{\pounds}_{v}\Omega^\Lambda, \mathop{\pounds}_{v}\partial_\mu\Omega^\Lambda). \tag{3.6}$$

Comparing (3.4) and (3.6), we have [1]

$$\underset{v}{\pounds}\, \partial_\mu \Omega^\Lambda = \partial_\mu \underset{v}{\pounds}\, \Omega^\Lambda. \tag{3.7}$$

Applying this formula, we can easily prove that [2]

$$\underset{v}{\pounds}\, \partial_{[\mu} w_{\lambda_1 \ldots \lambda_p]} = \partial_{[\mu} \underset{v}{\pounds}\, w_{\lambda_1 \ldots \lambda_p]} \tag{3.8}$$

$$\underset{v}{\pounds}\, \partial_\mu \mathfrak{w}^{\mu\varkappa_2 \ldots \varkappa_p} = \partial_\mu \underset{v}{\pounds}\, \mathfrak{w}^{\mu\varkappa_2 \ldots \varkappa_p}, \tag{3.9}$$

where $w^{\lambda_1 \ldots \lambda_p}$ is a p-vector and $\mathfrak{w}^{\mu\varkappa_2 \ldots \varkappa_p}$ a p-vector density.

We introduce now an anholonomic coordinate system (h) [3] defined by the fields $\underset{i}{e}^\varkappa$ and $\overset{h}{e}_\lambda$; $h, i, j, \ldots = 1, 2, \ldots, n$, and denote by $A_i^\varkappa$ and A_λ^h the intermediate components of the unit tensor:

$$A_i^\varkappa \overset{\text{def}}{=} \overset{h}{e}_i\, \underset{h}{e}^\varkappa \overset{*}{=} \underset{i}{e}^\varkappa; \qquad A_\lambda^h \overset{\text{def}}{=} e^h\, \overset{i}{\underset{i}{e}}_\lambda \overset{*}{=} \overset{h}{e}_\lambda. \tag{3.10}$$

The object of anholonomity is given by [4]

$$\Omega_{ji}^{\cdot\cdot h} \overset{\text{def}}{=} A_{ji}^{\mu\lambda}\, \partial_{[\mu} A_{\lambda]}^h. \tag{3.11}$$

Then from (3.10), (3.16) and (3.28) of Ch. I, we find

$$A_\varkappa^h(\underset{v}{\pounds}\, u^\varkappa) = v^j \partial_j u^h - u^j(\partial_j v^h - 2v^i \Omega_{ji}^{\cdot\cdot h}), \tag{3.12}$$

$$A_i^\lambda(\underset{v}{\pounds}\, w_\lambda) = v^j \partial_j w_i + w_h (\partial_i v^h - 2v^i \Omega_{ji}^{\cdot\cdot h}), \tag{3.13}$$

$$\begin{aligned} A_{\varkappa\lambda j}^{hi\mu}(\underset{v}{\pounds}\, \mathfrak{P}^{\varkappa\lambda}_{\cdot\cdot\mu}) = {} & v^k \partial_k \mathfrak{P}^{hi}_{\cdot\cdot j} - \mathfrak{P}^{ki}_{\cdot\cdot j}(\partial_k v^h - 2v^l \Omega_{kl}^{\cdot\cdot h}) \\ & - \mathfrak{P}^{hk}_{\cdot\cdot j}(\partial_k v^i - 2v^l \Omega_{kl}^{\cdot\cdot i}) + \mathfrak{P}^{hi}_{\cdot\cdot k}(\partial_j v^k - 2v^l \Omega_{jl}^{\cdot\cdot k}) \\ & + w\, \mathfrak{P}^{hi}_{\cdot\cdot j}(\partial_k v^k - 2v^l \Omega_{kl}^{\cdot\cdot k}) \end{aligned} \tag{3.14}$$

respectively, where u^h, w_i and $\mathfrak{P}^{hi}_{\cdot\cdot j}$ are respectively components of $u^\varkappa$, w_λ and $\mathfrak{P}^{\varkappa\lambda}_{\cdot\cdot\mu}$ with respect to the anholonomic coordinate system (h) and where $\partial_j = A_j^\varkappa \partial_\varkappa$.

From these equations we obtain the following formulas for the Lie

---

[1] Nijenhuis [2], p. 25; Schouten [8] p. 105,.

[2] Schouten [8], p. 110.

[3] Schouten [8] p. 99.

[4] Schouten [8] p. 100.

derivatives in anholonomic coordinate systems. [1]

$$\underset{v}{\pounds}\, u^h = v^j \partial_j u^h - u^j(\partial_j v^h - 2v^i \Omega_{ji}^h), \tag{3.15}$$

$$\underset{v}{\pounds}\, w_i = v^j \partial_j w_i + w_h(\partial_i v^h - 2v^j \Omega_{ji}^h), \tag{3.16}$$

$$\begin{aligned}\underset{v}{\pounds}\, \mathfrak{P}^{hi}_{..j} = v^k \partial_k \mathfrak{P}^{hi}_{..j} - \mathfrak{P}^{ki}_{..j}(\partial_k v^h - 2v^l \Omega_{kl}^h) - \mathfrak{P}^{hk}_{..j}(\partial_k v^i - 2v^l \Omega_{kl}^i) \\ + \mathfrak{P}^{hi}_{..k}(\partial_j v^k - 2v^l \Omega_{jl}^k) + w\, \mathfrak{P}^{hi}_{..j}(\partial_k v^k - 2v^l \Omega_{kl}^k)\end{aligned} \tag{3.17}$$

and

$$\underset{v}{\pounds}\, A_i^{\varkappa} = v^j \partial_j A_i^{\varkappa} - A_i^{\lambda} \partial_{\lambda} v^{\varkappa} + A_h^{\varkappa}(\partial_i v^h - 2v^l \Omega_{il}^h) = 0, \tag{3.18}$$

$$\underset{v}{\pounds}\, A_{\lambda}^h = v^j \partial_j A_{\lambda}^h + A_{\varkappa}^h \partial_{\lambda} v^{\varkappa} - A_{\lambda}^j(\partial_j v^h - 2v^l \Omega_{jl}^h) = 0. \tag{3.19}$$

Using (3.7) and (3.19), we find

$$\underset{v}{\pounds}\, \Omega_{ji}^h = 0. \tag{3.20}$$

Applying the formula (3.15), we easily get

$$\underset{j}{\pounds}\, \underset{i}{e}{}^h \overset{*}{=} -2\Omega_{ji}^h \tag{3.21}$$

from which

$$\underset{j}{\pounds}\, \underset{i}{e}{}^{\varkappa} \overset{*}{=} -2\Omega_{ji}^h A_h^{\varkappa}, \tag{3.22}$$

where $\underset{j}{\pounds}$ denotes the Lie derivative with respect to $\underset{j}{e}{}^{\varkappa}$.

§ 4. Some general formulas.

In this section we shall consider a linear geometric object Ω^{Λ}, that is, a geometric object whose transformation law is given by

$$\Omega^{\Lambda'} = F_{\Pi}^{\Lambda}(\xi^{\nu}, \xi^{\nu'})\Omega^{\Pi} + G^{\Lambda}(\xi^{\nu}, \xi^{\nu'}). \tag{4.1}$$

Since Ω^{Λ} is a geometric object, the functions $F_{\Pi}^{\Lambda}(\xi, \xi')$ and $G^{\Lambda}(\xi, \xi')$ appearing in formula (4.1) must satisfy the relations

$$\begin{cases} F_{\Pi}^{\Lambda}(\xi^{\nu'}, \xi^{\nu''})F_{\Sigma}^{\Pi}(\xi^{\nu}, \xi^{\nu'}) = F_{\Sigma}^{\Lambda}(\xi^{\nu}, \xi^{\nu''}), \\ F_{\Pi}^{\Lambda}(\xi^{\nu'}, \xi^{\nu''})G^{\Pi}(\xi^{\nu}, \xi^{\nu'}) + G^{\Lambda}(\xi^{\nu'}, \xi^{\nu''}) = G^{\Lambda}(\xi^{\nu}, \xi^{\nu''}) \end{cases} \tag{4.2}$$

and

$$F_{\Pi}^{\Lambda}(\xi^{\nu}, \xi^{\nu}) = \delta_{\Pi}^{\Lambda}, \qquad G^{\Lambda}(\xi^{\nu}, \xi^{\nu}) = 0. \tag{4.3}$$

We consider two infinitesimal point transformations

$$'\xi^{\varkappa} = \xi^{\varkappa} + \underset{1}{v}{}^{\varkappa}(\xi)dt, \qquad ''\xi^{\varkappa} = {}'\xi^{\varkappa} + \underset{2}{v}{}^{\varkappa}('\xi)du. \tag{4.4}$$

---

[1] Nijenhuis [2] p. 106; Schouten [8] p. 110.

For the coordinate transformations

$$\xi^{\kappa'} = \xi^{\kappa} + \underset{1}{v}^{\kappa}(\xi^{\nu})dt, \qquad \xi^{\kappa''} = \xi^{\kappa'} + \underset{2}{v}^{\kappa}(\xi^{\nu'})du \tag{4.5}$$

we have

$$\begin{aligned} &F^{\Lambda}_{\Pi}(\xi^{\nu'}, \xi^{\nu''})F^{\Pi}_{\Sigma}(\xi^{\nu}, \xi^{\nu'}) \\ &= [\delta^{\Lambda}_{\Pi} + \Sigma^{p}_{s=0}\,(F^{\kappa(s)}_{\rho}{}^{\Lambda}_{\Pi}\,\partial_{\kappa(s)}\underset{2}{v}^{\rho} + \underset{1}{v}^{\sigma}\,\partial_{\sigma}(F^{\kappa(s)}_{\rho}{}^{\Lambda}_{\Pi}\,\partial_{\kappa(s)}\underset{2}{v}^{\rho})du)dt \\ &+ \tfrac{1}{2}\Sigma^{p}_{s,t=0}\,F^{\lambda(t)}_{\sigma}{}^{\kappa(s)}_{\rho}{}^{\Lambda}_{\Pi}(\partial_{\lambda(t)}\underset{2}{v}^{\sigma})(\partial_{\kappa(s)}\underset{2}{v}^{\rho})du^{2}] \times \\ &\times [\delta^{\Pi}_{\Sigma} + \Sigma^{p}_{s=0}\,F^{\kappa(s)}_{\rho}{}^{\Pi}_{\Sigma}\,\partial_{\kappa(s)}\underset{1}{v}^{\rho}\,dt + \tfrac{1}{2}\Sigma^{p}_{s,t=0}\,F^{\lambda(t)}_{\sigma}{}^{\kappa(s)}_{\rho}{}^{\Pi}_{\Sigma}\,(\partial_{\lambda(t)}\underset{1}{v}^{\sigma})(\partial_{\kappa(s)}\underset{1}{v}^{\rho})dt^{2}] \\ &= \delta^{\Lambda}_{\Sigma} + \Sigma^{p}_{s=0}\,F^{\kappa(s)}_{\rho}{}^{\Lambda}_{\Sigma}\,\partial_{\kappa(s)}\underset{1}{v}^{\rho}\,dt + \Sigma^{p}_{s=0}\,F^{\kappa(s)}_{\rho}{}^{\Lambda}_{\Sigma}\,\partial_{\kappa(s)}\underset{2}{v}^{\rho}\,du \\ &+ \tfrac{1}{2}\Sigma^{p}_{s,t=0}\,F^{\lambda(t)}_{\sigma}{}^{\kappa(s)}_{\rho}{}^{\Lambda}_{\Sigma}(\partial_{\lambda(t)}\underset{1}{v}^{\sigma})(\partial_{\kappa(s)}\underset{1}{v}^{\rho})dt^{2} \\ &+ [\Sigma^{p}_{s,t=0}\,F^{\lambda(t)}_{\sigma}{}^{\Lambda}_{\Pi}\,F^{\kappa(s)}_{\rho}{}^{\Pi}_{\Sigma}(\partial_{\lambda(t)}\underset{2}{v}^{\sigma})(\partial_{\kappa(s)}\underset{1}{v}^{\rho}) + \Sigma^{p}_{s=0}\,\underset{1}{v}^{\sigma}\,\partial_{\sigma}(F^{\kappa(s)}_{\rho}{}^{\Lambda}_{\Sigma}\,\partial_{\kappa(s)}\underset{2}{v}^{\rho})]dtdu \\ &+ \tfrac{1}{2}\Sigma^{p}_{s,t=0}\,F^{\lambda(t)}_{\sigma}{}^{\kappa(s)}_{\rho}{}^{\Lambda}_{\Sigma}(\partial_{\lambda(t)}\underset{2}{v}^{\sigma})(\partial_{\kappa(s)}\underset{2}{v}^{\rho})du^{2} \end{aligned} \tag{4.6}$$

where

$$F^{\kappa(0)}_{\rho}{}^{\Lambda}_{\Pi} \overset{\text{def}}{=} \left[\frac{\partial F^{\Lambda}_{\Pi}}{\partial f^{\rho}}\right]_{\xi}, \qquad F^{\lambda(t)}_{\sigma}{}^{\kappa(s)}_{\rho}{}^{\Lambda}_{\Pi} \overset{\text{def}}{=} \left[\frac{\partial^{2} F^{\Lambda}_{\Pi}}{\partial(\partial_{\lambda(t)} f^{\sigma})\partial(\partial_{\kappa(s)} f^{\rho})}\right]_{\xi}. \tag{4.7}$$

On the other hand, on taking account of

$$\xi^{\kappa''} = \xi^{\kappa} + \underset{1}{v}^{\kappa}(\xi)dt + \underset{2}{v}^{\kappa}(\xi)du + \underset{1}{v}^{\rho}\,\partial_{\rho}\underset{2}{v}^{\kappa}dtdu,$$

we find

$$\begin{aligned} &F^{\Lambda}_{\Sigma}\,(\xi^{\nu}, \xi^{\nu''}) \\ &= \delta^{\Lambda}_{\Sigma} + \Sigma^{p}_{s=0}\,F^{\kappa(s)}_{\rho}{}^{\Lambda}_{\Sigma}[\partial_{\kappa(s)}\underset{1}{v}^{\rho}\,dt + \partial_{\kappa(s)}\underset{2}{v}^{\rho}\,du + \partial_{\kappa(s)}(\underset{1}{v}^{\sigma}\,\partial_{\sigma}\underset{2}{v}^{\rho})dtdu] \\ &+ \tfrac{1}{2}\Sigma^{p}_{s,t=0}\,F^{\lambda(t)}_{\sigma}{}^{\kappa(s)}_{\rho}{}^{\Lambda}_{\Sigma}(\partial_{\lambda(t)}\underset{1}{v}^{\sigma}dt + \partial_{\lambda(t)}\underset{2}{v}^{\sigma}\,du)(\partial_{\kappa(s)}\underset{1}{v}^{\rho}\,dt + \partial_{\kappa(s)}\underset{2}{v}^{\rho}\,du) \\ &= \delta^{\Lambda}_{\Sigma} + \Sigma^{p}_{s=0}\,F^{\kappa(s)}_{\rho}{}^{\Lambda}_{\Sigma}\,\partial_{\kappa(s)}\underset{1}{v}^{\rho}\,dt + \Sigma^{p}_{s=0}\,F^{\kappa(s)}_{\rho}{}^{\Lambda}_{\Sigma}\,\partial_{\kappa(s)}\underset{2}{v}^{\rho}\,du \\ &+ \tfrac{1}{2}\Sigma^{p}_{s,t=0}\,F^{\lambda(t)}_{\sigma}{}^{\kappa(s)}_{\rho}{}^{\Lambda}_{\Sigma}(\partial_{\lambda(t)}\underset{1}{v}^{\sigma})(\partial_{\kappa(s)}\underset{1}{v}^{\rho})dt^{2} \\ &+ [\Sigma^{p}_{s=0}\,F^{\kappa(s)}_{\rho}{}^{\Lambda}_{\Sigma}\,\partial_{\kappa(s)}(\underset{1}{v}^{\sigma}\,\partial_{\sigma}\underset{2}{v}^{\rho}) + \Sigma^{p}_{s,t=0}\,F^{\lambda(t)}_{\sigma}{}^{\kappa(s)}_{\rho}{}^{\Lambda}_{\Sigma}(\partial_{\lambda(t)}\underset{2}{v}^{\sigma})(\partial_{\kappa(s)}\underset{1}{v}^{\rho})]dtdu \\ &+ \tfrac{1}{2}\Sigma^{p}_{s,t=0}\,F^{\lambda(t)}_{\sigma}{}^{\kappa(s)}_{\rho}{}^{\Lambda}_{\Sigma}(\partial_{\lambda(t)}\underset{2}{v}^{\sigma})(\partial_{\kappa(s)}\underset{2}{v}^{\rho})du^{2}. \end{aligned} \tag{4.8}$$

Consequently, on comparing (4.6) and (4.8), we find

$$
\begin{aligned}
(4.9)\quad & \Sigma_{s,t=0}^{p} F_{\sigma}^{\lambda(t)\Lambda}{}_{\Pi} F_{\rho}^{\varkappa(s)\Pi}{}_{\Sigma} (\partial_{\lambda(t)} \underset{2}{v}^{\sigma})(\partial_{\varkappa(s)} \underset{1}{v}^{\rho}) + \Sigma_{s=0}^{p} \underset{1}{v}^{\sigma} \partial_{\sigma}(F_{\rho}^{\varkappa(s)\Lambda}{}_{\Sigma} \partial_{\varkappa(s)} \underset{2}{v}^{\rho}) \\
& = \Sigma_{s=0}^{p} F_{\rho}^{\varkappa(s)\Lambda}{}_{\Sigma} \partial_{\varkappa(s)}(\underset{1}{v}^{\sigma} \partial_{\sigma} \underset{2}{v}^{\rho}) + \Sigma_{s,t=0}^{p} F_{\sigma}^{\lambda(t)}{}_{\rho}^{\varkappa(s)\Lambda}{}_{\Sigma} (\partial_{\lambda(t)} \underset{2}{v}^{\rho})(\partial_{\varkappa(s)} \underset{1}{v}^{\rho}).
\end{aligned}
$$

Thus, if we put

$$
(4.10)\quad \begin{cases}
\{F, v\}_{\Pi}^{\Lambda} \stackrel{\text{def}}{=} \Sigma_{s=0}^{p} F_{\rho}^{\varkappa(s)\Lambda}{}_{\Pi} \partial_{\varkappa(s)} v^{\rho} \\
\{F, \underset{2}{v}, \underset{1}{v}\}_{\Pi}^{\Lambda} \stackrel{\text{def}}{=} \Sigma_{s,t=0}^{p} F_{\sigma}^{\lambda(t)}{}_{\rho}^{\varkappa(s)\Lambda}{}_{\Pi} (\partial_{\lambda(t)} \underset{2}{v}^{\sigma})(\partial_{\varkappa(s)} \underset{1}{v}^{\rho})
\end{cases}
$$

we obtain

$$
(4.11)\quad \{F, \underset{2}{v}\}_{\Pi}^{\Lambda} \{F, \underset{1}{v}\}_{\Sigma}^{\Pi} + \underset{1}{v}^{\rho} \partial_{\rho} \{F, \underset{2}{v}\}_{\Sigma}^{\Lambda} = \{F, \underset{1}{v}^{\rho} \partial_{\rho} \underset{2}{v}\}_{\Sigma}^{\Lambda} + \{F, \underset{2}{v}, \underset{1}{v}\}_{\Sigma}^{\Lambda}.
$$

Similarly we have

$$
\begin{aligned}
(4.12)\quad & F_{\Pi}^{\Lambda}(\xi^{\nu'}, \xi^{\nu''}) G^{\Pi}(\xi^{\nu}, \xi^{\nu'}) + G^{\Lambda}(\xi^{\nu'}, \xi^{\nu''}) \\
& = [\delta_{\Pi}^{\Lambda} + \Sigma_{s=0}^{p} (F_{\rho}^{\varkappa(s)\Lambda}{}_{\Pi} \partial_{\varkappa(s)} \underset{2}{v}^{\rho} + \underset{1}{v}^{\sigma} \partial_{\sigma}(F_{\rho}^{\varkappa(s)\Lambda}{}_{\Pi} \partial_{\varkappa(s)} \underset{2}{v}^{\rho}) dt) du \\
& + \tfrac{1}{2} \Sigma_{s,t=0}^{p} F_{\sigma}^{\lambda(t)}{}_{\rho}^{\varkappa(s)\Lambda}{}_{\Pi} (\partial_{\lambda(t)} \underset{2}{v}^{\sigma})(\partial_{\varkappa(s)} \underset{2}{v}^{\rho}) du^{2}] \times \\
& \times [\Sigma_{s=0}^{p} G_{\rho}^{\varkappa(s)\Pi} \partial_{\varkappa(s)} \underset{1}{v}^{\rho} dt + \tfrac{1}{2} \Sigma_{s,t=0}^{p} G_{\sigma}^{\lambda(t)}{}_{\rho}^{\varkappa(s)\Pi} (\partial_{\lambda(t)} \underset{1}{v}^{\sigma})(\partial_{\varkappa(s)} \underset{1}{v}^{\rho}) dt^{2}] \\
& + \Sigma_{s=0}^{p} (G_{\rho}^{\varkappa(s)\Lambda} \partial_{\varkappa(s)} \underset{2}{v}^{\rho} + \underset{1}{v}^{\sigma} \partial_{\sigma}(G_{\rho}^{\varkappa(s)\Lambda} \partial_{\varkappa(s)} \underset{2}{v}^{\rho}) dt) du \\
& + \tfrac{1}{2} \Sigma_{s,t=0}^{p} G_{\sigma}^{\lambda(t)}{}_{\rho}^{\varkappa(s)\Lambda} (\partial_{\lambda(t)} \underset{2}{v}^{\sigma})(\partial_{\varkappa(s)} \underset{2}{v}^{\rho}) du^{2} \\
& = \Sigma_{s=0}^{p} G_{\rho}^{\varkappa(s)\Lambda} \partial_{\varkappa(s)} \underset{1}{v}^{\rho} dt + \Sigma_{s=0}^{p} G_{\rho}^{\varkappa(s)\Lambda} \partial_{\varkappa(s)} \underset{2}{v}^{\rho} du \\
& + \tfrac{1}{2} \Sigma_{s,t=0}^{p} G_{\sigma}^{\lambda(t)}{}_{\rho}^{\varkappa(s)\Lambda} (\partial_{\lambda(t)} \underset{1}{v}^{\sigma})(\partial_{\varkappa(s)} \underset{1}{v}^{\rho}) dt^{2} \\
& + (\Sigma_{s,t=0}^{p} F_{\sigma}^{\lambda(t)\Lambda}{}_{\Pi} G_{\rho}^{\varkappa(s)\Pi} (\partial_{\lambda(t)} \underset{2}{v}^{\sigma})(\partial_{\varkappa(s)} \underset{1}{v}^{\rho}) + \Sigma_{s=0}^{p} \underset{1}{v}^{\sigma} \partial_{\sigma}(G_{\rho}^{\varkappa(s)\Lambda} \partial_{\varkappa(s)} \underset{2}{v}^{\rho})) dt du \\
& + \tfrac{1}{2} \Sigma_{s,t=0}^{p} G_{\sigma}^{\lambda(t)}{}_{\rho}^{\varkappa(s)\Lambda} (\partial_{\lambda(t)} \underset{2}{v}^{\sigma})(\partial_{\varkappa(s)} \underset{2}{v}^{\rho}) du^{2}
\end{aligned}
$$

and

$$
\begin{aligned}
(4.13)\quad & G^{\Lambda}(\xi^{\nu}, \xi^{\nu''}) \\
& = \Sigma_{s=0}^{p} G_{\rho}^{\varkappa(s)\Lambda} [\partial_{\varkappa(s)} \underset{1}{v}^{\rho} dt + \partial_{\varkappa(s)} \underset{2}{v}^{\rho} du + \partial_{\varkappa(s)} (\underset{1}{v}^{\sigma} \partial_{\sigma} \underset{2}{v}^{\rho}) dt du] \\
& + \tfrac{1}{2} \Sigma_{s,t=0}^{p} G_{\sigma}^{\lambda(t)}{}_{\rho}^{\varkappa(s)\Lambda} (\partial_{\lambda(t)} \underset{1}{v}^{\sigma} dt + \partial_{\lambda(t)} \underset{2}{v}^{\sigma} du)(\partial_{\varkappa(s)} \underset{1}{v}^{\rho} dt + \partial_{\varkappa(s)} \underset{2}{v}^{\rho} du)
\end{aligned}
$$

$$
\begin{aligned}
&= \Sigma_{s=0}^{p} G_{\rho}^{\varkappa(s)\Lambda} \partial_{\varkappa(s)} \underset{1}{v}^{\rho} dt + \Sigma_{s=0}^{p} G_{\rho}^{\varkappa(s)\Lambda} \partial_{\varkappa(s)} \underset{2}{v}^{\rho} du \\
&+ \tfrac{1}{2} \Sigma_{s,t=0}^{p} G_{\sigma\ \rho}^{\lambda(t)\varkappa(s)\Lambda} (\partial_{\lambda(t)} \underset{1}{v}^{\sigma})(\partial_{\varkappa(s)} \underset{1}{v}^{\rho}) dt^2 \\
&+ [\Sigma_{s=0}^{p} G_{\rho}^{\varkappa(s)\Lambda} \partial_{\varkappa(s)} (\underset{1}{v}^{\sigma} \partial_{\sigma} \underset{2}{v}^{\rho}) + \Sigma_{s,t=0}^{p} G_{\sigma\ \rho}^{\lambda(t)\varkappa(s)\Lambda} (\partial_{\lambda(t)} \underset{2}{v}^{\sigma})(\partial_{\lambda(s)} \underset{1}{v}^{\rho})] dt du \\
&+ \tfrac{1}{2} \Sigma_{s,t=0}^{p} G_{\sigma\ \rho}^{\lambda(t)\varkappa(s)\Lambda} (\partial_{\lambda(t)} \underset{2}{v}^{\sigma})(\partial_{\varkappa(s)} \underset{2}{v}^{\rho}) du^2,
\end{aligned}
$$

where

$$
(4.14) \qquad G_{\rho}^{\varkappa(s)\Lambda} = \left[\frac{\partial G^{\Lambda}}{\partial(\partial_{\varkappa(s)} f^{\rho})} \right]_{\xi}, \qquad G_{\sigma\ \rho}^{\lambda(t)\varkappa(s)\Lambda} = \left[\frac{\partial^2 G^{\Lambda}}{\partial(\partial_{\lambda(t)} f^{\sigma}) \partial(\partial_{\varkappa(s)} f^{\rho})} \right]_{\xi}.
$$

On comparing (4.12) and (4.13) we obtain

$$
(4.15) \qquad
\begin{aligned}
&\Sigma_{s,t=0}^{p} F_{\sigma\ \Pi}^{\lambda(t)\Lambda} G_{\rho}^{\varkappa(s)\Pi} (\partial_{\lambda(t)} \underset{2}{v}^{\sigma})(\partial_{\varkappa(s)} \underset{1}{v}^{\rho}) + \Sigma_{s=0}^{p} \underset{1}{v}^{\sigma} \partial_{\sigma} (G_{\rho}^{\varkappa(s)\Lambda} \partial_{\varkappa(s)} \underset{2}{v}^{\rho}) \\
&= \Sigma_{s=0}^{p} G_{\rho}^{\varkappa(s)\Lambda} \partial_{\varkappa(s)} (\underset{1}{v}^{\sigma} \partial_{\sigma} \underset{2}{v}^{\rho}) + \Sigma_{s,t=0}^{p} G_{\sigma\ \rho}^{\lambda(t)\varkappa(s)\Lambda} (\partial_{\lambda(t)} \underset{2}{v}^{\sigma})(\partial_{\varkappa(s)} \underset{1}{v}^{\rho}).
\end{aligned}
$$

Thus if we put

$$
(4.16) \qquad
\begin{cases}
\{G, v\}^{\Lambda} \overset{\text{def}}{=} \Sigma_{s=0}^{p} G_{\rho}^{\varkappa(s)\Lambda} \partial_{\varkappa(s)} v^{\rho} \\
\{G, \underset{2}{v}, \underset{1}{v}\}^{\Lambda} \overset{\text{def}}{=} G_{\sigma\ \rho}^{\lambda(t)\varkappa(s)\Lambda} (\partial_{\lambda(t)} \underset{2}{v}^{\sigma})(\partial_{\varkappa(s)} \underset{1}{v}^{\rho})
\end{cases}
$$

we obtain

$$
(4.17) \qquad \{F, \underset{2}{v}\}_{\Pi}^{\Lambda} \{G, \underset{1}{v}\}^{\Pi} + \underset{1}{v}^{\rho} \partial_{\rho} \{G, \underset{2}{v}\}^{\Lambda} = \{G, \underset{1}{v}^{\rho} \partial_{\rho} \underset{2}{v}\}^{\Lambda} + \{G, \underset{2}{v}, \underset{1}{v}\}^{\Lambda}.
$$

We now consider r infinitesimal point transformations

$$
(4.18) \qquad {}'\xi^{\varkappa} = \xi^{\varkappa} + \underset{a}{v}^{\varkappa}(\xi) dt \qquad (a, b, c, \ldots = \dot{1}, \dot{2}, \ldots, \dot{r}).
$$

Then, for two vectors $\underset{b}{v}^{\varkappa}$ and $\underset{c}{v}^{\varkappa}$, formula (4.11) gives

$$
(4.19) \qquad
\begin{aligned}
&\{F, \underset{c}{v}\}_{\Pi}^{\Lambda} \{F, \underset{b}{v}\}_{\Sigma}^{\Pi} - \{F, \underset{b}{v}\}_{\Pi}^{\Lambda} \{F, \underset{c}{v}\}_{\Sigma}^{\Pi} - \underset{c}{v}^{\rho} \partial_{\rho} \{F, \underset{b}{v}\}_{\Sigma}^{\Lambda} + \underset{b}{v}^{\rho} \partial_{\rho} \{F, \underset{c}{v}\}_{\Sigma}^{\Lambda} \\
&= - \{F, \underset{c}{v}^{\rho} \partial_{\rho} \underset{b}{v} - \underset{b}{v}^{\rho} \partial_{\rho} \underset{c}{v}\}_{\Sigma}^{\Lambda}
\end{aligned}
$$

and (4.17) gives

$$
(4.20) \qquad
\begin{aligned}
&\{F, \underset{c}{v}\}_{\Pi}^{\Lambda} \{G, \underset{b}{v}\}^{\Pi} - \{F, \underset{b}{v}\}_{\Pi}^{\Lambda} \{G, \underset{c}{v}\}^{\Pi} - \underset{c}{v}^{\rho} \partial_{\rho} \{G, \underset{b}{v}\}^{\Lambda} + \underset{b}{v}^{\rho} \partial_{\rho} \{G, \underset{c}{v}\}^{\Lambda} \\
&= - \{G, \underset{c}{v}^{\rho} \partial_{\rho} \underset{b}{v} - \underset{b}{v}^{\rho} \partial_{\rho} \underset{c}{v}\}^{\Lambda}.
\end{aligned}
$$

The Lie derivatives of a linear geometric object Ω^{Λ} with respect to $\underset{b}{v}^{\varkappa}$

are given by

$$\underset{b}{\pounds}\Omega^\Lambda = \underset{b}{v}^\rho \partial_\rho \Omega^\Lambda - [\{F, \underset{b}{v}\}^\Lambda_\Pi \Omega^\Pi + \{G, \underset{b}{v}\}^\Lambda],$$

where $\underset{b}{\pounds}$ represents the Lie derivatives with respect to $\underset{b}{v}^\varkappa$.

Since $\underset{b}{\pounds}\Omega^\Lambda$ is a linear homogeneous geometric object which has the transformation law

$$\underset{b}{\pounds}\Omega^{\Lambda'} = F^\Lambda_\Pi(\xi^\nu, \xi^{\nu'})\underset{b}{\pounds}\Omega^\Pi,$$

we obtain

$$(4.21) \qquad \underset{c}{\pounds}\underset{b}{\pounds}\Omega^\Lambda =$$

$$\underset{c}{v}^\sigma[\partial_\sigma \underset{b}{v}^\rho \partial_\rho \Omega^\Lambda + \underset{b}{v}^\rho \partial_\sigma \partial_\rho \Omega^\Lambda - (\partial_\sigma\{F, \underset{b}{v}\}^\Lambda_\Pi)\Omega^\Pi$$
$$- \{F, \underset{b}{v}\}^\Lambda_\Pi \partial_\sigma \Omega^\Pi - \partial_\sigma\{G, \underset{b}{v}\}^\Lambda]$$
$$- \{F, \underset{c}{v}\}^\Lambda_\Pi[\underset{b}{v}^\rho \partial_\rho \Omega^\Pi - \{F, \underset{b}{v}\}^\Pi_\Sigma \cdot \Omega^\Sigma + \{G, \underset{b}{v}\}^\Pi].$$

Consequently on taking account of (4.19) and (4.20), we obtain from equation (4.21)

$$(4.22) \qquad (\underset{c}{\pounds}\underset{b}{\pounds})\Omega^\Lambda \overset{\text{def}}{=} (\underset{c}{\pounds}\underset{b}{\pounds} - \underset{b}{\pounds}\underset{c}{\pounds})\Omega^\Lambda$$
$$= (\underset{c}{\pounds}\underset{b}{v}^\rho)\partial_\rho \Omega^\Lambda - \{F, \underset{c}{\pounds}\underset{b}{v}\}^\Lambda_\Pi \Omega^\Pi - \{G, \underset{c}{\pounds}\underset{b}{v}\}^\Lambda,$$
$$= \underset{cb}{\pounds}\Omega^\Lambda,$$

where

$$(4.23) \qquad \underset{cb}{v}^\varkappa = \underset{c}{\pounds}\underset{b}{v}^\varkappa = - \underset{b}{\pounds}\underset{c}{v}^\varkappa = \underset{c}{v}^\rho \partial_\rho \underset{b}{v}^\varkappa - \underset{b}{v}^\rho \partial_\rho \underset{c}{v}^\varkappa.$$

Thus we have

THEOREM 4.1. *Let $\underset{b}{\pounds} f = \underset{b}{v}^\varkappa \partial_\varkappa f$ be r infinitesimal operators and let Ω^Λ be N components of a linear geometric object. Then $(\underset{c}{\pounds}\underset{b}{\pounds})\Omega^\Lambda$ is equal to the Lie derivative of Ω^Λ with respect to the vector $\underset{c}{\pounds}\underset{b}{v}^\varkappa$.*

If $\underset{b}{\pounds} f$ $a, b, c, \ldots = 1, 2, \ldots, r$ are r infinitesimal operators of an r-parameter group G_r of transformations, then we have

$$(4.24) \qquad (\underset{c}{\pounds}\underset{b}{\pounds})f = c^a_{cb}\underset{a}{\pounds} f$$

or

$$\underset{c}{\pounds}\,\underset{b}{v}^{\varkappa} = c_{cb}^{a}\,\underset{a}{v}^{\varkappa} \tag{4.25}$$

where c_{cb}^{a} are the structural constants of the group G_r. Consequently we can state

THEOREM 4.2. *If* $\underset{b}{\pounds}f$ *are* r *infinitesimal operators of an* r*-parameter group* G_r *of transformations, then we have the formula*

$$(\underset{c}{\pounds}\,\underset{b}{\pounds})\Omega^{\Lambda} = c_{cb}^{a}\,\underset{a}{\pounds}\,\Omega^{\Lambda} \tag{4.26}$$

for any linear geometric object Ω^{Λ}.

Chapter III

GROUPS OF TRANSFORMATIONS LEAVING A GEOMETRIC OBJECT INVARIANT

§ 1. Projective and conformal motions.

Let us first consider an A_n with a symmetric linear connexion $\Gamma^{\varkappa}_{\mu\lambda}$. The geodesics of the space are given by

$$\frac{d^2\xi^{\varkappa}}{dt^2} + \Gamma^{\varkappa}_{\mu\lambda}\frac{d\xi^{\mu}}{dt}\frac{d\xi^{\lambda}}{dt} = \alpha(t)\frac{d\xi^{\varkappa}}{dt}. \tag{1.1}$$

When a point transformation

$$'\xi^{\varkappa} = f^{\varkappa}(\xi^{\nu}) \tag{1.2}$$

transforms the system of geodesics into the same system, (1.2) is called a *projective motion* in A_n. The necessary and sufficient condition that (1.2) be a projective motion in A_n is that the Lie difference of $\Gamma^{\varkappa}_{\mu\lambda}$ with respect to (1.2) has the form [1]

$$'\Gamma^{\varkappa}_{\mu\lambda} - \Gamma^{\varkappa}_{\mu\lambda} = A^{\varkappa}_{\mu}p_{\lambda} + A^{\varkappa}_{\lambda}p_{\mu}, \tag{1.3}$$

where p_{λ} is a covariant vector.

When (1.2) is an infinitesimal transformation:

$$'\xi^{\varkappa} = \xi^{\varkappa} + v^{\varkappa}(\xi)dt, \tag{1.4}$$

the condition is

$$\boxed{\underset{v}{\pounds}\Gamma^{\varkappa}_{\mu\lambda} = A^{\varkappa}_{\mu}p_{\lambda} + A^{\varkappa}_{\lambda}p_{\mu}.} \tag{1.5}$$

Thus we have

Theorem 1.1. *A necessary and sufficient condition that* (1.4) *be a projective motion in an A_n is that the Lie derivative of $\Gamma^{\varkappa}_{\mu\lambda}$ has the form* (1.5).

[1] The necessary and sufficient condition that two linear connexions $'\Gamma^{\varkappa}_{\mu\lambda}$ and $\Gamma^{\varkappa}_{\mu\lambda}$ give the same system of geodesics is that $'\Gamma^{\varkappa}_{\mu\lambda} = \Gamma^{\varkappa}_{\mu\lambda} + A^{\varkappa}_{\mu}p_{\lambda} + A^{\varkappa}_{\lambda}p_{\mu}$ for a certain covariant vector p_{λ}. Cf. Schouten [8], p. 156; p. 287.

From (1.5), we have

$$\underset{v}{\pounds}\Gamma^{\rho}_{\mu\rho} = (n+1)p_{\mu}. \tag{1.6}$$

Eliminating p_μ from (1.5) and (1.6) we find

$$\boxed{\underset{v}{\pounds}\overset{p}{\Gamma}{}^{\varkappa}_{\mu\lambda} = 0,} \tag{1.7}$$

where

$$\overset{p}{\Gamma}{}^{\varkappa}_{\mu\lambda} \overset{\text{def}}{=} \Gamma^{\varkappa}_{\mu\lambda} - \frac{1}{n+1}(A^{\varkappa}_{\mu}\Gamma^{\rho}_{\lambda\rho} + A^{\varkappa}_{\lambda}\Gamma^{\rho}_{\mu\rho}) \tag{1.8}$$

are the well-known projective parameters introduced by T. Y. Thomas.[1]

If we write out $\underset{v}{\pounds}\overset{p}{\Gamma}{}^{\varkappa}_{\mu\lambda}$ explicitly, we get

$$\underset{v}{\pounds}\overset{p}{\Gamma}{}^{\varkappa}_{\mu\lambda} = \partial_\mu\partial_\lambda v^\varkappa + v^\rho\partial_\rho\overset{p}{\Gamma}{}^{\varkappa}_{\mu\lambda} - \overset{p}{\Gamma}{}^{\rho}_{\mu\lambda}\partial_\rho v^\varkappa + \overset{p}{\Gamma}{}^{\varkappa}_{\rho\lambda}\partial_\mu v^\rho + \overset{p}{\Gamma}{}^{\varkappa}_{\mu\rho}\partial_\lambda v^\rho$$
$$- \frac{1}{n+1}(A^{\varkappa}_{\mu}\partial_\lambda\partial_\rho v^\rho + A^{\varkappa}_{\lambda}\partial_\mu\partial_\rho v^\rho). \tag{1.9}$$

Conversely, if (1.7) holds, it can easily be proved that the Lie derivative $\underset{v}{\pounds}\Gamma^{\varkappa}_{\mu\lambda}$ of $\Gamma^{\varkappa}_{\mu\lambda}$ has the form (1.5).

Thus we have

THEOREM 1.2. *A necessary and sufficient condition that* (1.4) *be a projective motion in an* A_n *is that the Lie derivative of* $\overset{p}{\Gamma}{}^{\varkappa}_{\mu\lambda}$ *vanish.*

Let us next consider a V_n with the fundamental tensor $g_{\lambda\varkappa}$. When a point transformation (1.2) does not change the angle between two directions at a point, (1.2) is called a *conformal motion* in the V_n. The necessary and sufficient condition that (1.2) be a conformal motion in a V_n is that the Lie difference of $g_{\lambda\varkappa}$ with respect to (1.2) be proportional to $g_{\lambda\varkappa}$:[2]

$$'g_{\lambda\varkappa} - g_{\lambda\varkappa} = 2\phi g_{\lambda\varkappa}. \tag{1.10}$$

where ϕ is a scalar.

---

[1] T. Y. THOMAS [3]; cf. SCHOUTEN [8], p. 300.
[2] SCHOUTEN [8] p. 304.

When (1.2) is infinitesimal, the condition is

$$\boxed{\underset{v}{\pounds} g_{\lambda\varkappa} = 2\phi g_{\lambda\varkappa}.} \tag{1.11}$$

Thus we have

THEOREM 1.3. *A necessary and sufficient condition that* (1.4) *be a conformal motion in a* V_n *is that the Lie derivative of* $g_{\lambda\varkappa}$ *be a multiple of* $g_{\lambda\varkappa}$.

From (1.11) we find

$$\underset{v}{\pounds}\, \mathfrak{g}^{\frac{1}{n}} = 2\phi; \quad \mathfrak{g} \overset{\text{def}}{=} |\mathrm{Det}(g_{\lambda\varkappa})|. \tag{1.12}$$

Eliminating ϕ from (1.11) and (1.12), we obtain

$$\underset{v}{\pounds}\, \mathfrak{G}_{\lambda\varkappa} = 0, \tag{1.13}$$

where

$$\mathfrak{G}_{\lambda\varkappa} \overset{\text{def}}{=} \mathfrak{g}^{-\frac{1}{n}} g_{\lambda\varkappa}.\ [1] \tag{1.14}$$

If we write out $\underset{v}{\pounds}\, \mathfrak{G}_{\lambda\varkappa}$ explicitly, we get

$$\underset{v}{\pounds}\, \mathfrak{G}_{\lambda\varkappa} = v^{\mu}\partial_{\mu}\mathfrak{G}_{\lambda\varkappa} + \mathfrak{G}_{\rho\varkappa}\partial_{\lambda}v^{\rho} + \mathfrak{G}_{\lambda\rho}\partial_{\varkappa}v^{\rho} - \frac{2}{n}\mathfrak{G}_{\lambda\varkappa}\partial_{\rho}v^{\rho}. \tag{1.15}$$

Conversely, if (1.13) holds, then it can easily be seen that the Lie derivative $\underset{v}{\pounds} g_{\lambda\varkappa}$ of $g_{\lambda\varkappa}$ is proportional to $g_{\lambda\varkappa}$. Thus we have

THEOREM 1.4. *A necessary and sufficient condition that* (1.4) *be a conformal motion in a* V_n *is that the Lie derivative of* $\mathfrak{G}_{\lambda\varkappa}$ *vanish.* [2]

Kobayashi [3] proved

THEOREM 1.5. *The group of affine, projective or conformal motions in a space is a Lie group.*

§ 2. Invariance group of a geometric object. [4]

Let us consider a geometric object $\Omega^{\Lambda}(\xi)$ in an X_n of class C^{ω} and an

[1] SCHOUTEN [8], p. 315.

[2] The projective and conformal motions will be studied in detail in Ch. VI and VII.

[3] KOBAYASHI [2, 3].

[4] NIJENHUIS [2]; TASHIRO [1].

r-parameter group G_r of transformations:

(2.1) $$'\xi^{\varkappa} = f^{\varkappa}(\xi^1, \xi^2, \ldots, \xi^{\varkappa}; \eta^1, \eta^2, \ldots, \eta^r).$$

If the group has the property

(2.2) $$'\Omega^{\Lambda} - \Omega^{\Lambda} = 0,$$

we call G_r *an invariance group* of the geometric object Ω^{Λ}. Groups of motions in V_n, of affine motions in L_n, of projective motions in A_n, or of conformal motions in V_n are invariance groups of $g_{\lambda\varkappa}$, of $\Gamma^{\varkappa}_{\mu\lambda}$, of $\overset{p}{\Gamma}{}^{\varkappa}_{\mu\lambda}$ or of $\mathfrak{G}_{\lambda\varkappa}$ respectively.

We now suppose that the space admits an infinitesimal point transformation (1.4) for which the Lie derivative of a *linear differential* geometric object Ω^{Λ} vanishes:

(2.3) $$\underset{v}{\pounds}\, \Omega^{\Lambda} = v^{\mu}\, \partial_{\mu}\, \Omega^{\Lambda} - \{F, v\}^{\Lambda}_{\Pi}\, \Omega^{\Pi} - \{G, v\}^{\Lambda} = 0.$$

We know on the one hand that if the transformation law of Ω^{Λ} is given by

(2.4) $$\Omega^{\Lambda'} = F^{\Lambda}_{\Pi}(\xi^{\nu}, \xi^{\nu'})\Omega^{\Pi} + G^{\Lambda}(\xi^{\nu}, \xi^{\nu'}),$$

then that of $\underset{v}{\pounds}\Omega^{\Lambda}$ is given by

(2.5) $$\underset{v}{\pounds}\, \Omega^{\Lambda'} = F^{\Lambda}_{\Pi}(\xi^{\nu}, \xi^{\nu'})\underset{v}{\pounds}\, \Omega^{\Pi}.$$

Thus we see that the equation

(2.6) $$\underset{v}{\pounds}\Omega^{\Lambda} = 0$$

has a meaning which does not depend on the choice of coordinate systems.

We know on the other hand that if a contravariant vector field $v^{\varkappa}(\xi)$ is given, we can choose a coordinate system $(\varkappa)$ in a suitable neighbourhood of a regular point [1] of $v^{\varkappa}$ such that, [2] in this neighbourhood,

(2.7) $$v^{\varkappa} = \underset{1}{e}{}^{\varkappa}.$$

In this coordinate system, the infinitesimal point transformation

(2.8) $$'\xi^{\varkappa} = \xi^{\varkappa} + \underset{1}{e}{}^{\varkappa}\, dt$$

---

[1] By a regular point, we mean a point at which $v^{\varkappa} \neq 0$.

[2] GOURSAT [1] p. 117; cf. SCHOUTEN [8] p. 83.

generates a finite point transformation

$$'\xi^{\varkappa} = \xi^{\varkappa} + t \cdot \underset{1}{e^{\varkappa}}. \tag{2.9}$$

For (2.8), the equation (2.6) becomes

$$\underset{v}{\pounds}\,\Omega^{\Lambda} = \partial_1 \Omega^{\Lambda} = 0, \tag{2.10}$$

which means that the components Ω^{Λ} of the geometric object are independent of the variable ξ^1. Consequently for the finite point transformation (2.9), we have

$$'\Omega^{\Lambda} = F^{\Lambda}(\Omega^{\Sigma}('\xi), \xi^{\nu'}, \xi^{\nu}) = \Omega^{\Lambda} \tag{2.11}$$

because $\Omega^{\Sigma}('\xi) = \Omega^{\Sigma}(\xi)$ and

$$F^{\Lambda}(\Omega^{\Sigma}, \xi^{\nu'}, \xi^{\nu}) = F^{\Lambda}(\Omega^{\Sigma}, \xi^{\nu}, \xi^{\nu}) = \Omega^{\Lambda},$$

the object being linear differential.

This can also be derived in the following way. Because of

$$\underset{v}{\pounds}\Omega^{\Lambda} = \partial_1 \Omega^{\Lambda},$$

we have

$$'\Omega^{\Lambda} = e^{t\underset{v}{\pounds}}\Omega^{\Lambda} = \Omega^{\Lambda} + t\underset{v}{\pounds}\,\Omega^{\Lambda} + \frac{t^2}{2!}\cdot \underset{v}{\pounds^2}\,\Omega^{\Lambda} + \dots \tag{2.12}$$

in the case where Ω^{Λ} is of class C^{ω}. The equation (2.12) shows that (2.6) implies (2.11). Gathering these results we can state

THEOREM 2.1. *If a space admits an infinitesimal point transformation with respect to which the Lie derivative of a linear differential geometric object vanishes, then it admits also a one-parameter invariance group of this geometric object.*

THEOREM 2.2. *In order that a space admit a one-parameter invariance group of a linear differential geometric object, it is necessary and sufficient that there exist a coordinate system with respect to which the components of the geometric object are independent of one of the coordinates.*

Suppose that r contravariant vectors $\underset{a}{v^{\varkappa}}$; $a, b, c, \ldots = 1, 2, \ldots, r$, define r one-parameter invariance groups of a same linear differential geometric object Ω^{Λ}, then we have

$$\underset{a}{\pounds}\,\Omega^{\Lambda} = \underset{a}{v^{\mu}}\,\partial_{\mu}\Omega^{\Lambda} - \{F, \underset{a}{v}\}^{\Lambda}_{\Pi}\,\Omega^{\Pi} - \{G, \underset{a}{v}\}^{\Lambda} = 0, \tag{2.13}$$

from which

$$(2.14) \qquad c^a \underset{a}{\pounds}\, \Omega^\Lambda = (c^a \underset{a}{v^\mu})\partial_\mu \Omega^\Lambda - \{F, c^a \underset{a}{v}\}^\Lambda_\Pi \Omega^\Pi - \{G, c^a \underset{a}{v}\}^\Lambda = 0,$$

c^a being r constants. Thus we have

THEOREM 2.3. *If each of r contravariant vectors generates a one-parameter invariance group of a same linear differential geometric object, then a linear combination of these contravariant vectors with constant coefficients generates also a one-parameter invariance group of the same geometric object.*

THEOREM 2.4. *If each of r infinitesimal operators of an r-parameter group of transformations generates a one-parameter invariance group of a same linear differential geometric object, then all the transformations of the r-parameter group leave invariant the geometric object, that is, the group is an invariance group of the geometric object.*

Moreover, according to Theorem 4.1 of Ch. II, we get

THEOREM 2.5. *If each of r vectors $\underset{a}{v^\varkappa}$ defines a one-parameter invariance group of the same linear differential geometric object, then each of the vectors $\underset{c}{\pounds}\underset{b}{v^\varkappa}$ defines also a one-parameter invariance group of this geometric object.*

Suppose that each of r linearly independent [1] vectors $\underset{a}{v^\varkappa}$ defines a one-parameter invariance group of the same linear differential geometric object. If any vector $v^\varkappa$ which defines a one-parameter invariance group of this geometric object is a linear combination of $\underset{a}{v^\varkappa}$ with constant coefficients, then the set of vectors $\underset{a}{v^\varkappa}$ is said to be *complete*.

Now if r vectors $\underset{a}{v^\varkappa}$ form a complete set of vectors defining r one-parameter invariance groups of the same linear differential geometric object, then since the $\underset{c}{\pounds}\underset{b}{v^\varkappa}$ are also vectors defining an invariance group, we must have

$$(2.15) \qquad \underset{c}{\pounds}\, \underset{b}{v^\varkappa} = c_{cb}^{a}\, \underset{a}{v^\varkappa}$$

where the c_{cb}^a are constants. The equation (2.15) shows that the $\underset{a}{\pounds}f =$

[1] This means: whenever equations $c^a \underset{a}{v^\varkappa} = 0$ ($c^a =$ constants) hold, then $c^a = 0$ (cf. SCHOUTEN [8], p. 203).

$\underset{a}{v}^{\varkappa}\partial_{\varkappa}f$ are r infinitesimal operators of an r-parameter group. [1] Thus we have

THEOREM 2.6. *If r vectors $\underset{a}{v}^{\varkappa}$ form a complete set of vectors defining r one-parameter invariance groups of the same linear differential geometric object, then the $\underset{a}{\pounds}f = \underset{a}{v}^{\varkappa}\partial_{\varkappa}f$ are r infinitesimal operators of an r-parameter invariance group of the object.*

§ 3. A group as invariance group of a geometric object.

We shall consider in this section the following problem. Given an r-parameter group of transformations in an n-dimensional space, does there exist a linear geometric object Ω^{Λ} with a given manner of transformation such that the given group is an invariance group of the object? In other words, if r vectors $\underset{a}{v}^{\varkappa}$ define an r-parameter group of transformations and if Ω^{Λ} are components of a linear geometric object whose transformation law is

$$\Omega^{\Lambda'} = F^{\Lambda}_{\Pi}(\xi^{\nu}, \xi^{\nu'})\Omega^{\Pi} + G^{\Lambda}(\xi^{\nu}, \xi^{\nu'}), \tag{3.1}$$

is the system of partial differential equations

$$\underset{a}{\pounds}\,\Omega^{\Lambda} = \underset{a}{v}^{\mu}\,\partial_{\mu}\Omega^{\Lambda} - \{F, \underset{a}{v}\}^{\Lambda}_{\Pi}\Omega^{\Pi} - \{G, \underset{a}{v}\}^{\Lambda} = 0 \tag{3.2}$$

integrable?

First of all we notice that the functions $\{F, v\}^{\Lambda}_{\Pi}$ and $\{G, \underset{a}{v}\}^{\Lambda}$ satisfy (4.19) and (4.20) of Ch. II, from which

$$\{F, \underset{c}{v}\}^{\Lambda}_{\Pi}\{F, \underset{b}{v}\}^{\Pi}_{\Sigma} - \{F, \underset{b}{v}\}^{\Lambda}_{\Pi}\{F, \underset{c}{v}\}^{\Pi}_{\Sigma} - \underset{c}{v}^{\rho}\,\partial_{\rho}\{F, \underset{b}{v}\}^{\Lambda}_{\Sigma} + \underset{b}{v}^{\rho}\,\partial_{\rho}\{F, \underset{c}{v}\}^{\Lambda}_{\Sigma} = -\,c^{a}_{cb}\{F, \underset{a}{v}\}^{\Lambda}_{\Sigma}, \tag{3.3}$$

$$\{F, \underset{c}{v}\}^{\Lambda}_{\Pi}\{G, \underset{b}{v}\}^{\Pi} - \{F, \underset{b}{v}\}^{\Lambda}_{\Pi}\{G, \underset{c}{v}\}^{\Pi} - \underset{c}{v}^{\rho}\,\partial_{\rho}\{G, \underset{b}{v}\}^{\Lambda} + \underset{b}{v}^{\rho}\,\partial_{\rho}\{G, \underset{c}{v}\}^{\Lambda} = -\,c^{a}_{cb}\{G, \underset{a}{v}\}^{\Lambda}, \tag{3.4}$$

because of the relations

$$\underset{c}{v}^{\rho}\,\partial_{\rho}\underset{b}{v}^{\varkappa} - \underset{b}{v}^{\rho}\,\partial_{\rho}\underset{c}{v}^{\varkappa} = c^{a}_{cb}\,\underset{a}{v}^{\varkappa}. \tag{3.5}$$

---

[2] This is the second fundamental theorem of Lie. Cf. EISENHART [4], p. 54; SCHOUTEN [8], p. 206ff.

We shall first consider the case in which the rank of $\underset{a}{v}^{\varkappa}$ in a certain neighbourhood is equal to $r \leq n$.

In this case, we can take a coordinate system $(\varkappa)$ with respect to which the components of the vectors $\underset{a}{v}^{\varkappa}$ satisfy the following conditions[1]:

$$\text{(3.6)} \qquad \operatorname{Det}(\underset{a}{v}^{\alpha}) \neq 0, \ \underset{a}{v}^{\xi} = 0; \quad a, b, c = 1, 2, \ldots, r,$$
$$\alpha, \beta, \gamma = 1, 2, \ldots, r; \ \xi, \eta, \zeta = r+1, \ldots, n.$$

In such a coordinate system the equation (3.2) takes the form

$$\text{(3.7)} \qquad \underset{a}{\pounds}\, \Omega^{\Lambda} = \underset{a}{v}^{\alpha} \partial_{\alpha} \Omega^{\Lambda} - \{F, \underset{a}{v}\}^{\Lambda}_{\Pi} \Omega^{\Pi} - \{G, \underset{a}{v}\}^{\Lambda} = 0,$$

and consequently if we define the functions $\Theta^{\Lambda}_{\alpha}(\Omega, \xi)$ by

$$\text{(3.8)} \qquad \underset{a}{v}^{\alpha} \Theta^{\Lambda}_{\alpha}(\Omega, \xi) \overset{\text{def}}{=} \{F, \underset{a}{v}\}^{\Lambda}_{\Pi} \Omega^{\Pi} + \{G, \underset{a}{v}\}^{\Lambda},$$

we get from (3.7)

$$\text{(3.9)} \qquad \underset{a}{\pounds}\, \Omega^{\Lambda} = \underset{a}{v}^{\alpha}[\partial_{\alpha} \Omega^{\Lambda} - \Theta^{\Lambda}_{\alpha}(\Omega, \xi)] = 0$$

or

$$\text{(3.10)} \qquad \partial_{\alpha} \Omega^{\Lambda} = \Theta^{\Lambda}_{\alpha}(\Omega, \xi).$$

We shall examine the integrability conditions of this system of partial differential equations. From (3.8) we get

$$\text{(3.11)} \qquad \underset{b}{v}^{\beta} \partial_{\Pi} \Theta^{\Lambda}_{\beta} = \{F, \underset{b}{v}\}^{\Lambda}_{\Pi}$$

and consequently

$$\underset{c}{v}^{\gamma} \underset{b}{v}^{\beta} \Theta^{\Pi}_{\gamma} \partial_{\Pi} \Theta^{\Lambda}_{\beta} = \{F, \underset{b}{v}\}^{\Lambda}_{\Pi} \{F, \underset{c}{v}\}^{\Pi}_{\Sigma} \Omega^{\Sigma} + \{F, \underset{b}{v}\}^{\Lambda}_{\Pi} \{G, \underset{c}{v}\}^{\Pi}$$

from which

$$\underset{c}{v}^{\gamma} \underset{b}{v}^{\beta} [\Theta^{\Pi}_{\gamma} \partial_{\Pi} \Theta^{\Pi}_{\beta} - \Theta^{\Pi}_{\beta} \partial_{\Pi} \Theta^{\Pi}_{\gamma}]$$
$$= [\{F, \underset{b}{v}\}^{\Lambda}_{\Pi} \{F, \underset{c}{v}\}^{\Pi}_{\Sigma} - \{F, \underset{c}{v}\}^{\Lambda}_{\Pi} \{F, \underset{b}{v}\}^{\Pi}_{\Sigma}] \Omega^{\Sigma}$$
$$+ \{F, \underset{b}{v}\}^{\Lambda}_{\Pi} \{G, \underset{c}{v}\}^{\Pi} - \{F, \underset{c}{v}\}^{\Lambda}_{\Pi} \{G, \underset{b}{v}\}^{\Pi}.$$

---

[1] Cf. EISENHART [5], p. 74.

Because of the equations (3.3) and (3.4), the above equation becomes

$$\begin{aligned}(3.12)\qquad & \underset{c}{v}^\gamma \underset{b}{v}^\beta[\Theta^\Pi_\gamma \partial_\Pi \Theta^\Lambda_\beta - \Theta^\Pi_\beta \partial_\Pi \Theta^\Lambda_\gamma] \\ & = [c^a_{cb}\{F, \underset{a}{v}\}^\Lambda_\Pi + \underset{b}{v}^\gamma \partial_\gamma\{F, \underset{c}{v}\}^\Lambda_\Pi - \underset{c}{v}^\gamma \partial_\gamma\{F, \underset{b}{v}\}^\Lambda_\Pi]\Omega^\Pi \\ & + [c^a_{cb}\{G, \underset{a}{v}\}^\Lambda + \underset{b}{v}^\gamma \partial_\gamma\{G, \underset{c}{v}\}^\Lambda - \underset{c}{v}^\gamma \partial_\gamma\{G, \underset{b}{v}\}^\Lambda].\end{aligned}$$

Also from (3.8) we find

$$(\underset{c}{v}^\gamma \partial_\gamma \underset{b}{v}^\beta)\Theta^\Lambda_\beta + \underset{c}{v}^\gamma \underset{b}{v}^\beta \partial_\gamma \Theta^\Lambda_\beta = (\underset{c}{v}^\gamma \partial_\gamma\{F, \underset{b}{v}\}^\Lambda_\Pi)\Omega^\Pi + \underset{c}{v}^\gamma \partial_\gamma\{G, \underset{b}{v}\}^\Lambda,$$

from which

$$\begin{aligned}& (\underset{c}{v}^\gamma \partial_\gamma \underset{b}{v}^\beta - \underset{b}{v}^\gamma \partial_\gamma \underset{c}{v}^\beta)\Theta^\Lambda_\beta + \underset{c}{v}^\gamma \underset{b}{v}^\beta[\partial_\gamma \Theta^\Lambda_\beta - \partial_\beta \Theta^\Lambda_\gamma] \\ & = [\underset{c}{v}^\gamma \partial_\gamma\{F, \underset{b}{v}\}^\Lambda_\Pi - \underset{b}{v}^\gamma \partial_\gamma\{F, \underset{c}{v}\}^\Lambda_\Pi]\Omega^\Pi \\ & + [\underset{c}{v}^\gamma \partial_\gamma\{G, \underset{b}{v}\}^\Lambda - \underset{b}{v}^\gamma \partial_\gamma\{G, \underset{c}{v}\}^\Lambda]\end{aligned}$$

or

$$\begin{aligned}(3.13)\qquad & \underset{c}{v}^\gamma \underset{b}{v}^\beta[\partial_\gamma \Theta^\Lambda_\beta - \partial_\gamma \Theta^\Lambda_\gamma] \\ & = - c^a_{cb}\underset{a}{v}^\alpha \Theta^\Lambda_\alpha + [\underset{c}{v}^\gamma \partial_\gamma\{F, \underset{b}{v}\}^\Lambda_\Pi - \underset{b}{v}^\gamma \partial_\gamma\{F, \underset{c}{v}\}^\Lambda_\Pi]\Omega^\Pi \\ & + [\underset{c}{v}^\gamma \partial_\gamma\{G, \underset{b}{v}\}^\Lambda - \underset{b}{v}^\gamma \partial_\gamma\{G, \underset{c}{v}\}^\Lambda]\end{aligned}$$

because of (3.5). Comparing the two equations (3.12) and (3.13) and taking account of (3.8), we find

$$\underset{c}{v}^\gamma \underset{b}{v}^\gamma[\Theta^\Pi_\gamma \partial_\Pi \Theta^\Lambda_\beta - \Theta^\Pi_\beta \partial_\Pi \Theta^\Lambda_\gamma + \partial_\gamma \Theta^\Lambda_\beta - \partial_\beta \Theta^\Lambda_\gamma] = 0,$$

from which

$$(3.14)\qquad \Theta^\Pi_\gamma \partial_\Pi \Theta^\Lambda_\beta + \partial_\gamma \Theta^\Lambda_\beta = \Theta^\Pi_\beta \partial_\Pi \Theta^\Lambda_\gamma + \partial_\beta \Theta^\Lambda_\gamma$$

by virtue of $\mathrm{Det}(\underset{a}{v}^\alpha) \neq 0$. The equation (3.14) means that the system (3.10) of partial differential equations is completely integrable. Thus we have

THEOREM 3.1. *If there is given an r-parameter group of transformations in a space of $n(\geq r)$ dimensions such that the rank of $\underset{a}{v}^\kappa$ in a neighbourhood is r, the group can be regarded as an invariance group of a linear geometric object.*

We next consider the case in which the group is intransitive and the rank q of $\underset{a}{v}^{\varkappa}$ in a neighbourhood is less than r.

In this case, we can take a coordinate system $(\varkappa)$ with respect to which the components of the vectors $\underset{a}{v}^{\varkappa}$ satisfy the following relations: [1]

$$\text{(3.15)} \qquad \operatorname{Det}(\underset{i}{v}^{\alpha}) \neq 0, \quad \underset{i}{v}^{\xi} = 0, \quad \underset{u}{v}^{\varkappa} = \varphi_u^i(\xi^\nu)\underset{i}{v}^{\varkappa}$$

$$a, b, c = 1, 2, \ldots, \mathrm{n},$$

$$i, j, k = 1, 2, \ldots, \mathrm{q};\ u, v, w = \mathrm{q} + 1, \ldots, \mathrm{n},$$

$$\varkappa, \lambda, \mu, \nu = 1, 2, \ldots, n,$$

$$\alpha, \beta, \gamma = 1, 2, \ldots, q;\ \xi, \eta, \zeta = q + 1, \ldots, n.$$

In such a coordinate system, the equation (3.2) takes the form

$$\text{(3.16)} \quad \begin{cases} \text{(a)}\ \underset{h}{\pounds}\,\Omega^\Lambda = \underset{h}{v}^{\alpha}\partial_\alpha\Omega^\Lambda - \{F, \underset{h}{v}\}^\Lambda_\Pi\,\Omega^\Pi - \{G, \underset{h}{v}\}^\Lambda = 0, \\ \text{(b)}\ \underset{u}{\pounds}\,\Omega^\Lambda = \varphi_u^h\underset{h}{v}^{\alpha}\partial_\alpha\Omega^\Lambda - \{F, \varphi_u^h\underset{h}{v}\}^\Lambda_\Pi\,\Omega^\Pi - \{G, \varphi_u^h\underset{h}{v}\}^\Lambda = 0. \end{cases}$$

If we define the functions $\Theta_\alpha^\Lambda(\Omega, \xi)$ and $\Xi_u^\Lambda(\Omega, \xi)$ by

$$\text{(3.17)} \qquad \underset{h}{v}^{\alpha}\,\Theta_\alpha^\Lambda(\Omega, \xi) \overset{\text{def}}{=} \{F, \underset{h}{v}\}^\Lambda_\Pi\,\Omega^\Pi + \{G, \underset{h}{v}\}^\Lambda$$

and

$$\text{(3.18)} \qquad \Xi_u^\Lambda(\Omega, \xi) \overset{\text{def}}{=} \varphi_u^h\,[\{F, \underset{h}{v}\}^\Lambda_\Pi\,\Omega^\Pi + \{G, \underset{h}{v}\}^\Lambda] - \{F, \varphi_u^h\underset{h}{v}\}^\Lambda_\Pi\,\Omega^\Pi - \{G, \varphi_u^h\underset{h}{v}\}^\Lambda$$

respectively, we can write (3.16) in the form

$$\text{(3.19)} \quad \begin{cases} \text{(a)}\ \underset{h}{\pounds}\,\Omega^\Lambda = \underset{h}{v}^{\alpha}[\partial_\alpha\Omega^\Lambda - \Theta_\alpha^\Lambda(\Omega, \xi)] = 0, \\ \text{(b)}\ \underset{u}{\pounds}\,\Omega^\Lambda = \varphi_u^h\underset{h}{\pounds}\,\Omega^\Lambda + \Xi_u^\Lambda(\Omega, \xi) = 0. \end{cases}$$

These equations are equivalent to

$$\text{(3.20)} \qquad \partial_\alpha\Omega^\Lambda = \Theta_\alpha^\Lambda(\Omega, \xi), \quad \Xi_u^\Lambda(\Omega, \xi) = 0.$$

We shall examine the integrability conditions of this mixed system of partial differential equations.

---

[1] Cf. Eisenhart [5], p. 74.

From (3.17) we find

$$\underset{i}{v}^{\beta}\partial_{\Pi}\Theta^{\Lambda}_{\beta} = \{F, \underset{i}{v}\}^{\Lambda}_{\Pi}$$

and consequently

$$\underset{j}{v}^{\gamma}\underset{i}{v}^{\beta}\Theta^{\Pi}_{\gamma}\partial_{\Pi}\Theta^{\Lambda}_{\beta} = \{F, \underset{i}{v}\}^{\Lambda}_{\Pi}\{F, \underset{j}{v}\}^{\Pi}_{\Sigma}\Omega^{\Sigma} + \{F, \underset{i}{v}\}^{\Lambda}_{\Pi}\{G, \underset{j}{v}\}^{\Pi},$$

from which

$$\begin{aligned}(3.21)\quad & \underset{j}{v}^{\gamma}\underset{i}{v}^{\beta}[\Theta^{\Pi}_{\gamma}\partial_{\Pi}\Theta^{\Lambda}_{\beta} - \Theta^{\Pi}_{\beta}\partial_{\Pi}\Theta^{\Lambda}_{\gamma}] \\ &= [\{F, \underset{i}{v}\}^{\Lambda}_{\Pi}\{F, \underset{j}{v}\}^{\Pi}_{\Sigma} - \{F, \underset{j}{v}\}^{\Lambda}_{\Pi}\{F, \underset{i}{v}\}^{\Pi}_{\Sigma}]\Omega^{\Sigma} \\ &\quad + \{F, \underset{i}{v}\}^{\Lambda}_{\Pi}\{G, \underset{j}{v}\}^{\Pi} - \{F, \underset{j}{v}\}^{\Lambda}_{\Pi}\{G, \underset{i}{v}\}^{\Pi}.\end{aligned}$$

Because of the equations (3.3) and (3.4), the equation (3.21) becomes

$$\begin{aligned}(3.22)\quad & \underset{j}{v}^{\gamma}\underset{i}{v}^{\beta}[\Theta^{\Pi}_{\gamma}\partial_{\Pi}\Theta^{\Lambda}_{\beta} - \Theta^{\Pi}_{\beta}\partial_{\Pi}\Theta^{\Lambda}_{\gamma}] \\ &= [c^{h}_{ji}\{F, \underset{h}{v}\}^{\Lambda}_{\Sigma} + c^{u}_{ji}\{F, \varphi^{h}_{u}\underset{h}{v}\}^{\Lambda}_{\Sigma} + \underset{i}{v}^{\gamma}\partial_{\gamma}\{F, \underset{j}{v}\}^{\Lambda}_{\Sigma} - \underset{j}{v}^{\gamma}\partial_{\gamma}\{F, \underset{i}{v}\}^{\Lambda}_{\Sigma}]\Omega^{\Sigma} \\ &\quad + [c^{h}_{ji}\{G, \underset{h}{v}\}^{\Lambda}_{\Sigma} + c^{u}_{ji}\{G, \varphi^{h}_{u}\underset{h}{v}\}^{\Lambda} + \underset{i}{v}^{\gamma}\partial_{\gamma}\{G, \underset{j}{v}\}^{\Lambda} - \underset{j}{v}^{\gamma}\partial_{\gamma}\{G, \underset{i}{v}\}^{\Lambda}].\end{aligned}$$

Also from (3.17) we find

$$(\underset{j}{v}^{\gamma}\partial_{\gamma}\underset{i}{v}^{\beta})\Theta^{\Lambda}_{\beta} + \underset{j}{v}^{\gamma}\underset{i}{v}^{\beta}\partial_{\gamma}\Theta^{\Lambda}_{\beta} = (\underset{j}{v}^{\gamma}\partial_{\gamma}\{F, \underset{i}{v}\}^{\Lambda}_{\Pi})\Omega^{\Pi} + \underset{j}{v}^{\gamma}\partial_{\gamma}\{G, \underset{i}{v}\}^{\Lambda},$$

from which

$$\begin{aligned}& (\underset{j}{v}^{\gamma}\partial_{\gamma}\underset{i}{v}^{\beta} - \underset{i}{v}^{\gamma}\partial_{\gamma}\underset{j}{v}^{\beta})\Theta^{\Lambda}_{\beta} + \underset{j}{v}^{\gamma}\underset{i}{v}^{\beta}[\partial_{\gamma}\Theta^{\Lambda}_{\beta} - \partial_{\beta}\Theta^{\Lambda}_{\gamma}] \\ &\qquad = [\underset{j}{v}^{\gamma}\partial_{\gamma}\{F, \underset{i}{v}\}^{\Lambda}_{\Pi} - \underset{i}{v}^{\gamma}\partial_{\gamma}\{F, \underset{j}{v}\}^{\Lambda}_{\Pi}]\Omega^{\Pi} \\ &\qquad\quad + [\underset{j}{v}^{\gamma}\partial_{\gamma}\{G, \underset{i}{v}\}^{\Lambda} - \underset{i}{v}^{\gamma}\partial_{\gamma}\{G, \underset{j}{v}\}^{\Lambda}],\end{aligned}$$

or

$$\begin{aligned}(3.23)\quad & \underset{j}{v}^{\gamma}\underset{i}{v}^{\beta}[\partial_{\gamma}\Theta^{\Lambda}_{\beta} - \partial_{\beta}\Theta^{\Lambda}_{\gamma}] \\ &= -c^{h}_{ji}\underset{h}{v}^{\alpha}\Theta^{\Lambda}_{\alpha} - c^{u}_{ji}\varphi^{h}_{u}\underset{h}{v}^{\alpha}\Theta^{\Lambda}_{\alpha} + [\underset{j}{v}^{\gamma}\partial_{\gamma}\{F, \underset{i}{v}\}^{\Lambda}_{\Pi} - \underset{i}{v}^{\gamma}\partial_{\gamma}\{F, \underset{j}{v}\}^{\Lambda}_{\Pi}]\Omega^{\Pi} \\ &\quad + [\underset{j}{v}^{\gamma}\partial_{\gamma}\{G, \underset{i}{v}\}^{\Lambda} - \underset{i}{v}^{\gamma}\partial_{\gamma}\{G, \underset{j}{v}\}^{\Lambda}].\end{aligned}$$

Comparing (3.22) and (3.23) and taking account of (3.17) and (3.18), we obtain

$$(3.24)\quad \underset{j}{v}^{\gamma}\underset{i}{v}^{\beta}[\Theta^{\Pi}_{\gamma}\partial_{\Pi}\Theta^{\Lambda}_{\beta} - \Theta^{\Pi}_{\beta}\partial_{\Pi}\Theta^{\Lambda}_{\gamma} + \partial_{\gamma}\Theta^{\Lambda}_{\beta} - \partial_{\beta}\Theta^{\Lambda}_{\gamma}] + c^{u}_{ji}\Xi^{\Lambda}_{u} = 0.$$

The equation (3.24) shows that, if we take account of $\Xi^{\Lambda}_{u}(\Omega, \xi) = 0$, then we have

$$(3.25)\quad \Theta^{\Pi}_{\gamma}\partial_{\Pi}\Theta^{\Lambda}_{\beta} + \partial_{\gamma}\Theta^{\Lambda}_{\beta} = \Theta^{\Pi}_{\beta}\partial_{\Pi}\Theta^{\Lambda}_{\gamma} + \partial_{\beta}\Theta^{\Lambda}_{\gamma}.$$

From (3.18) we get

$$\partial_{\Pi}\Xi^{\Lambda}_{u} = \varphi^{h}_{u}\{F, \underset{h}{v}\}^{\Lambda}_{\Pi} - \{F, \varphi^{h}_{u}\underset{h}{v}\}^{\Lambda}_{\Pi}$$

and consequently

$$(3.26)\quad \underset{i}{v}^{\alpha}\Theta^{\Pi}_{\alpha}\partial_{\Pi}\Xi^{\Lambda}_{u} = \varphi^{h}_{u}\{F, \underset{h}{v}\}^{\Lambda}_{\Pi}\{F, \underset{i}{v}\}^{\Pi}_{\Sigma}\Omega^{\Sigma} + \varphi^{h}_{u}\{F, \underset{h}{v}\}^{\Lambda}_{\Pi}\{G, \underset{i}{v}\}^{\Pi}$$
$$- \{F, \varphi^{h}_{u}\underset{h}{v}\}^{\Lambda}_{\Pi}\{F, \underset{i}{v}\}^{\Pi}_{\Sigma}\Omega^{\Sigma} - \{F, \varphi^{h}_{u}\underset{h}{v}\}^{\Lambda}_{\Pi}\{G, \underset{i}{v}\}^{\Pi}.$$

Also from (3.18) we obtain

$$(3.27)\quad \underset{i}{v}^{\alpha}\partial_{\alpha}\Xi^{\Lambda}_{u} = (\underset{i}{v}^{\alpha}\partial_{\alpha}\varphi^{h}_{u})[\{F, \underset{h}{v}\}^{\Lambda}_{\Pi}\Omega^{\Pi} + \{G, \underset{h}{v}\}^{\Lambda}]$$
$$+ \varphi^{h}_{u}[(\underset{i}{v}^{\alpha}\partial_{\alpha}\{F, \underset{h}{v}\}^{\Lambda}_{\Pi})\Omega^{\Pi} + \underset{i}{v}^{\alpha}\partial_{\alpha}\{G, \underset{h}{v}\}^{\Lambda}]$$
$$- (\underset{i}{v}^{\alpha}\partial_{\alpha}\{F, \varphi^{h}_{u}\underset{h}{v}\}^{\Lambda}_{\Pi})\Omega^{\Pi} - \underset{i}{v}^{\alpha}\partial_{\alpha}\{G, \varphi^{h}_{u}\underset{h}{v}\}^{\Lambda}.$$

Adding (3.26) and (3.27) and taking account of (3.3) and (3.4), we find

$$(3.28)\quad \underset{i}{v}^{\alpha}[\Theta^{\Pi}_{\alpha}\partial_{\Pi}\Xi^{\Lambda}_{u} + \partial_{\alpha}\Xi^{\Lambda}_{u}]$$
$$= \varphi^{h}_{u}[\{F, \underset{i}{v}\}^{\Lambda}_{\Pi}\{F, \underset{h}{v}\}^{\Pi}_{\Sigma} + \underset{h}{v}^{\alpha}\partial_{\alpha}\{F, \underset{i}{v}\}^{\Lambda}_{\Sigma} - c^{j}_{hi}\{F, \underset{j}{v}\}^{\Lambda}_{\Sigma} - c^{v}_{hi}\{F, \varphi^{l}_{v}\underset{l}{v}\}^{\Lambda}_{\Sigma}]\Omega^{\Sigma}$$
$$+ \varphi^{h}_{u}[\{F, \underset{i}{v}\}^{\Lambda}_{\Pi}\{G, \underset{h}{v}\}^{\Pi} + \underset{h}{v}^{\alpha}\partial_{\alpha}\{G, \underset{i}{v}\}^{\Lambda} - c^{j}_{hi}\{G, \underset{j}{v}\}^{\Lambda} - c^{v}_{hi}\{G, \varphi^{l}_{v}\underset{l}{v}\}^{\Lambda}]$$
$$- [\{F, \underset{i}{v}\}^{\Lambda}_{\Pi}\{F, \varphi^{l}_{u}\underset{l}{v}\}^{\Pi}_{\Sigma} + \varphi^{l}_{u}\underset{l}{v}^{\alpha}\partial_{\alpha}\{F, \underset{i}{v}\}^{\Lambda}_{\Sigma} + c^{j}_{iu}\{F, \underset{j}{v}\}^{\Lambda}_{\Sigma} + c^{v}_{iu}\{F, \varphi^{l}_{v}\underset{l}{v}\}^{\Lambda}_{\Sigma}]\Omega^{\Sigma}$$
$$- [\{F, \underset{i}{v}\}^{\Lambda}_{\Pi}\{G, \varphi^{l}_{u}\underset{l}{v}\}^{\Pi} + \varphi^{l}_{u}\underset{l}{v}^{\alpha}\partial_{\alpha}\{G, \underset{i}{v}\}^{\Lambda} + c^{j}_{iu}\{G, \underset{j}{v}\}^{\Lambda} + c^{v}_{iu}\{G, \varphi^{l}_{v}\underset{l}{v}\}^{\Lambda}]$$
$$+ (\underset{i}{v}^{\alpha}\partial_{\alpha}\varphi^{h}_{u})[\{F, \underset{h}{v}\}^{\Lambda}_{\Pi}\Omega^{\Pi} + \{G, \underset{h}{v}\}^{\Lambda}].$$

On the other hand, putting $c = i$, $b = u$, $\varkappa = \alpha$ in (3.5), we find

$$\text{(3.29)} \qquad \underset{i}{v}^{\gamma}(\partial_\gamma \varphi^h_u)\underset{h}{v}^{\alpha} = \varphi^h_u c^j_{hi} \underset{j}{v}^{\alpha} + \varphi^h_u c^v_{hi} \varphi^l_v \underset{l}{v}^{\alpha} + c^j_{iu} \underset{j}{v}^{\alpha} + c^v_{iu} \varphi^l_v \underset{l}{v}^{\alpha}.$$

Thus from (3.28), we obtain

$$\text{(3.30)} \qquad \underset{i}{v}^{\alpha}[\Theta^{\Pi}_{\alpha} \partial_{\Pi} \Xi^{\Lambda}_{u} + \partial_\alpha \Xi^{\Lambda}_{u}] = \{F, \underset{i}{v}\}^{\Lambda}_{\Pi} \Xi^{\Pi}_{u} + \varphi^h_u c^v_{hi} \Xi^{\Lambda}_{v} + c^v_{iu} \Xi^{\Lambda}_{v}.$$

The equation (3.30) shows that if we take account of $\Xi^{\Lambda}_{u} = 0$ we have

$$\text{(3.31)} \qquad \Theta^{\Pi}_{\alpha} \partial_{\Pi} \Xi^{\Lambda}_{u} + \partial_\alpha \Xi^{\Lambda}_{u} = 0.$$

The equations (3.25) and (3.31) show that the mixed system (3.21) of partial differential equations is completely integrable. We thus have

THEOREM 3.2. *Let there be given an r-parameter intransitive group of transformations in a space of n dimensions such that the rank q of* $\underset{a}{v}^{\varkappa}$ *in a certain neighbourhood is less than r. We choose a coordinate system with respect to which the vectors* $\underset{a}{v}^{\varkappa}$ *have the components* (3.15). *If the equations* $\Xi^{\Lambda}_{u}(\Omega, \xi) = 0$ *are consistent in* Ω^{Λ} *at a point of the space, then we can find a linear geometric object* Ω^{Λ} *of the given kind which has the given group as invariance group.*

We consider finally the case in which the group is multiply transitive, that is, $q = n < r$. The above discussions are valid also in this case if the indices take the values

$$i, j, k = 1, 2, \ldots, \mathrm{n}; \; u, v = \mathrm{n} + 1, \ldots, \mathrm{n};$$
$$\alpha, \beta, \gamma = 1, 2, \ldots, n,$$

and we have

THEOREM 3.3. *If there is given an r-parameter multiply transitive group of transformations in a space of n dimensions and if* $\Xi^{\Lambda}_{u}(\Omega, \xi) = 0$ *are consistent in* Ω^{Λ} *at a point of the space, then we can find a linear geometric object of the given kind which has the given group as invariance group.*

§ 4. Generalizations of the preceding theorems.[1]

Let us consider an r-parameter group of transformations generated by r infinitesimal operators $\underset{a}{\pounds} f = \underset{a}{v}^{\varkappa} \partial_\varkappa f$ in a space of n dimensions. Then

[1] YANO and TASHIRO [1].

for any linear geometric object Ω^Λ we have the formula

$$(\underset{c}{\pounds}\underset{b}{\pounds})\Omega^\Lambda = c_{cb}^{a}\underset{a}{\pounds}\Omega^\Lambda. \tag{4.1}$$

If the transformation law of Ω^Λ is given by (2.4) then that of $\underset{a}{\pounds}\Omega^\Lambda = \underset{a}{\Phi}^\Lambda$ is given by (2.5). Substituting $\underset{a}{\pounds}\Omega^\Lambda = \underset{a}{\Phi}^\Lambda$ in (4.1), we find

$$\underset{c}{\pounds}\underset{b}{\Phi}^\Lambda - \underset{b}{\pounds}\underset{c}{\Phi}^\Lambda = c_{cb}^{a}\underset{a}{\Phi}^\Lambda. \tag{4.2}$$

If r linear homogeneous geometric objects $\underset{a}{\Phi}^\Lambda$ satisfy (4.2) we say that they from a *complete system* with respect to the given r-parameter group.

We shall consider in this section the following problem. If $\underset{a}{v}^\varkappa$ are r vectors defining an r-parameter group of transformations in a space of n dimensions and if $\underset{a}{\Phi}^\Lambda$ are r linear homogeneous geometric objects of the same type forming a complete system with respect to the given group, is the system of partial differential equations

$$\underset{a}{\pounds}\Omega^\Lambda = \underset{a}{v}^\mu \partial_\mu \Omega^\Lambda - \{F, \underset{a}{v}\}_\Pi^\Lambda \Omega^\Pi - \{G, \underset{a}{v}\}^\Lambda = \underset{a}{\Phi}^\Lambda \tag{4.3}$$

integrable?

We shall first consider the case in which the rank of $\underset{a}{v}^\varkappa$ in a certain neighbourhood is equal to $r \leq n$. In this case we can choose a coordinate system with respect to which the components of the vectors $\underset{a}{v}^\varkappa$ satisfy (3.6). In such a coordinate system the equations (4.3) take the form

$$\underset{a}{\pounds}\Omega^\Lambda = \underset{a}{v}^\alpha \partial_\alpha \Omega^\Lambda - \{F, \underset{a}{v}\}_\Pi^\Lambda \Omega^\Pi - \{G, \underset{a}{v}\}^\Lambda = \underset{a}{\Phi}^\Lambda,$$

consequently if we define the functions $\Theta_\alpha^\Lambda(\Omega, \Phi, \xi)$ by

$$\underset{a}{v}^\alpha \Theta_\alpha^\Lambda(\Omega, \Phi, \xi) \stackrel{\text{def}}{=} \{F, \underset{a}{v}\}_\Pi^\Lambda \Omega^\Pi + \{G, \underset{a}{v}\}^\Lambda + \underset{a}{\Phi}^\Lambda, \tag{4.4}$$

we get

$$\underset{a}{\pounds}\Omega^\Lambda - \underset{a}{\Phi}^\Lambda = \underset{a}{v}^\alpha[\partial_\alpha \Omega^\Lambda - \Theta_\alpha^\Lambda(\Omega, \Phi, \xi)] = 0 \tag{4.5}$$

or

$$\partial_\alpha \Omega^\Lambda = \Theta_\alpha^\Lambda(\Omega, \Phi, \xi). \tag{4.6}$$

By the same method as was used in § 3, we can prove

$$\begin{aligned} &\Theta_\gamma^\Pi \partial_\Pi \Theta_\beta^\Lambda + (\partial_\gamma \underset{a}{\Phi}^\Pi)(\partial_\Pi^a \Theta_\beta^\Lambda) + \partial_\gamma \Theta_\beta^\Lambda \\ &\qquad = \Theta_\beta^\Pi \partial_\Pi \Theta_\gamma^\Lambda + (\partial_\beta \underset{a}{\Phi}^\Pi)(\partial_\Pi^a \Theta_\gamma^\Lambda) + \partial_\beta \Theta_\gamma^\Lambda;\ \partial_\Pi^a = \frac{\partial}{\partial \underset{a}{\Phi}^\Pi}. \end{aligned} \tag{4.7}$$

This equation means that the system (4.6) of partial differential equations is completely integrable. Thus we have

THEOREM 4.1. *If r linear homogeneous geometric objects $\underset{a}{\Phi}^{\Lambda}$ of the same type form a complete system with respect to an r-parameter group of transformations in a space of n dimensions and the group is such that the rank of $\underset{a}{v}^{\varkappa}$ in a neighbourhood is $r \leq n$, then the system of partial differential equations $\underset{a}{\pounds}\Omega^{\Lambda} = \underset{a}{\Phi}^{\Lambda}$ for a linear geometric object Ω^{Λ} is completely integrable.*

We shall next consider the case in which the group is intransitive and the rank q of $\underset{a}{v}^{\varkappa}$ in a neighbourhood is less than r. In this case, we can take a coordinate system with respect to which the components of $\underset{a}{v}^{\varkappa}$ satisfy (3.15). In such a coordinate system, the equations (4.3) take the form

$$
(4.8) \quad \begin{cases} \text{(a)} \ \underset{h}{\pounds}\Omega^{\Lambda} = \underset{h}{v}^{\alpha}\partial_{\alpha}\Omega^{\Lambda} - \{F, \underset{h}{v}\}^{\Lambda}_{\Pi}\Omega^{\Pi} - \{G, \underset{h}{v}\}^{\Lambda} = \underset{h}{\Phi}^{\Lambda}, \\ \text{(b)} \ \underset{u}{\pounds}\Omega^{\Lambda} = \varphi^{h}_{u}\underset{h}{v}^{\alpha}\partial_{\alpha}\Omega^{\Lambda} - \{F, \varphi^{h}_{u}\underset{h}{v}\}^{\Lambda}_{\Pi}\Omega^{\Pi} - \{G, \varphi^{h}_{u}\underset{h}{v}\}^{\Lambda} = \underset{u}{\Phi}^{\Lambda}. \end{cases}
$$

If we define the functions $\Theta^{\Lambda}_{\alpha}(\Omega, \Phi, \xi)$ and $\Xi^{\Lambda}_{u}(\Omega, \Phi, \xi)$ by

$$
(4.9) \qquad \underset{h}{v}^{\alpha}\Theta^{\Lambda}_{\alpha}(\Omega, \Phi, \xi) \overset{\text{def}}{=} \{F, \underset{h}{v}\}^{\Lambda}_{\Pi}\Omega^{\Pi} + \{G, \underset{h}{v}\}^{\Lambda} + \underset{h}{\Phi}^{\Lambda}
$$

and

$$
(4.10) \quad \Xi^{\Lambda}_{u}(\Omega, \Phi, \xi) \overset{\text{def}}{=} \varphi^{h}_{u}[\{F, \underset{h}{v}\}^{\Lambda}_{\Pi}\Omega^{\Pi} + \{G, \underset{h}{v}\}^{\Lambda}] - \{F, \varphi^{h}_{u}\underset{h}{v}\}^{\Lambda}_{\Pi}\Omega^{\Pi} - \{G, \varphi^{h}_{u}\underset{h}{v}\}^{\Lambda} - \underset{u}{\Phi}^{\Lambda}
$$

respectively, we can write (4.9) in the form

$$
(4.11) \quad \begin{cases} \text{(a)} \ \underset{h}{\pounds}\Omega^{\Lambda} - \underset{h}{\Phi}^{\Lambda} = \underset{h}{v}^{\alpha}[\partial_{\alpha}\Omega^{\Lambda} - \Theta^{\Lambda}_{\alpha}(\Omega, \Phi, \xi)] = 0, \\ \text{(b)} \ \underset{u}{\pounds}\Omega^{\Lambda} - \underset{u}{\Phi}^{\Lambda} = \varphi^{h}_{u}[\underset{h}{\pounds}\Omega^{\Lambda} - \underset{h}{\Phi}^{\Lambda}] + \Xi^{\Lambda}_{u}(\Omega, \Phi, \xi) = 0. \end{cases}
$$

These equations are equivalent to

$$
(4.12) \qquad \partial_{\alpha}\Omega^{\Lambda} = \Theta^{\Lambda}_{\alpha}(\Omega, \Phi, \xi), \ \ \Xi^{\Lambda}_{u}(\Omega, \Phi, \xi) = 0.
$$

By the same method as was used in § 3, we can prove

$$
\underset{j}{v}^{\gamma}\underset{i}{v}^{\beta}[\Theta^{\Pi}_{\gamma}\partial_{\Pi}\Theta^{\Lambda}_{\beta} + (\partial_{\gamma}\underset{a}{\Phi}^{\Pi})(\partial^{a}_{\Pi}\Theta^{\Lambda}_{\beta}) + \partial_{\gamma}\Theta^{\Lambda}_{\beta} - \Theta^{\Pi}_{\beta}\partial_{\Pi}\Theta^{\Lambda}_{\gamma} - (\partial_{\beta}\underset{a}{\Phi}^{\Pi})(\partial^{a}_{\Pi}\Theta^{\Lambda}_{\gamma}) - \partial_{\beta}\Theta^{\Lambda}_{\gamma}] + c^{u}_{ji}\Xi^{\Lambda}_{u} = 0,
$$

which shows that if we take account of $\Xi^{\Lambda}_{u} = 0$, we have

$$(4.13) \qquad \Theta^{\Pi}_{\gamma} \partial_{\Pi} \Theta^{\Lambda}_{\beta} + (\partial_{\gamma} \underset{a}{\Phi}^{\Pi})(\partial^{a}_{\Pi} \Theta^{\Lambda}_{\beta}) + \partial_{\gamma} \Theta^{\Lambda}_{\beta} = \Theta^{\Pi}_{\beta} \partial_{\Pi} \Theta^{\Lambda}_{\gamma} + (\partial_{\beta} \underset{a}{\Phi}^{\Pi})(\partial^{a}_{\Pi} \Theta^{\Lambda}_{\gamma}) + \partial_{\beta} \Theta^{\Lambda}_{\gamma}.$$

We can also prove that

$$(4.14) \qquad \underset{i}{v}^{\alpha}[\Theta^{\Pi}_{\alpha} \partial_{\Pi} \Xi^{\Lambda}_{u} + (\partial_{\alpha} \underset{a}{\Phi}^{\Pi})(\partial^{a}_{\Pi} \Xi^{\Lambda}_{u}) + \partial_{\alpha} \Xi^{\Lambda}_{u}] = \{F, \underset{i}{v}\}^{\Lambda}_{\Pi} \Xi^{\Pi}_{u} + \varphi^{h}_{u} c^{v}_{hi} \Xi^{\Lambda}_{v} + c^{v}_{iu} \Xi^{\Lambda}_{v},$$

which shows that if we take account of $\Xi^{\Lambda}_{u} = 0$, we have

$$(4.15) \qquad \Theta^{\Pi}_{\alpha} \partial_{\Pi} \Xi^{\Lambda}_{u} + (\partial_{\alpha} \underset{a}{\Phi}^{\Pi})(\partial^{a}_{\Pi} \Xi^{\Lambda}_{u}) + \partial_{\alpha} \Xi^{\Lambda}_{u} = 0.$$

The equations (4.13) and (4.15) show that the mixed system (4.12) of partial differential equations is completely integrable. We thus have

THEOREM 4.2. *If r linear homogeneous geometric objects $\underset{a}{\Phi}^{\Lambda}$ of the same type form a complete system with respect to an intransitive r-parameter group of transformations in a space of n dimensions, if the rank q of $\underset{a}{v}^{\varkappa}$ defining the group is less than r and if the equations $\Xi^{\Lambda}_{u}(\Omega, \Phi, \xi) = 0$ are consistent in Ω^{Λ} at a point of the space, then the mixed system of partial differential equations $\underset{a}{\pounds}\Omega^{\Lambda} = \underset{a}{\Phi}^{\Lambda}$ is completely integrable.*

Similarly we have

THEOREM 4.3. *If r linear homogeneous geometric objects $\underset{a}{\Phi}^{\Lambda}$ of the same type form a complete system with respect to a multiply transitive r-parameter group of transformations in a space of n dimensions and if the equations $\Xi^{\Lambda}_{u}(\Omega, \Phi, \xi) = 0$ are consistent in Ω^{Λ} at a point of the space, then the mixed system of partial differential equations $\underset{a}{\pounds}\Omega^{\Lambda} = \underset{a}{\Phi}^{\Lambda}$ is completely integrable.*

§ 5. Some applications.

We shall state in this section some applications of the theorems proved in the preceding sections. But for the sake of simplicity, we shall state them only for the case in which the rank of $\underset{a}{v}^{\varkappa}$ is equal to $r \leq n$.

THEOREM 5.1. *Consider an n-dimensional space with a linear homogeneous geometric object Ω^{Λ} and suppose that the field admits an r-para-*

meter group of transformations such that the rank of $\underset{a}{v}^{\varkappa}$ *in a neighbourhood is* $r \leq n$ *and that*

$$\underset{a}{\pounds}\Omega^{\Lambda} = \underset{a}{\rho}\Omega^{\Lambda}, \tag{5.1}$$

where $\underset{a}{\rho}$ *are* r *scalars. Then we can find a scalar* ρ *such that the group is an invariance group of the geometric object* $\rho\Omega^{\Lambda}$.

On substituting (5.1) in the identities

$$\underset{c}{\pounds}\underset{b}{\pounds}\Omega^{\Lambda} - \underset{b}{\pounds}\underset{c}{\pounds}\Omega^{\Lambda} = c_{cb}^{a}\underset{a}{\pounds}\Omega^{\Lambda}, \tag{5.2}$$

we find

$$\underset{c}{\pounds}\underset{b}{\rho} - \underset{b}{\pounds}\underset{c}{\rho} = c_{cb}^{a}\underset{a}{\rho}, \tag{5.3}$$

which shows that the r scalars $\underset{a}{\rho}$ form a complete system with respect to the given group.

To prove the theorem we have now only to show that there exists a scalar ρ such that

$$\underset{a}{\pounds}(\rho\Omega^{\Lambda}) = 0$$

or

$$\underset{a}{\pounds}\log\rho = -\underset{a}{\rho}. \tag{5.4}$$

But since the $\underset{a}{\rho}$ form a complete system with respect to the group, according to Theorem 4.1 this system of partial differential equations is completely integrable. The theorem is thus proved.

THEOREM 5.2. *Consider an n-dimensional space with a linear geometric object* Ω^{Λ} *and suppose that the space admits an r-parameter group of transformations such that the rank of* $\underset{a}{v}^{\varkappa}$ *in a neighbourhood is* $r \leq n$ *and that*

$$\underset{a}{\pounds}\Omega^{\Lambda} = \underset{a}{\Phi}^{\Lambda}, \tag{5.5}$$

where $\underset{a}{\Phi}^{\Lambda}$ *are* r *linear homogeneous geometric objects. Then we can always find a linear homogeneous geometric object* Φ^{Λ} *such that the group is an invariance group of the geometric object* $\Omega^{\Lambda} + \Phi^{\Lambda}$.

On substituting (5.5) into (5.2), we find

$$\underset{c}{\pounds}\underset{b}{\Phi}^{\Lambda} - \underset{b}{\pounds}\underset{c}{\Phi}^{\Lambda} = c_{cb}^{a}\underset{a}{\Phi}^{\Lambda} \tag{5.6}$$

which shows that the functions $\underset{a}{\Phi}^{\Lambda}$ form a complete system with respect to the given group.

To prove the theorem, we have only to show the existence of a geometric object Φ^{Λ} such that

$$\underset{a}{\pounds}(\Omega^{\Lambda} + \Phi^{\Lambda}) = 0.$$

But this equation can be written as

$$\underset{a}{\pounds}\Phi^{\Lambda} = -\underset{a}{\Phi}^{\Lambda}, \tag{5.7}$$

and since the functions $\underset{a}{\Phi}^{\Lambda}$ form a complete system with respect to the group, according to Theorem 4.1, the system (5.7) is completely integrable. Thus the theorem is proved.

CHAPTER IV

GROUPS OF MOTIONS IN V_n

§ 1. Groups of motions.

Consider an n-dimensional Riemannian space V_n with the fundamental quadratic differential form

$$ds^2 = g_{\lambda\varkappa} d\xi^\lambda d\xi^\varkappa. \tag{1.1}$$

In order that an infinitesimal point transformation

$$'\xi^\varkappa = \xi^\varkappa + v^\varkappa(\xi)dt \tag{1.2}$$

be a motion in the V_n, it is necessary and sufficient that the Lie derivative of $g_{\lambda\varkappa}$ with respect to (1.2) vanish:

$$\underset{v}{\pounds} g_{\lambda\varkappa} = 2\nabla_{(\lambda} v_{\varkappa)} = 0. \tag{1.3}$$

Now take a geodesic $\xi^\varkappa(s)$ in the V_n and consider the inner product of a Killing vector $v_\varkappa$ and a unit tangent $\frac{d\xi^\varkappa}{ds}$ to the geodesic. Then along the geodesic we have

$$\frac{\delta}{ds}\left(v_\varkappa \frac{d\xi^\varkappa}{ds}\right) = \nabla_{(\lambda} v_{\varkappa)} \frac{d\xi^\lambda}{ds}\frac{d\xi^\varkappa}{ds} = 0, \tag{1.4}$$

which shows that the inner product is constant along the geodesic.

Conversely, if the inner product is constant along any geodesic, then we have (1.4) for any $d\xi^\varkappa/ds$ and consequently we have (1.3). Thus we have

THEOREM 1.1.[1] *In order that a vector field $v^\varkappa(\xi)$ define an infinitesimal motion in a V_n, it is necessary and sufficient that the inner product of $v^\varkappa$ and a unit tangent to an arbitrary geodesic be constant along this geodesic.*

In order that, $\rho v^\varkappa$ define a motion always if $v^\varkappa$ defines a motion, it is necessary and sufficient that

$$\nabla_\lambda(\rho v_\varkappa) + \nabla_\varkappa(\rho v_\lambda) = 0,$$

---

[1] Cf. EISENHART [4], p. 212; K. YANO [13], p. 30. This book will be referred to as G. T.

or

$$(\nabla_\lambda \rho)v_\varkappa + (\nabla_\varkappa \rho)v_\lambda = 0,$$

from which we conclude $\nabla_\lambda \rho = 0$ and consequently $\rho =$ constant. Thus we have

THEOREM 1.2.[1] *Two infinitesimal motions cannot have the same trajectories.* [2]

Now since the tensor $g_{\lambda\varkappa}$ is a linear homogeneous differential geometric object, according to Theorems 2.1 and 2.2 of Ch. III, we have respectively

THEOREM 1.3. [3] *If a V_n admits an infinitesimal motion, it admits also a one-parameter group of motions generated by this infinitesimal motion.*

THEOREM 1.4.[4] *In order that a V_n admit a one-parameter group of motions it is necessary and sufficient that there exist a coordinate system in which the components of the fundamental tensor are independent of one of the coordinates.*

If we choose a coordinate system [5] in which $v^\varkappa = \xi^\varkappa$, the Killing equation becomes

$$\underset{v}{\pounds} g_{\lambda\varkappa} = \xi^\mu \partial_\mu g_{\lambda\varkappa} + 2g_{\lambda\varkappa} = 0,$$

from which

THEOREM 1.5. [6] *In order that a V_n admit a one-parameter group of motions, it is necessary and sufficient that there exist a coordinate system with respect to which the components of the fundamental tensor are homogeneous functions of degree -2 of the coordinates.*

---

[1] Cf. EISENHART [4], p. 210; G. T. p. 31.

[2] The trajectores of an infinitesimal transformation $'\xi^\varkappa = \xi^\varkappa + v^\varkappa dt$ are the curves defined by $\dfrac{d\xi^\varkappa}{dt} = v^\varkappa(\xi)$.

[3] Cf. EISENHART [4], p. 209; G. T., p. 31. S. Kobayashi proved that in a complete V_n, this theorem is globally true.

[4] Cf. EISENHART [4], p. 209; G.T., p. 31.

[5] Such a coordinate system is obtained in the following way. Take a coordinate system $(\varkappa)$ in which $v^\varkappa = \underset{1}{e^\varkappa}$. If we effect a coordinate transformation of the form

$$\xi^{\varkappa'} = e^{\xi^1} f^\varkappa(\xi^2, \xi^3, \ldots, \xi^u),$$

we have

$$v^{\varkappa'} = (\partial_\varkappa \xi^{\varkappa'})\underset{1}{e^\varkappa} = \xi^{\varkappa'}.$$

[6] Cf. G. T., p. 31.

Moreover from Theorems 2.3, 2.4, 2.5 and 2.6 of Ch. III, we have respectively

THEOREM 1.6.[1] *If each of r vectors generates a one-parameter group of motions in a V_n, then a linear combination of these vectors with constant coefficients generates also a one-parameter group of motions.*

THEOREM 1.7.[2] *If each of r infinitesimal operators of an r-parameter group generates a one-parameter group of motions in a V_n, then the group contains only motions.*

THEOREM 1.8.[3] *If each of r vectors $\overset{a}{v}{}^{\kappa}$ defines a one-parameter group of motions in a V_n, then the vector $\underset{c}{\pounds}\overset{}{\underset{b}{v}}{}^{\kappa}$ defines also a one-parameter group of motions.*

THEOREM 1.9.[4] *If r infinitesimal operators $\underset{a}{\pounds}f$ form a complete system of r one-parameter groups of motions, then the operators $\underset{a}{\pounds}f$ are generators of an r-parameter group of motions.*

§ 2. Groups of translations.

If the trajectories of a motion are geodesics, the motion is called a *translation*. In order that (1.2) be a translation, it is necessary and sufficient that

$$\underset{v}{\pounds} g_{\lambda\kappa} = 2\nabla_{(\lambda} v_{\kappa)} = 0, \quad v^{\lambda}\nabla_{\lambda} v_{\kappa} = \alpha v_{\kappa}. \tag{2.1}$$

On transvecting the second equation of (2.1) with v^{κ} ,we find

$$v^{\lambda} v^{\kappa} \nabla_{(\lambda} v_{\kappa)} = \alpha v^{\kappa} v_{\kappa},$$

from which, using the first equation of (2.1), we get $\alpha = 0$. Consequently, transvecting the first equation of (2.1) with v^{λ}, we find

$$\tfrac{1}{2}\nabla_{\kappa}(v^{\lambda} v_{\lambda}) = 0,$$

from which it follows that $v^{\lambda}v_{\lambda} =$ constant.

Conversely, if a vector v^{κ} satisfying $v^{\lambda}v_{\lambda} =$ constant defines a motion, transvecting the first equation of (2.1) with v^{λ}, we find the second equation with $\alpha = 0$. Thus we have

---

[1] Cf. EISENHART [4], p. 210; G. T., p. 31.
[2] Cf. EISENHART [4], p. 210; G. T., p. 31.
[3] Cf. EISENHART [4], p. 216; G. T., p. 33.
[4] Cf. EISENHART [4], p. 216; G. T., p. 33.

THEOREM 2.1.[1] *In order that* (1.2) *be a translation in a* V_n, *it is necessary and sufficient that* $\underset{v}{\pounds} g_{\lambda\varkappa} = 0$ *and* $g_{\lambda\varkappa}v^\lambda v^\varkappa =$ *constant. In this case every point is moved over the same distance.*

Now according to Theorems 1.1 and 2.1, we have

THEOREM 2.2. *In order that a vector field of constant length define an infinitesimal translation, it is necessary and sufficient that along every geodesic it make the same angle with the tangent.*

If we take a coordinate system with respect to which $v^\varkappa = \underset{1}{e}^\varkappa$, we have from Theorem 2.1

$$\partial_1 g_{\lambda\varkappa} = 0, \quad g_{11} = \text{constant}. \tag{2.2}$$

It is evident that if (2.2) holds in some coordinate system, the group of transformations

$$'\xi^\varkappa = \xi^\varkappa + \underset{1}{e}^\varkappa t$$

is that of translations. Thus Theorem 1.3 is true also for the group of translations, and corresponding to 1.4, we have

THEOREM 2.3. *In order that a* V_n *admit a one-parameter group of translations it is necessary and sufficient that there exists a coordinate system with respect to which the components* $g_{\lambda\varkappa}$ *of the fundamental tensor are independent of* ξ^1 *and* g_{11} *is a constant.* [2]

Also Theorems 1.6 and 1.7 are true for translations as can be proved easily.

§ 3. Motions and affine motions.

The following theorem is geometrically evident.

THEOREM 3.1. [3] *A motion in a* V_n *is an affine motion.*

---

[1] Cf. EISENHART [4], p. 212; G. T., p. 32.

[2] Such a space has been used to construct a 5-dimensional unified field theory of gravitation and electromagnetism. The ds^2 of such a space is of the form

$$ds^2 = (d\xi^1 + g_{1\rho}d\xi^\rho)^2 + (g_{\sigma\rho} - g_{1\sigma}g_{1\rho})d\xi^\sigma d\xi^\rho, \quad \rho, \sigma = 2, 3, 4, 5.$$

In this equation, $g_{1\rho}$ is identified with the electromagnetic potential and $g_{\sigma\rho} - g_{1\sigma}g_{1\rho}$ with the gravitational potential. See for example KALUZA [1], KLEIN [1], YANO [3].

[3] Cf. EISENHART [4], p. 210; G. T., p. 34.

To prove this, we apply the formula (4.9) of Ch. I to the fundamental tensor $g_{\lambda\kappa}$:

$$\underset{v}{\pounds}(\nabla_\mu g_{\lambda\kappa}) - \nabla_\mu(\underset{v}{\pounds} g_{\lambda\kappa}) = -(\underset{v}{\pounds}\{{}^{\rho}_{\mu\lambda}\})g_{\rho\kappa} - (\underset{v}{\pounds}\{{}^{\rho}_{\mu\kappa}\})g_{\lambda\rho},$$

from which

$$\underset{v}{\pounds}\{{}^{\kappa}_{\mu\lambda}\} = \tfrac{1}{2}g^{\kappa\rho}[\nabla_\mu \underset{v}{\pounds} g_{\lambda\rho} + \nabla_\lambda \underset{v}{\pounds} g_{\mu\rho} - \nabla_\rho \underset{v}{\pounds} g_{\mu\lambda}]. \tag{3.1}$$

This equation shows that $\underset{v}{\pounds} g_{\lambda\kappa} = 0$ implies $\underset{v}{\pounds}\{{}^{\kappa}_{\mu\lambda}\} = 0$. [1]

The following theorem is also geometrically evident.

THEOREM 3.2. [2] *For a motion in a V_n the Lie derivatives of the curvature tensor and its successive covariant derivatives vanish.*

We prove this by applying the formula (4.14) of CH. I to the Christoffel symbol:

$$\nabla_\nu \underset{v}{\pounds}\{{}^{\kappa}_{\mu\lambda}\} - \nabla_\mu \underset{v}{\pounds}\{{}^{\kappa}_{\nu\lambda}\} = \underset{v}{\pounds} K_{\nu\mu\lambda}{}^{\kappa}, \tag{3.2}$$

where $K_{\nu\mu\lambda}{}^{\kappa}$ is the curvature tensor of V_n. Thus for a motion we have

$$\underset{v}{\pounds} K_{\nu\mu\lambda}{}^{\kappa} = 0. \tag{3.3}$$

On the other hand since a motion is an affine motion, the covariant derivation and the Lie derivation are commutative. Thus from (3.3) we obtain

$$\underset{v}{\pounds}\nabla_\omega K_{\nu\mu\lambda}{}^{\kappa} = 0, \quad \underset{v}{\pounds}\nabla_{\omega_2}\nabla_{\omega_1} K_{\nu\mu\lambda}{}^{\kappa} = 0, \quad \ldots . \tag{3.4}$$

§ 4. Some theorems on projectively or conformally related spaces

Consider two Riemannian spaces V_n and $'V_n$ which are in geodesic correspondence. Then denoting the Christoffel symbols of them by $\{{}^{\kappa}_{\mu\lambda}\}$ and $'\{{}^{\kappa}_{\mu\lambda}\}$ respectively, we have

$$'\{{}^{\kappa}_{\mu\lambda}\} = \{{}^{\kappa}_{\mu\lambda}\} + A^{\kappa}_{\mu} p_\lambda + A^{\kappa}_{\lambda} p_\mu$$

But since V_n and $'V_n$ are both Riemannian, the vector p_λ should be a gradient. [3] Thus putting $p_\lambda = \frac{1}{2}\partial_\lambda \log \phi$, we get

$$'\{{}^{\kappa}_{\mu\lambda}\} = \{{}^{\kappa}_{\mu\lambda}\} + \tfrac{1}{2}A^{\kappa}_{\mu}\partial_\lambda \log\phi + \tfrac{1}{2}A^{\kappa}_{\lambda}\partial_\mu \log\phi. \tag{4.1}$$

We now assume that the V_n admits a motion with symbol $\underset{v}{\pounds} f$. Then

[1] Under some global conditions $\underset{v}{\pounds}\{{}^{\kappa}_{\mu\lambda}\} = 0$ implies $\underset{v}{\pounds} g_{\lambda\kappa} = 0$. See Ch. IX.

[2] Cf. EISENHART [4], p. 213; G. T., p. 37.

[3] SCHOUTEN [8], p. 292.

we have

$$\underset{v}{\pounds} g_{\lambda\varkappa} = \nabla_\lambda v_\varkappa + \nabla_\varkappa v_\lambda = \partial_\lambda v_\varkappa + \partial_\varkappa v_\lambda - 2\{^{\ \rho}_{\lambda\varkappa}\} v_\rho = 0.$$

Consequently on utilising (4.1) we have

$$\pounds g_{\lambda\varkappa} = \partial_\lambda v_\varkappa + \partial_\varkappa v_\lambda - 2[{}'\{^{\ \rho}_{\lambda\varkappa}\} - \tfrac{1}{2}A^\rho_\lambda \partial_\varkappa \log \phi - \tfrac{1}{2}A^\rho_\varkappa \partial_\lambda \log \phi] v_\rho$$
$$= \phi^{-1}[\partial_\lambda(\phi v_\varkappa) + \partial_\varkappa(\phi v_\lambda) - 2'\{^{\ \rho}_{\lambda\varkappa}\}\phi v_\rho]$$

Thus denoting by $'g_{\lambda\varkappa}$ the fundamental tensor of $'V_n$ and by $'\pounds f$ the symbol defined by $\phi v_\varkappa$ in $'V_n$, we have

$$\pounds g_{\lambda\varkappa} = \phi^{-1\prime}\pounds' g_{\lambda\varkappa}.$$

Thus we have

THEOREM 4.1.[1] *If two Riemannian spaces V_n and $'V_n$ are in geodesic correspondence and if V_n admits a group of motions, $'V_n$ also admits a group of motions.*

Consider a V_n which admits an r-parameter group G_r of motions such that the rank of $\underset{a}{v}{}^\varkappa$ is in a certain neighbourhood equal to $r < n$. Then we have $\underset{a}{\pounds} g_{\lambda\varkappa} = 0$. In order that a space $'V_n$ conformal to V_n admit the same group G_r as a group of motions, it is necessary and sufficient that there exist a function ρ^2 such that $\underset{a}{\pounds}(\rho^2 g_{\lambda\varkappa}) = 0$ or $\underset{a}{\pounds}\rho^2 = 0$. But on the other hand

$$(\underset{c}{\pounds}\underset{b}{\pounds})\rho^2 = c^a_{cb}\underset{a}{\pounds}\rho^2$$

and consequently $\underset{a}{\pounds}\rho^2 = 0$ admits $n - r$ independent solutions. Thus we have

THEOREM 4.2.[2] *If a V_n admits a G_r of motions such that the rank of $\underset{a}{v}{}^\varkappa$ in a neighbourhood is equal to $r < n$, then there exist $n - r$ V_n's, corresponding to $n - r$ independent solutions of $\underset{a}{\pounds}\rho^2 = 0$, which are conformal to the given V_n and admit the same group as a group of motions.*

As an application of Theorem 5.1 of Ch. III, we have

THEOREM 4.3. *If a V_n admits a G_r of conformal motions such that the rank of $\underset{a}{v}{}^\varkappa$ in a neighbourhood is $r \leqq n$, then there exists a $'V_n$ which is conformal to V_n and which admits the G_r as a group of motions.*

---

[1] KNEBELMAN [5].

[2] KNEBELMAN [4].

Using this theorem we can prove the following theorem which generalizes a theorem of J. Levine. [1]

THEOREM 4.4.[2] *In order that a G_r in X_n such that the rank of $\underset{a}{v}^\varkappa$ in a neighbourhood is equal to $r \leq n$, can be regarded as a group of motions in a C_n,* [3] *it is necessary and sufficient that the group be a subgroup of a group of conformal transformations.*

The necessity is evident. Conversely, if the group is a subgroup of a group of conformal transformations, it is a group of conformal motions in a C_n. Consequently according to Theorem 4.3, there exists a V_n which is conformal to C_n and is itself a C_n which admits the group as a group of motions.

§ 5. A theorem of Knebelman. [4]

If a V_n admits an r-parameter group G_r of affine motions and if $\underset{a}{\pounds} f = \underset{a}{v}^\varkappa \partial_\varkappa f$ denotes r linearly independent infinitesimal operators of the group, then we have $(\underset{c}{\pounds}\underset{b}{\pounds})f = c^a_{cb}\underset{a}{\pounds} f$ and

$$\underset{a}{\pounds}\{^{\ \varkappa}_{\mu\lambda}\} = 0. \tag{5.1}$$

We ask for a necessary and sufficient condition that the group of affine motions contain a subgroup of motions.

In order that this be the case, it is necessary that there exist r constants c^a, which are not all zero, and such that $\pounds f = c^a \underset{a}{\pounds} f$ is a motion. But this means that $c^a \underset{a}{\pounds} g_{\lambda\varkappa} = 0$ and this is only possible if the a-rank [5] of $\underset{a}{\pounds} g_{\lambda\varkappa}$ is $< r$.

Conversely if the a-rank s of $\underset{a}{\pounds} g_{\lambda\varkappa}$ is $< r$, there are $r - s$ linearly independent solutions $\phi^a_u(\xi^\nu)$; $u, v, w, \ldots = 1, 2, \ldots, r - s$, of the equations

$$\phi^a_u(\xi^\nu)\underset{a}{\pounds} g_{\lambda\varkappa} = 0. \tag{5.2}$$

---

[1] LEVINE [2, 3].

[2] YANO and TASHIRO [1].

[3] C_n stands for a conformally Euclidean space, cf. SCHOUTEN [8], p. 305ff.

[4] KNEBELMAN [7]; cf. G. T., p. 43.

[5] The *a-rank* of the $\underset{a}{\pounds} g_{\lambda\varkappa}$ is the rank of the matrix $\underset{a}{\pounds} g_{\lambda\varkappa}$ where a denotes the rows and $\lambda\varkappa$ denotes the columns. Cf. SCHOUTEN [8], p. 20.

By covariant differentiation we get

$$(\nabla_\mu \phi^a_u) \underset{a}{\pounds} g_{\lambda\varkappa} + \phi^a_u \nabla_\mu \underset{a}{\pounds} g_{\lambda\varkappa} = 0. \tag{5.3}$$

But on the other hand, since $\underset{a}{\pounds} f$ are operators of a group of affine motions, (5.1) holds and consequently according to the formula

$$\underset{a}{\pounds}(\nabla_\mu g_{\lambda\varkappa}) - \nabla_\mu(\underset{a}{\pounds} g_{\lambda\varkappa}) = -(\underset{a}{\pounds}\{{}^{\rho}_{\mu\lambda}\}) g_{\rho\varkappa} - (\underset{a}{\pounds}\{{}^{\rho}_{\mu\varkappa}\}) g_{\lambda\rho},$$

which is a special case of (4.9) of Chapter I, we find

$$\nabla_\mu(\underset{a}{\pounds} g_{\lambda\varkappa}) = 0, \tag{5.4}$$

and consequently from (5.3)

$$(\nabla_\mu \phi^a_u) \underset{a}{\pounds} g_{\lambda\varkappa} = 0. \tag{5.5}$$

But since the ϕ^a_u are $r - s$ linearly independent solutions of $\phi^a \underset{a}{\pounds} g_{\lambda\varkappa} = 0$, we have $\nabla_\mu \phi^a_v = \Lambda^u_{\mu v} \phi^a_u$, where the $\Lambda^u_{\mu v}$ are functions of $\xi^\varkappa$. The integrability conditions of these equations are

$$\nabla_\nu \Lambda^u_{\mu v} - \nabla_\mu \Lambda^u_{\nu v} + \Lambda^u_{\nu w} \Lambda^w_{\mu v} - \Lambda^u_{\mu w} \Lambda^w_{\nu v} = 0. \tag{5.6}$$

Now if there exist $r - s$ functions $f^u(\xi)$ such that the transvections

$$c^a = f^u(\xi) \phi^a_u(\xi) \tag{5.7}$$

are constants, the group of affine motions contains a subgroup of motions. In order that the c^a in (5.7) be constants, the functions $f^u(\xi)$ should satisfy the equations

$$0 = (\nabla_\mu f^u) \phi^a_u + f^u(\nabla_\mu \phi^a_u),$$
$$0 = (\nabla_\mu f^u + \Lambda^u_{\mu v} f^v) \phi^a_u,$$

from which

$$\nabla_\mu f^u + \Lambda^u_{\mu v} f^v = 0. \tag{5.8}$$

But the integrability conditions of (5.8) are exactly given by (5.6). Thus we have proved the existence of functions $f^u(\xi)$ such that the c^a given by (5.7) are constants. This proves:

THEOREM 5.1. *In order that a G_r of affine motions in a V_n contains a subgroup of motions, it is necessary and sufficient that the a-rank of $\underset{a}{\pounds} g_{\lambda\varkappa}$ be less than r.*

§ 6. Integrability conditions of Killing's equation. [1]

The integrability conditions of Killing's equation

$$(6.1) \qquad \underset{v}{\pounds} g_{\lambda\varkappa} = \nabla_\lambda v_\varkappa + \nabla_\varkappa v_\lambda = 0$$

can be deduced from it, considering first the equation

$$(6.2) \qquad \underset{v}{\pounds}\{^{\varkappa}_{\mu\lambda}\} = \nabla_\mu \nabla_\lambda v^\varkappa + K_{\nu\mu\lambda}^{\cdot\cdot\cdot\varkappa} v^\nu = 0$$

and next the mixed system of partial differential equations

$$(6.3) \qquad \begin{cases} v_{\lambda\varkappa} + v_{\varkappa\lambda} = 0 & (v_{\lambda\varkappa} = v_{\lambda}^{\cdot\rho} g_{\rho\varkappa}) \\ \nabla_\lambda v^\varkappa = v_\lambda^{\cdot\varkappa}, & \nabla_\mu v_\lambda^{\cdot\varkappa} = - K_{\nu\mu\lambda}^{\cdot\cdot\cdot\varkappa} v^\nu. \end{cases}$$

We know that the equations (6.1) and (6.2) or the equations (6.3) have for integrability conditions

$$(6.4) \qquad \underset{v}{\pounds} K_{\nu\mu\lambda}^{\cdot\cdot\cdot\varkappa} = 0, \ \underset{v}{\pounds}\nabla_\omega K_{\nu\mu\lambda}^{\cdot\cdot\cdot\varkappa} = 0, \ \underset{v}{\pounds}\nabla_{\omega_2}\nabla_{\omega_1} K_{\nu\mu\lambda}^{\cdot\cdot\cdot\varkappa} = 0, \ \ldots.$$

Thus we have by a theorem [2] on partial differential equations

THEOREM 6.1. *In order that a V_n admit a group of motions, it is necessary and sufficient that there exist a positive integer N such that the first N sets of equations*

$$\underset{v}{\pounds} g_{\lambda\varkappa} = 0, \ \underset{v}{\pounds} K_{\nu\mu\lambda}^{\cdot\cdot\cdot\varkappa} = 0, \ \underset{v}{\pounds}\nabla_\omega K_{\nu\mu\lambda}^{\cdot\cdot\cdot\varkappa} = 0, \ \ldots$$

are compatible in $v^\varkappa$ and $v_\lambda^{\cdot\varkappa}$ and that the $v^\varkappa$ and $v_\lambda^{\cdot\varkappa}$ satisfying these equations satisfy the $(N+1)$st set of equations.

When there exist $\frac{1}{2}n(n+1) - r$ linearly independent equations in the first N sets except the first, the space admits a G_r of motions.

We shall examine the case in which the conditions

$$(6.5) \qquad \underset{v}{\pounds} K_{\nu\mu\lambda}^{\cdot\cdot\cdot\varkappa} = v^\rho \nabla_\rho K_{\nu\mu\lambda}^{\cdot\cdot\cdot\varkappa} - K_{\nu\mu\lambda}^{\cdot\cdot\cdot\rho} \nabla_\rho v^\varkappa + K_{\rho\mu\lambda}^{\cdot\cdot\cdot\varkappa} \nabla_\nu v^\rho + K_{\nu\rho\lambda}^{\cdot\cdot\cdot\varkappa} \nabla_\mu v^\rho + K_{\nu\mu\rho}^{\cdot\cdot\cdot\varkappa} \nabla_\lambda v^\rho = 0$$

are identically satisfied by arbitrary $v_\varkappa$ and $\nabla_\lambda v^\varkappa$ such that

$$\nabla_\lambda v_\varkappa + \nabla_\varkappa v_\lambda = 0.$$

[1] Cf. EISENHART [4], p. 214; SCHOUTEN [8], p. 350; G. T., p. 34.

[2] See for example, EISENHART [4], p. 1; T. Y. THOMAS [3]; VEBLEN [1], p. 73.

The equation (6.5) can be written as

$$v^\rho \nabla_\rho K_{\nu\mu\lambda\varkappa} - K_{\nu\mu\lambda}^{\cdot\cdot\cdot\rho} \nabla_\rho v_\varkappa + K_{\rho\mu\lambda\varkappa} \nabla_\nu v^\rho + K_{\nu\rho\lambda\varkappa} \nabla_\mu v^\rho + K_{\nu\mu\rho\varkappa} \nabla_\lambda v^\rho = 0$$

from which

$$v^\rho \nabla_\rho K_{\nu\mu\lambda\varkappa} - (K_{\nu\mu\lambda}^{\cdot\cdot\cdot\sigma} A_\varkappa^\rho - K_{\varkappa\lambda\mu}^{\cdot\cdot\cdot\rho} A_\nu^\sigma - K_{\lambda\varkappa\nu}^{\cdot\cdot\cdot\rho} A_\mu^\sigma + K_{\nu\mu\varkappa}^{\cdot\cdot\cdot\rho} A_\lambda^\sigma) \nabla_\sigma v_\rho = 0.$$

Since here v^ρ and $\nabla_\sigma v_\rho - \nabla_\rho v_\sigma$ are arbitrary, we get

$$\nabla_\rho K_{\nu\mu\lambda\varkappa} = 0$$

and

$$K_{\nu\mu\lambda}^{\cdot\cdot\cdot\sigma} A_\varkappa^\rho - K_{\varkappa\lambda\mu}^{\cdot\cdot\cdot\rho} A_\nu^\sigma - K_{\lambda\varkappa\nu}^{\cdot\cdot\cdot\rho} A_\mu^\sigma + K_{\nu\mu\varkappa}^{\cdot\cdot\cdot\rho} A_\lambda^\sigma$$
$$= K_{\nu\mu\lambda}^{\cdot\cdot\cdot\rho} A_\varkappa^\sigma - K_{\varkappa\lambda\mu}^{\cdot\cdot\cdot\sigma} A_\nu^\rho - K_{\lambda\varkappa\nu}^{\cdot\cdot\cdot\sigma} A_\mu^\rho + K_{\nu\mu\varkappa}^{\cdot\cdot\cdot\sigma} A_\lambda^\rho,$$

from which

THEOREM 6.2.[1] *In order that a V_n admit a group of motions of the maximum order $\frac{1}{2}n(n+1)$, it is necessary and sufficient that V_n be an S_n.*[2]

§ 7. A group as group of motions.

We shall consider in this section some applications of the Theorems 3.1, 3.2 and 3.3 of Ch. III.

We consider first a G_r in an X_n and suppose that the rank of $\underset{a}{v}^\varkappa$ in a neighbourhood is equal to $r \leq n$. In this case we can choose a coordinate system in which

$$\text{Det}\,(\underset{a}{v}^\alpha) \neq 0, \quad \underset{a}{v}^\xi = 0, \qquad \alpha, \beta, \gamma = 1, 2, \ldots, r;$$
$$\xi, \eta, \zeta = r+1, \ldots, n.$$

In this coordinate system Killing's equations take the form

$$\underset{a}{\pounds} g_{\lambda\varkappa} = \underset{a}{v}^\alpha \partial_\alpha g_{\lambda\varkappa} + g_{\alpha\varkappa} \partial_\lambda \underset{a}{v}^\alpha + g_{\lambda\alpha} \partial_\varkappa \underset{a}{v}^\alpha = 0. \tag{7.1}$$

Thus, defining the functions $\Theta_{\alpha\lambda\varkappa}(g, \xi)$ by

$$\underset{a}{v}^\alpha \Theta_{\alpha\lambda\varkappa}(g, \xi) \overset{\text{def}}{=} - g_{\alpha\varkappa} \partial_\lambda \underset{a}{v}^\alpha - g_{\lambda\alpha} \partial_\varkappa \underset{a}{v}^\alpha, \tag{7.2}$$

we get from (7.1)

$$\underset{a}{\pounds} g_{\lambda\varkappa} = \underset{a}{v}^\alpha [\partial_\alpha g_{\lambda\varkappa} - \Theta_{\alpha\lambda\varkappa}(g, \xi)] = 0,$$

---

[1] EISENHART [4], p. 215; SCHOUTEN [8], p. 350; G. T., p. 36.

[2] S_n stands for a Riemannian space of constant curvature. Cf .SCHOUTEN [8], p. 148.

from which

$$\partial_\alpha g_{\lambda\varkappa} = \Theta_{\alpha\lambda\varkappa}(g, \xi). \tag{7.3}$$

On using the method of § 3 of Ch. III, we find

$$\Theta_{\gamma\nu\mu}\frac{\partial\Theta_{\beta\lambda\varkappa}}{\partial g_{\nu\mu}} + \partial_\gamma\Theta_{\beta\lambda\varkappa} = \Theta_{\beta\nu\mu}\frac{\partial\Theta_{\gamma\lambda\varkappa}}{\partial g_{\nu\mu}} + \partial_\beta\Theta_{\gamma\lambda\varkappa}. \tag{7.4}$$

Moreover, as is easily to be seen from (7.2), the functions $\Theta_{\alpha\lambda\varkappa}$ satisfy the relations

$$\underset{a}{v^\alpha}\Theta_{\alpha[\lambda\varkappa]} = -g_{[\alpha\varkappa]}\partial_\lambda \underset{a}{v^\alpha} - g_{[\lambda\alpha]}\partial_\varkappa \underset{a}{v^\alpha}. \tag{7.5}$$

Thus, if the initial conditions of $g_{\lambda\varkappa}$ satisfy $g_{[\lambda\varkappa]} = 0$, then the solutions of (7.3), satisfy also $g_{[\lambda\varkappa]} = 0$.

Thus the equations (7.3) are completely integrable and the solutions $g_{\lambda\varkappa}(\xi)$ are determined by $\frac{1}{2}n(n+1)$ initial values of $g_{\lambda\varkappa}(\xi)$, which, in the case of $r < n$, can be arbitrary functions of $\xi^{r+1}, \ldots, \xi^n$. Thus we have

THEOREM 7.1.[1] *A G_r in X_n such that the rank of $\underset{a}{v^\varkappa}$ in a neighbourhood is $r \leq n$ can be regarded as a group of motions in V_n whose fundamental tensor contains $\frac{1}{2}n(n+1)$ arbitrary functions for $r < n$ and $\frac{1}{2}n(n+1)$ constants for $r = n$.*

We next consider a G_r in X_n and suppose that the rank of $\underset{a}{v^\varkappa}$ in a neighbourhood is $q < r$. In this case we can choose a coordinate system in which

$$\text{Det}\,(\underset{i}{v^\alpha}) \neq 0,\quad \underset{i}{v^\xi} = 0,\quad \underset{u}{v^\varkappa} = \varphi^i_u \underset{i}{v^\varkappa} \tag{7.5}$$

$$\alpha, \beta, \gamma, \ldots = 1, 2, \ldots, r; \qquad \xi, \eta, \zeta = r+1, \ldots, n,$$
$$i, j, k, \ldots = 1, 2, \ldots, q; \qquad u, v, w = q+1, \ldots, r.$$

Then Killing's equations take the form

$$\begin{cases} \underset{i}{\pounds} g_{\lambda\varkappa} = \underset{i}{v^\alpha}\partial_\alpha g_{\lambda\varkappa} + g_{\alpha\varkappa}\partial_\lambda \underset{i}{v^\alpha} + g_{\lambda\alpha}\partial_\varkappa \underset{i}{v^\alpha} = 0, \\ \underset{u}{\pounds} g_{\lambda\varkappa} = \varphi^i_u \underset{i}{\pounds} g_{\lambda\varkappa} + g_{\alpha\varkappa}(\partial_\lambda\varphi^i_u)\underset{i}{v^\alpha} + g_{\lambda\alpha}(\partial_\varkappa\varphi^i_u)\underset{i}{v^\alpha} = 0, \end{cases} \tag{7.7}$$

[1] EISENHART [4], p. 218; G. T., p. 44.

and consequently defining the functions $\Theta_{\alpha\lambda\varkappa}(g, \xi)$ and $\Xi_{u\lambda\varkappa}(g, \xi)$ by

$$\underset{i}{v}^{\alpha}\Theta_{\alpha\lambda\varkappa}(g, \xi) \overset{\text{def}}{=} -g_{\alpha\varkappa}\partial_{\lambda}\underset{i}{v}^{\alpha} - g_{\lambda\alpha}\partial_{\varkappa}\underset{i}{v}^{\alpha}, \tag{7.8}$$

and

$$\Xi_{u\lambda\varkappa}(g, \xi) \overset{\text{def}}{=} (g_{\alpha\varkappa}\partial_{\lambda}\varphi^{i}_{u} + g_{\lambda\alpha}\partial_{\varkappa}\varphi^{i}_{u})\underset{i}{v}^{\alpha} \tag{7.9}$$

respectively, the equations (7.7) give

$$\begin{cases} \underset{i}{\pounds}g_{\lambda\varkappa} = \underset{i}{v}^{\alpha}[\partial_{\alpha}g_{\lambda\varkappa} - \Theta_{\alpha\lambda\varkappa}(g, \xi)] = 0, \\ \underset{u}{\pounds}g_{\lambda\varkappa} = \varphi^{i}_{u}\underset{i}{\pounds}g_{\lambda\varkappa} + \Xi_{u\lambda\varkappa}(g, \xi) = 0, \end{cases} \tag{7.10}$$

from which

$$\partial_{\alpha}g_{\lambda\varkappa} = \Theta_{\alpha\lambda\varkappa}(g, \xi), \quad \Xi_{u\lambda\varkappa}(g, \xi) = 0. \tag{7.11}$$

On using the method of § 3 of Ch. III, we can prove that, if we take account of the second equations of (7.11), we have

$$\Theta_{\gamma\nu\mu}\frac{\partial\Theta_{\beta\lambda\varkappa}}{\partial g_{\nu\mu}} + \partial_{\gamma}\Theta_{\beta\lambda\varkappa} = \Theta_{\beta\nu\mu}\frac{\partial\Theta_{\gamma\lambda\varkappa}}{\partial g_{\nu\mu}} + \partial_{\beta}\Theta_{\gamma\lambda\varkappa} \tag{7.12}$$

and

$$\Theta_{\alpha\nu\mu}\frac{\partial\Xi_{u\lambda\varkappa}}{\partial g_{\nu\mu}} + \partial_{\alpha}\Xi_{u\lambda\varkappa} = 0. \tag{7.13}$$

Moreover, we have from (7.8)

$$\underset{i}{v}^{\alpha}\Theta_{\alpha[\lambda\varkappa]} = -g_{[\alpha\varkappa]}\partial_{\lambda}\underset{i}{v}^{\alpha} - g_{[\lambda\alpha]}\partial_{\varkappa}\underset{i}{v}^{\alpha}. \tag{7.14}$$

Thus if the equations $\Xi_{u\lambda\varkappa}(g, \xi) = 0$ are compatible in $g_{\lambda\varkappa}$ such that $g_{\lambda\varkappa} = g_{\varkappa\lambda}$ and $\det(g_{\lambda\varkappa}) \neq 0$ at a fixed point of the space, the mixed system of partial differential equations (7.11) is completely integrable and the solutions $g_{\lambda\varkappa}(\xi)$ are determined by the initial values of $g_{\lambda\varkappa}$ satisfying $\Xi_{u\lambda\varkappa}(g, \xi) = 0$, $g_{[\lambda\varkappa]} = 0$. But since $\Xi_{u\lambda\varkappa}(g, \xi)$ do not contain the $g_{\eta\zeta}$, these $\frac{1}{2}(n-q)(n-q+1)$ components of $g_{\lambda\alpha}$ can be taken as constants. Thus we have

THEOREM 7.2.[1] *Consider an intransitive G_r in an X_n such that the rank of $\underset{a}{v}^{\varkappa}$ in a neighbourhood is $q < r$. If, in a coordinate system where*

[1] EISENHART [4], p. 221; G. T., p. 45.

(7.5) is valid, the equations $\Xi_{u\lambda\varkappa}(g, \xi) = 0$, $g_{[\lambda\varkappa]} = 0$ are compatible for $g_{\lambda\varkappa}$ such that $\det(g_{\lambda\varkappa}) \neq 0$ at a fixed point of the space, then the group can be regarded as a group of motions in a V_n whose fundamental tensor depends on at least $\frac{1}{2}(n-q)(n-q+1)$ arbitrary constants.

We consider finally a multiply transitive G_r in an X_n. Then the rank of $\underset{a}{v}^\varkappa$ in a neighbourhood is $n < r$. If we put

$$\mathrm{Det}\,(\underset{a}{v}^\varkappa) \neq 0, \quad \underset{u}{v}^\varkappa = \varphi^a_u \underset{a}{v}^\varkappa$$

$$a, b, c = 1, 2, \ldots, \mathrm{n}; \; u = \mathrm{n}+1, \ldots, \mathrm{r},$$

according to the preceding arguments, we have

THEOREM 7.3.[1] *If, for a multiply transitive G_r in an X_n, the equations $\Xi_{u\lambda\varkappa}(g, \xi) = 0$, $g_{[\lambda\varkappa]} = 0$ are compatible in $g_{\lambda\varkappa}$ such that $\mathrm{Det}\,(g_{\lambda\varkappa}) \neq 0$ at a fixed point of the space, then the group can be regarded as a group of motions in a V_n.*

§ 8. A theorem of Wang.

Consider a V_n with positive definite fundamental tensor and suppose that the V_n admits a G_r of motions

$$(8.1) \qquad '\xi^\varkappa = f^\varkappa(\xi^1, \xi^2, \ldots, \xi^n; \eta^1, \eta^2, \ldots, \eta^r).$$

Take a point $P(\underset{0}{\xi}^\varkappa)$ in the space and consider all the motions of G_r which fix this point. The set of such motions form a subgroup $G(P)$ of G_r. If we denote the equations of motions belonging to this subgroup by

$$(8.2) \qquad T_\zeta\colon\; '\xi^\varkappa = h^\varkappa(\xi^1, \xi^2, \ldots, \xi^n; \zeta^1, \ldots, \zeta^{r_0}),$$

then we have

$$(8.3) \qquad \underset{0}{\xi}^\varkappa = h^\varkappa(\underset{0}{\xi}; \zeta)$$

for any ζ. It is well-known that[2] the subgroup $G(P)$ depends on $r_0 \geq r - n$ parameters if $r > n$. It is called the *group of stability* or the *isotropy group* of the V_n at P.

Now to each motion T_ζ there corresponds a linear homogeneous

[1] EISENHART [4], p. 221.
[2] EISENHART [4], p. 65.

transformation $\tilde{T}_\zeta$ at $\underset{0}{\xi}^\kappa$ given by

$$\tilde{T}_\zeta\colon\ d'\xi^\kappa = h^\kappa_\lambda(\zeta)d\xi^\lambda, \tag{8.4}$$

where

$$h^\kappa_\lambda(\zeta) \overset{\text{def}}{=} \partial_\lambda h^\kappa(\underset{0}{\xi};\zeta).$$

Consider now two transformations $T_{\underset{1}{\zeta}}$ and $T_{\underset{2}{\zeta}}$ of $G(P)$, then their product is given by

$$T_{\underset{2}{\zeta}}T_{\underset{1}{\zeta}}\colon\ {}''\xi^\kappa = h^\kappa(h(\xi;\underset{1}{\zeta}):\underset{2}{\zeta}),$$

from which

$$d''\xi^\kappa = h^\kappa_\lambda(\underset{2}{\zeta})h^\lambda_\mu(\underset{1}{\zeta})d\xi^\mu.$$

This equation shows that if $\tilde{T}_{\underset{1}{\zeta}}$ belongs to $T_{\underset{1}{\zeta}}$ and $\tilde{T}_{\underset{2}{\zeta}}$ belongs to $T_{\underset{2}{\zeta}}$, then $\tilde{T}_{\underset{2}{\zeta}}\tilde{T}_{\underset{1}{\zeta}}$ belongs to $T_{\underset{2}{\zeta}}T_{\underset{1}{\zeta}}$. Consequently all the $\tilde{T}_\zeta$ form a linear group $\tilde{G}(P)$ and the correspondence from T_ζ to $\tilde{T}_\zeta$ is a homomorphism. We shall now consider the kernel of this homomorphism. Suppose that T_ζ corresponds to the identity of $\tilde{G}(P)$. Geometrically this means that each direction $d\xi^\kappa$ issuing from the point P is invariant by T_ζ. Since T_ζ changes a geodesic into a geodesic and does not change any direction issuing from P, T_ζ does not change any geodesic issuing from P. On the other hand, since T_ζ is a motion which leaves invariant the point P, T_ζ must leave invariant all the points on the geodesics issuing from the point P. Thus T_ζ is the identity of $G(P)$. Consequently the two groups $G(P)$ and $\tilde{G}(P)$ are isomorphic. Furthermore, since the correspondence is continuous, they are isomorphic in the sense of the theory of topological groups. Thus $\tilde{G}(P)$ depends on r_0 $(\geq r - n)$ parameters if $G(P)$ does.

Now since T_ζ is a motion, we must have

$$g_{\lambda\kappa}('\xi)d'\xi^\lambda d'\xi^\kappa = g_{\lambda\kappa}(\xi)d\xi^\lambda d\xi^\kappa$$

and consequently

$$g_{\nu\mu}(\underset{0}{\xi})h^\nu_\lambda(\zeta)h^\mu_\kappa(\zeta)d\xi^\lambda d\xi^\kappa = g_{\lambda\kappa}(\underset{0}{\xi})d\xi^\lambda d\xi^\kappa$$

at the point $P(\underset{0}{\xi})$. Thus the group $\tilde{G}(P)$ is a subgroup of the rotation group in the tangent Euclidean space at P.

Suppose that the V_n admits a G_r of motions with $r > \frac{1}{2}n(n-1)+1$ parameters. Then the group $\tilde{G}(P)$ depends on

$$r_0 \geq r - n > \tfrac{1}{2}(n-1)(n-2)$$

parameters.

But on the other hand we have the following theorem of Montgomery and Samelson [1].

THEOREM 8.1. *In a Euclidean space of $n(\neq 4)$ dimensions there is no proper subgroup of the rotation group of an order greater than $\frac{1}{2}(n-1)$ $(n-2)$.*

Thus if $n \neq 4$, we must have

$$r_0 = \tfrac{1}{2}n(n-1), \tag{8.5}$$

and the group $G(P)$ coincides with the rotation group.

Consequently the group $G(P)$ contains a motion which changes any direction at P into any direction at P, the point P being arbitrary.

Take now two points P_1 and P_2 in V_n in such a way that they are sufficiently near to each other to be joined by a geodesic. We consider the midpoint M of the geodesic segment P_1P_2. Then in the group of stability $G(M)$ at M, there exists a motion which changes the direction of the tangent to the geodesic at M into the opposite direction. Since this motion changes a geodesic into a geodesic and does not change the length of a geodesic, it carries the point P_1 into the point P_2.

If there are two points A and B at a finite distance in V_n, we join A and B by an arbitrary curve and take a series of points $P_1, P_2, \ldots, P_N$ on the curve in such a way that A and P_1, P_1 and P_2, $\ldots$, P_N and B can respectively be joined by geodesics. We denote by $M_0, M_1, \ldots, M_N$ the midpoints of the geodesic segments $AP_1, P_1P_2, \ldots, P_NB$ respectively. Then applying appropriate motions belonging to $G(M_0)$, $G(M_1)$, $\ldots$, $G(M_N)$ successively we can carry A into B by a product of motions of G_r. Since A and B are two arbitrary points in V_n, this shows that G_r is transitive and consequently that

$$r = r_0 + n = \tfrac{1}{2}n(n+1).$$

Thus according to Theorem 6.2, we have the following theorem due to Wang: [2]

---

[1] MONTGOMERY and SAMELSON [1].

[2] WANG [1].

THEOREM 8.2. *If a* V_n *for* $n > 2$, $n \neq 4$ *admits a* G_r *of motions of order greater than* $\frac{1}{2}n(n-1)+1$, then the V_n is an S_n.

The same argument gives

THEOREM 8.3.[1] *In a* V_n *for* $n \neq 4$, *there does not exist a* G_r *of motions such that*

$$\tfrac{1}{2}n(n+1) > r > \tfrac{1}{2}n(n-1)+1.$$

§ 9. Two theorems of Egorov.

In 1903, G. Fubini[2] proved

THEOREM 9.1. *A* V_n, $n > 2$, *cannot admit a complete group of motions of order* $\frac{1}{2}n(n+1)-1$.

Generalizing this result, I. P. Egorov[3] has proved in 1949 the two following theorems.

THEOREM 9.2. *The maximum order of the complete group of motions in those* V_n*'s which are not Einstein spaces is* $\frac{1}{2}n(n-1)+1$.

In fact, if the operator $\underset{v}{\pounds}f$ is that of a motion, we have

$$\underset{v}{\pounds}K_{\mu\lambda} = v^\rho \nabla_\rho K_{\mu\lambda} + K_{\rho\lambda}\nabla_\mu v^\rho + K_{\mu\rho}\nabla_\lambda v^\rho = 0$$

or

$$\underset{v}{\pounds}K_{\mu\lambda} = v^\rho \nabla_\rho K_{\mu\lambda} + (A_\mu^\sigma K_\lambda^{\cdot\rho} + A_\lambda^\sigma K_\mu^{\cdot\rho})\nabla_\sigma v_\rho = 0. \tag{9.1}$$

Since $\nabla_\sigma v_\rho$ must satisfy $\nabla_{(\sigma}v_{\rho)} = 0$, we can write (9.1) also in the form

$$\underset{v}{\pounds}K_{\mu\lambda} = v^\rho \nabla_\rho K_{\mu\lambda} + \Sigma^{1\ldots\ldots n}_{\alpha_j>\alpha_i} T_{\mu\lambda}^{\cdot\cdot\alpha_j\alpha_i}\nabla_{\alpha_j}v_{\alpha_i} = 0,\text{[4]} \tag{9.2}$$

where

$$T_{\mu\lambda}^{\cdot\cdot\alpha_j\alpha_i} = 4A_{(\mu}^{[\alpha_j}K_{\lambda)}^{\cdot\alpha_i]}. \tag{9.3}$$

We now consider a matrix by letting the two lower indices denote the rows and the two upper indices the columns. The rank of this matrix is what is called the $\mu\lambda$-rank (or $\alpha_j\alpha_i$-rank) of $T_{\mu\lambda}^{\cdot\cdot\alpha_j\alpha_i}$.[5]

---

[1] YANO [19].

[2] FUBINI [1].

[3] EGOROV [4].

[4] Here we have introduced a new kind of indices α_i, α_j also running from 1 to n for which we assume that $\alpha_i \neq \alpha_j$ for $i \neq j$ and that the summation convention does not hold. So summation over these indices must always be denoted by a sign Σ.

[5] Cf. SCHOUTEN [8], p. 20.

$\mu\lambda \backslash \alpha_j\alpha_i$	$\alpha_2\alpha_1$	$\alpha_r\alpha_2$
$\alpha_1\alpha_1$	$-2K_{\alpha_1}^{\cdot\alpha_2}$	0
$\alpha_s\alpha_1$	*	$\delta_s^r K_{\alpha_1}^{\cdot\alpha_2}$

$(r, s = 3, \ldots, n)$

If the space admits a complete group of motions of order greater than $\frac{1}{2}n(n-1)+1$, the $\mu\lambda$-rank of $T_{\mu\lambda}^{\cdot\cdot\alpha_j\alpha_i}$ must be less than

$$\tfrac{1}{2}n(n+1) - [\tfrac{1}{2}n(n-1)+1] = n-1.$$

We consider the $(n-1)$-rowed determinant formed by the above elements of this matrix. Since this determinant vanishes we have

$$K_{\alpha_1}^{\cdot\alpha_2} = 0. \tag{9.4}$$

On taking account of (9.4), we next consider the $(n-1)$-rowed determinant formed by the following elements:

$\mu\lambda \backslash \alpha_j\alpha_i$	$\alpha_{j_1}\alpha_{i_1}$	$\alpha_{j_2}\alpha_{i_2}$	.....	$\alpha_{j_n}\alpha_{i_n}$
$\alpha_{j_1}\alpha_{i_1}$	$K_{\alpha_{i_1}}^{\cdot\alpha_{i_1}} - K_{\alpha_{j_1}}^{\cdot\alpha_{j_1}}$	0	.....	0
$\alpha_{j_2}\alpha_{i_2}$	0	$K_{\alpha_{i_2}}^{\cdot\alpha_{i_2}} - K_{\alpha_{j_2}}^{\cdot\alpha_{j_2}}$	.....	0
⋮	⋮		.....	⋮
			.....	
$\alpha_{j_n}\alpha_{i_n}$	0	0	.....	$K_{\alpha_{i_n}}^{\cdot\alpha_{i_n}} - K_{\alpha_{j_n}}^{\cdot\alpha_{j_n}}$

Since all such determinants vanish, the number of the differences $K_{\alpha_i}^{\cdot\alpha_i} - K_{\alpha_j}^{\cdot\alpha_j}$ which are not zero is at most $n-2$, from which

$$K_1^{\cdot 1} = K_2^{\cdot 2} = \ldots = K_n^{\cdot n}. \tag{9.5}$$

The equations (9.4) and (9.5) give

$$K_\mu^{\cdot\lambda} = \frac{1}{n} K A_\mu^\lambda \tag{9.6}$$

which shows that the space is an Einstein space.

Thus the order of the complete group of motions in a V_n which is not an Einstein space is at most $\frac{1}{2}n(n-1)+1$.

On the other hand a V_n with the metric $ds^2 = (d\xi^1)^2 + d\sigma^2$ where $d\sigma^2$ is the fundamental form of an S_{n-1} with non-vanishing curvature and with coordinates $\xi^2, \ldots, \xi^n$, is not an Einstein space and it obviously admits a group of motions of order $\frac{1}{2}n(n-1)+1$. This proves Theorem 9.2.

THEOREM 9.3.[1] *The order of the complete group of motions in a V_n which is not an S_n is at most $\frac{1}{2}n(n-1)+2$.*

In fact, if the operator $\underset{v}{\pounds} f$ is that of a motion, we have

$$\underset{v}{\pounds} K_{\nu\mu\lambda\varkappa} = v^\rho \nabla_\rho K_{\nu\mu\lambda\varkappa} + \Sigma^{1\cdots n}_{\alpha_j>\alpha_i} T_{\nu\mu\lambda\varkappa}^{\cdot\cdot\cdot\cdot\alpha_j\alpha_i} \nabla_{\alpha_j} v_{\alpha_i} = 0, \tag{9.7}$$

where

$$T_{\nu\mu\lambda\varkappa}^{\cdot\cdot\cdot\cdot\alpha_j\alpha_i} = 2(A_\nu^{[\alpha_j} K_{\varkappa\lambda\mu}^{\cdot\cdot\cdot\alpha_i]} - A_\mu^{[\alpha_j} K_{\varkappa\lambda\nu}^{\cdot\cdot\cdot\alpha_i]} - A_\lambda^{[\alpha_j} K_{\nu\mu\varkappa}^{\cdot\cdot\cdot\alpha_i]} + A_\varkappa^{[\alpha_j} K_{\nu\mu\lambda}^{\cdot\cdot\cdot\alpha_i]}). \tag{9.8}$$

We now consider a matrix by letting denote $\nu\mu\lambda\varkappa$ the rows and $\alpha_j\alpha_i$ the columns. The rank of this matrix is what is called the $\nu\mu\lambda\varkappa$-rank (or $\alpha_j\alpha_i$-rank) of $T_{\nu\mu\lambda\varkappa}^{\cdot\cdot\cdot\cdot\alpha_j\alpha_i}$.

If the space admits a complete group of motions of order greater than $\frac{1}{2}n(n-1)+2$, the rank of this matrix must be less than

$$\tfrac{1}{2}n(n+1) - [\tfrac{1}{2}n(n-1)+2] = n-2.$$

$\nu\mu\lambda\varkappa$ \ $\alpha_j\alpha_i$	$\alpha_2\alpha_3$	$\alpha_t\alpha_3$
$\alpha_1\alpha_2\alpha_2\alpha_1$	$2K_{\alpha_2\alpha_1\alpha_1}^{\cdot\cdot\cdot\alpha_3}$	0
$\alpha_1\alpha_2\alpha_u\alpha_1$	*	$\delta_u^t K_{\alpha_2\alpha_1\alpha_1}^{\cdot\cdot\cdot\alpha_3}$
	$(t, u = 4, 5, \ldots, n)$	

We consider the $(n-2)$-rowed determinant formed by the elements of this matrix in the table. Since the determinant vanishes, we have

$$K_{\alpha_2\alpha_1\alpha_1}^{\cdot\cdot\cdot\alpha_3} = 0. \tag{9.9}$$

On taking account of (9.9) we next consider the following submatrix

---

[1] For $n \neq 4$, this is a special case of Theorem 8.2 of Wang. For $n = 4$,

$$\tfrac{1}{2}n(n+1) = 10,\ \tfrac{1}{2}n(n+1) - 1 = 9,\ \tfrac{1}{2}n(n-1)+2 = 8$$

so that it is a consequence of Theorem 9.1 of Fubini. But we shall give the proof of this theorem to illustrate Egorov's method.

with $n-1$ rows and $n-1$ columns:

$\nu\mu\lambda\varkappa$ \ $\alpha_j\alpha_i$	$\alpha_1\alpha_3$	$\alpha_2\alpha_3$	$\alpha_4\alpha_3$	$\alpha_v\alpha_3$
$\alpha_1\alpha_2\alpha_1\alpha_4$	$K^{\cdot\cdot\cdot\alpha_3}_{\alpha_4\alpha_1\alpha_2} - K^{\cdot\cdot\cdot\alpha_3}_{\alpha_1\alpha_2\alpha_4}$	0	0	0
$\alpha_1\alpha_2\alpha_2\alpha_4$	*	$K^{\cdot\cdot\cdot\alpha_3}_{\alpha_2\alpha_4\alpha_1} - K^{\cdot\cdot\cdot\alpha_3}_{\alpha_1\alpha_2\alpha_4}$	0	0
$\alpha_1\alpha_4\alpha_2\alpha_4$	*	*	$K^{\cdot\cdot\cdot\alpha_3}_{\alpha_2\alpha_4\alpha_1} - K^{\cdot\cdot\cdot\alpha_3}_{\alpha_4\alpha_1\alpha_2}$	0
$\alpha_1\alpha_w\alpha_2\alpha_4$	*	*	*	$\delta^v_w K^{\cdot\cdot\cdot\alpha_3}_{\alpha_2\alpha_4\alpha_1}$

$$(v, w = 5, 6, \ldots, n)$$

Since the rank of this matrix is less than $n-2$, if $K^{\cdot\cdot\cdot\alpha_3}_{\alpha_2\alpha_4\alpha_1}$ were not zero, two of the differences

$$K^{\cdot\cdot\cdot\alpha_3}_{\alpha_4\alpha_1\alpha_2} - K^{\cdot\cdot\cdot\alpha_3}_{\alpha_1\alpha_2\alpha_4}, \quad K^{\cdot\cdot\cdot\alpha_3}_{\alpha_2\alpha_4\alpha_1} - K^{\cdot\cdot\cdot\alpha_3}_{\alpha_2\alpha_1\alpha_4}, \quad K^{\cdot\cdot\cdot\alpha_3}_{\alpha_2\alpha_4\alpha_1} - K^{\cdot\cdot\cdot\alpha_3}_{\alpha_4\alpha_1\alpha_2}$$

should be zero. But the sum of the first and the third is equal to the second, so all three vanish. Then using

$$K^{\cdot\cdot\cdot\alpha_3}_{\alpha_4\alpha_1\alpha_2} + K^{\cdot\cdot\cdot\alpha_3}_{\alpha_1\alpha_2\alpha_4} + K^{\cdot\cdot\cdot\alpha_3}_{\alpha_2\alpha_4\alpha_1} = 0,$$

we have

$$K^{\cdot\cdot\cdot\alpha_3}_{\alpha_4\alpha_1\alpha_2} = K^{\cdot\cdot\cdot\alpha_3}_{\alpha_1\alpha_2\alpha_4} = K^{\cdot\cdot\cdot\alpha_3}_{\alpha_2\alpha_4\alpha_1} = 0.$$

This proves that

$$K^{\cdot\cdot\cdot\alpha_3}_{\alpha_2\alpha_1\alpha_4} = 0. \tag{9.10}$$

The equations (9.9) and (9.10) show that

$$K^{\cdot\cdot\cdot\varkappa}_{\nu\mu\lambda} = 0 \quad \text{for } \varkappa \neq \nu, \mu, \lambda. \tag{9.11}$$

On the other hand, it is known that[1] an A_n, $n > 2$, is a D_n (= projective Euclidean space) if and only if the equation (9.11) holds for every coordinate system at every point. Since the V_n is a D_n, V_n is an S_n.[2] This proves Theorem 9.3.

---

[1] SCHOUTEN [8], p. 290.
[2] SCHOUTEN [8], p. 294.

§ 10. V_n's admitting a group G_r of motions of order $r = \frac{1}{2}n(n-1)+1$.[1]

Let $\underset{a}{\pounds} f = \underset{a}{v}^{\varkappa}\partial_{\varkappa} f$ denote r infinitesimal operators of G_r and be q the rank of $\underset{a}{v}^{\varkappa}$ in a certain neighbourhood. Then we have $n \geq q$ and the group of stability $G(P)$ at a point $P(\xi)$ is of order $r - q = \frac{1}{2}n(n-1) + + 1 - q$.

If $n > q$ then we have

$$r - q > \tfrac{1}{2}n(n-1) + 1 - n = \tfrac{1}{2}(n-1)(n-2)$$

and consequently for $n \neq 4$ we can conclude from Theorem 8.1 that $\tilde{G}(P)$ coincides with the rotation group. Thus, as is to be seen from the proof of Theorem 8.2, the group G_r is transitive and consequently we should have $n = q$, which is a contradiction.

This proves that $n = q$, hence:

THEOREM 10.1. *If a V_n, $n \neq 4$, admits a group of motions of order $\frac{1}{2}n(n-1) + 1$, then the group is transitive.*

Since G_r is transitive, the group of stability $G(P)$ and consequently $\tilde{G}(P)$ is of order

$$r - n = \tfrac{1}{2}n(n-1) + 1 - n = \tfrac{1}{2}(n-1)(n-2).$$

On the other hand, we have a theorem of D. Montgomery and H. Samelson [2]:

THEOREM 10.2. *In a Euclidean R_n, $n \neq 4$, $n \neq 8$, a subgroup of order $\frac{1}{2}(n-1)(n-2)$ of the rotation group leaves invariant one and only one direction.*

In the following we shall assume that $n \neq 4$, $n \neq 8$.

Following Theorem 10.2, $\tilde{G}(P)$ leaves invariant one and only one direction at P. We denote this direction by $u(P)$. Now take two arbitrary points P and Q in V_n. Since the group is transitive, there exists a motion T which carries P into Q. If we denote by $T(Q)$ an arbitrary motion which leaves invariant Q, then the motion $T^{-1}T(Q)T$ leaves invariant P.

---

[1] YANO [19].

[2] MONTGOMERY and SAMELSON [1].

Consequently on applying this motion to $u(P)$ we get

$$T^{-1}T(Q)Tu(P) = u(P),$$

from which

$$T(Q)Tu(P) = Tu(P),$$

which shows that $Tu(P)$ is invariant for any motion which leaves invariant Q. Hence

THEOREM 10.3. *If a V_n, $n \neq 4$, $n \neq 8$ admits a G_r of order $r = \frac{1}{2}n(n-1) + 1$ there exists a field of directions such that the direction $u(P)$ at P is carried into $u(Q)$ at Q by any motion of the group which carries P into Q.*

Consider now a geodesic which passes through a point P and which is tangent to the direction $u(P)$. Since the group of stability $G(P)$ at P leaves invariant P and $u(P)$, it leaves invariant not only this geodesic but also all the points on the geodesic. Thus, if we take a point Q different from P on the geodesic, then $G(P)$ leaves invariant Q.

Consider next an orthogonal frame $\underset{i}{e}^{\varkappa}(P)$ at P whose first axis $\underset{1}{e}^{\varkappa}(P)$ is taken along $u(P)$ and displace this frame parallelly from P to Q along the geodesic. Then we obtain an orthogonal frame $\underset{i}{e}^{\varkappa}(Q)$ at Q whose first axis is tangent to the geodesic. Now, if we apply a motion T of $G(P)$, we get

$$T\underset{i}{e}^{\varkappa}(P) \overset{\text{def}}{=} {}'\underset{i}{e}^{\varkappa}(P), \quad T\underset{i}{e}^{\varkappa}(Q) \overset{\text{def}}{=} {}'\underset{i}{e}^{\varkappa}(Q).$$

Since the parallel displacement is preserved by a motion, by displacing ${}'\underset{i}{e}^{\varkappa}(P)$ from P to Q along the geodesic, we obtain ${}'\underset{i}{e}^{\varkappa}(Q)$. Thus the mutual position between $\underset{i}{e}^{\varkappa}(P)$ and ${}'\underset{i}{e}^{\varkappa}(P)$ and that between $\underset{i}{e}^{\varkappa}(Q)$ and ${}'\underset{i}{e}^{\varkappa}(Q)$ are exactly the same. This shows that $G(P)$ at P acts, at Q, as a group of motions which leaves invariant Q and is of order $\frac{1}{2}(n-1)(n-2)$. Thus we can conclude $G(P) = G(Q)$.

Since the group $G(Q)$ fixes the tangent to the geodesic and the direction $u(Q)$ at the same time, the tangent must coincide with the direction $u(Q)$ and consequently we can say that the geodesic is a streamline[1] of the field u.

---

[1] The *streamlines* of a vector field $u^{\varkappa}(\xi)$ are the curves defined by the differential equations

$$\frac{d\xi^{\varkappa}}{dt} = u^{\varkappa}(\xi).$$

Since there is one and only one streamline passing through a point, these streamlines depend on $n - 1$ parameters and they are transformed one into the other by a motion of G_r. Thus we have

THEOREM 10.4. *If a V_n, $n \neq 4$, $n \neq 8$, admits a group G_r of motions of order $r = \frac{1}{2}n(n-1) + 1$, there exists a family of geodesics such that there is one and only one geodesic of the family passing through each point and a geodesic passing through a point P is transformed into a geodesic passing through a point Q by a motion of G_r which carries P into Q.*

With any point $\underset{0}{\xi^\varkappa}$ of V_n, there is now associated a direction $u^\varkappa(\underset{0}{\xi})$. We attach to $\underset{0}{\xi^\varkappa}$ a unit orthogonal frame $\underset{h}{e^\varkappa}(\underset{0}{\xi})$ in such a way that the first axis $\underset{1}{e^\varkappa}(\underset{0}{\xi})$ is in the direction $u^\varkappa(\underset{0}{\xi})$, and we consider all the frames obtainable from $\underset{h}{e^\varkappa}(\underset{0}{\xi})$ by applying to it all the motions of G_r. Such a family of frames is said to *be adapted* to the group of motions under consideration.

The frames $\underset{h}{e^\varkappa}$; $h, i, j, \ldots = 1, 2, \ldots, n$, thus attached to the different points of the space depend on $\frac{1}{2}n(n-1) + 1$ parameters, the first n of which are the coordinates $\xi^\varkappa$ of the origin and the other $\frac{1}{2}(n-1)(n-2)$ are parameters η^α $(\alpha = 1, 2, \ldots, \frac{1}{2}(n-1)(n-2))$ which fix the direction of the vectors $\underset{2}{e^\varkappa}, \underset{3}{e^\varkappa}, \ldots, \underset{n}{e^\varkappa}$. Thus the $\underset{h}{e^\varkappa}$ are functions of the $\xi^\varkappa$ and the η^α.

Now for a variation of coordinates, we have[1]

$$d\xi^\varkappa = A^h \underset{h}{e^\varkappa}, \tag{10.1}$$

where

$$A^h \overset{\text{def}}{=} \overset{h}{e_\varkappa}(\xi, \eta) d\xi^\varkappa = (d\xi)^h. \tag{10.2}$$

Since $g_{\lambda\varkappa} \underset{i}{e^\lambda} \underset{h}{e^\varkappa} = \delta_{ih}$, we have from (10.1)

$$ds^2 = g_{\lambda\varkappa} d\xi^\lambda d\xi^\varkappa = \sum_h A^h A^h = \sum_h (d\xi)^h (d\xi)^h. \tag{10.3}$$

The equation (10.1) represents the relative position of $d\xi^\varkappa$ and $\underset{h}{e^\varkappa}$. Since this relative position is preserved by any motion of the group, the Pfaffian forms A^h are invariant for any motion of the group G_r.

---

[1] SCHOUTEN [8], p. 172.

For a variation of coordinates and parameters, we have

$$d\underset{h}{e}^{\varkappa} = d\xi^{\mu}\partial_{\mu}\underset{h}{e}^{\varkappa} + d\eta^{\alpha}\partial_{\alpha}\underset{h}{e}^{\varkappa}; \quad \partial_{\alpha} = \partial/\partial\eta^{\alpha}.$$

If $\underset{h}{e}^{\varkappa} + d\underset{h}{e}^{\varkappa}$ is displaced parallelly from $\xi^{\varkappa} + d\xi^{\varkappa}$ to $\xi^{\varkappa}$, we get $\underset{h}{e}^{\varkappa} + \delta\underset{h}{e}^{\varkappa}$ at $\xi^{\varkappa}$, where

$$\delta\underset{h}{e}^{\varkappa} \overset{\text{def}}{=} d\underset{h}{e}^{\varkappa} + \{^{\,\varkappa}_{\mu\lambda}\}\underset{h}{e}^{\lambda}d\xi^{\mu}. \tag{10.4}$$

From (10.4), we find[1]

$$\delta\underset{i}{e}^{\varkappa} = \Gamma^h_i\underset{h}{e}^{\varkappa}, \tag{10.5}$$

where

$$\Gamma^h_i = \overset{h}{e}_{\varkappa}[d\underset{i}{e}^{\varkappa} + \{^{\,\varkappa}_{\mu\lambda}\}\underset{i}{e}^{\lambda}d\xi^{\mu}] \tag{10.6}$$

are Pfaffian forms with respect to $\xi^{\varkappa}$ and η^{α}. The equation (10.5) represents the Riemannian connexion with respect to the frames $\underset{h}{e}^{\varkappa}$. Because of the same reason as for (10.1), the Γ^h_i are invariant for any motion of the group.

Since $\delta(g_{\lambda\varkappa}\underset{i}{e}^{\lambda}\underset{h}{e}^{\varkappa}) = 0$, we have from (10.5)

$$\Gamma^h_i + \Gamma^i_h = 0. \tag{10.7}$$

As we see from (10.2) the A^h are linear homogeneous with respect to $d\xi^{\varkappa}$. On the other hand, since the vector $\underset{1}{e}^{\varkappa}$ has the definite direction $u^{\varkappa}(\xi)$ at each point $\xi^{\varkappa}$, it does not depend on the parameters η^{α}. consequently we see from (10.6) that Γ^h_1 are also linear homogeneous with respect to $d\xi^{\varkappa}$. Thus putting

$$\Gamma^h_1 = f^h_{\varkappa}(\xi, \eta)d\xi^{\varkappa}, \tag{10.8}$$

we obtain from (10.2) and (10.8)

$$\Gamma^h_1 = c^h_i A^i, \tag{10.9}$$

where

$$c^h_i = f^h_{\varkappa}\underset{i}{e}^{\varkappa} \tag{10.10}$$

are functions of $\xi^{\varkappa}$ and η^{α}.

---

[1] SCHOUTEN [8], p. 177. The forms A^h and Γ^h_i were denoted by ω^h and ω^h_i respectively in E. CARTAN's papers [6, 7, 9, 10, 11].

The Pfaffian forms Γ^h_1 and A^h are invariant for any motion T of the group. If we denote by $'c^h_i$ the transform of c^h_i by T, we have

$$\Gamma^h_1 = {'c^h_i} A^i. \tag{10.11}$$

From (10.9) and (10.11), we find

$$('c^h_i - c^h_i)A^i = 0.$$

But the A^i are n linear independent Pfaffian forms and consequently we have

$$'c^h_i = c^h_i.$$

Since c^h_i are functions of $\xi^\varkappa$ and η^α, this equation shows that c^h_i are constants.

To find the values of these constants we apply a method of E. Cartan.[1]

At two points $\xi^\varkappa$ and $\xi^\varkappa + d\xi^\varkappa$ we consider the frames $\underset{h}{e}^\varkappa(\xi)$ and $\underset{h}{e}^\varkappa(\xi+d\xi)$ both adapted to the given group of motions. We effect to $\underset{h}{e}^\varkappa(\xi)$ and $\underset{h}{e}^\varkappa(\xi + d\xi)$ the same infinitesimal rotation around the first axes. This rotation can be represented by the formulas

$$\begin{cases} '\underset{h}{e}^\varkappa(\xi) = \underset{h}{e}^\varkappa(\xi) + k^i_h dt \underset{i}{e}^\varkappa(\xi), \\ '\underset{h}{e}^\varkappa(\xi + d\xi) = \underset{h}{e}^\varkappa(\xi + d\xi) + k^i_h dt \underset{i}{e}^\varkappa(\xi + d\xi) \end{cases} \tag{10.12}$$

using the same infinitesimal constants $k^i_h dt$ which satisfy

$$k^i_h + k^h_i = 0 \tag{10.13}$$

and

$$k^i_1 = - k^1_i = 0. \tag{10.14}$$

Now the figure composed of $\underset{h}{e}^\varkappa(\xi)$ and $'\underset{h}{e}^\varkappa(\xi)$ is congruent to that composed of $\underset{h}{e}^\varkappa(\xi + d\xi)$ and $'\underset{h}{e}^\varkappa(\xi + d\xi)$ in the sense that there exists a motion which carries $\underset{h}{e}^\varkappa(\xi)$ into $\underset{h}{e}^\varkappa(\xi + d\xi)$ and at the same time $'\underset{h}{e}^\varkappa(\xi)$ into $'\underset{h}{e}^\varkappa(\xi + d\xi)$. This motion can be represented analytically by A^h and Γ^h_i with respect to the frame $\underset{h}{e}^\varkappa$. But during the orthogonal transformation

[1] E. Cartan [6, 11], Ch. XII, XIII.

of the frames which carries $\underset{h}{e}^{\varkappa}(\xi)$ into $'\underset{h}{e}^{\varkappa}(\xi)$, the components A^h and Γ^h_i receive the variations

$$\Delta A^h = -k^h_i dt A^i,\ \Delta\Gamma^h_j = -k^i_l dt \Gamma^l_j + \Gamma^i_l k^l_j dt. \tag{10.15}$$

On the other hand we have from (10.9)

$$\Delta\Gamma^h_1 = c^h_i \Delta A^i. \tag{10.16}$$

On substituting (10.15) in (10.16) and using (10.9), we find

$$(k^h_l c^l_i - c^h_l k^l_i)A^i = 0,$$

from which

$$k^h_l c^l_i - c^h_l k^l_i = 0 \tag{10.17}$$

because the A^i are linearly independent.

First of all, putting $h = 1$ in (10.9) and taking account of $\Gamma^1_1 = 0$, we obtain

$$c^1_i = 0. \tag{10.18}$$

Next putting $i = 1$ in (10.17) and taking account of $k^l_1 = -k^1_l = 0$, we find

$$k^h_l c^l_1 = 0,$$

which should be satisfied for any k^h_l satisfying $k^h_1 = -k^1_h = 0$ and $k^h_i + k^i_h = 0$. Thus we get

$$c^1_l = 0. \tag{10.19}$$

Thus the equation (10.17) becomes

$$k^r_s c^s_t - c^r_s k^s_t = 0 \tag{10.20}$$

where $r, s, t, u, v = 2, 3, \ldots, n$. This equation can be written also in the form

$$(\delta^r_v c^u_t - c^r_v \delta^u_t)k^v_u = 0 \tag{10.21}$$

and should be satisfied by any k^v_u satisfying $k^v_u + k^u_v = 0$, from which

$$(\delta^r_v c^u_t - c^r_v \delta^u_t) - (\delta^r_u c^v_t - c^r_u \delta^v_t) = 0.$$

On contracting this equation with respect to r and v, we find

$$(n-2)(c^u_t + c^t_u) = c^v_v \delta^u_t. \tag{10.22}$$

Since the cases $n = 3$ and $n = 4$ are exceptional, we shall hereafter assume $n > 4$, $n \neq 8$.

Taking the antisymmetric part of (10.22), we find

$$(n-3)(c_t^u - c_u^t) = 0$$

from which $c_t^u = c_u^t$ and consequently from (10.22)

$$c_t^u = \frac{1}{n-1} c_v^v \delta_t^u.$$

Thus the matrix c_i^h has the form

$$(10.23) \qquad (c_i^h) = \begin{pmatrix} 0 & 0 & 0 & \dots & 0 \\ 0 & c & 0 & \dots & 0 \\ 0 & 0 & c & \dots & 0 \\ & \cdot & \cdot & \cdot & \cdot \\ 0 & 0 & 0 & \dots & c \end{pmatrix}$$

thus, on account of (10.9)

$$(10.24) \qquad \Gamma_1^r = cA^r.$$

Thus from the equations of structure [1]

$$[dA^h] = [A^i \Gamma_i^h],$$

we get

$$[dA^1] = 0,$$

which shows that the form is exact:

$$(10.25) \qquad A^1 = df(\xi).$$

Thus there exists a family of hypersurfaces $f(\xi) =$ constants along which

$$A^1 = 0 \text{ or}$$

$$d\xi^\varkappa = A^2 \underset{2}{e}^\varkappa + \dots + A^n \underset{n}{e}^\varkappa.$$

Since the vectors $\underset{2}{e}^\varkappa, \dots, \underset{n}{e}^\varkappa$ are always tangent to one of these hypersurfaces, we can see that these hypersurfaces admit groups of motions of the maximum order. Consequently these hypersurfaces regarded as V_{n-1}'s are all S_{n-1}'s.

It is clear that the orthogonal trajectories of these hypersurfaces are the geodesics that appeared in Theorem 10.4.

---

[1] SCHOUTEN [8], p. 177, (10.27) with $S^h = 0$.

Since there is always a motion of the transitive group G_r which transforms a given orthogonal trajectory into another given trajectory, it follows that there also must be always such a motion transforming a given hypersurface of the kind considered into another given hypersurface of the same kind. Consequently the hypersurfaces are all of the *same* constant curvature.

Now we distinguish two cases (I) $c = 0$ and (II) $c \neq 0$.

In case I, $c = 0$, we have from (10.24) $\Gamma^r_1 = 0$ and consequently

$$\delta \underset{1}{e}^{\varkappa} = 0, \tag{10.26}$$

which shows that $\underset{1}{e}^{\varkappa}$ is a covariant constant vector field. Since the normals to theh ypersurfaces $f(\xi) =$ constant are parallel, the hypersurfaces are all geodesic.

In case II, $c \neq 0$, we have from (10.24)

$$A^r = \frac{1}{c} \Gamma^r_1 \tag{10.27}$$

and consequently

$$d\xi^{\varkappa} = A^1 \underset{1}{e}^{\varkappa} + A^r \underset{r}{e}^{\varkappa} = A^1 \underset{1}{e}^{\varkappa} + \frac{1}{c} \Gamma^r_1 \underset{1}{e}^{\varkappa} = A^1 \underset{1}{e}^{\varkappa} + \frac{1}{c} \delta \underset{1}{e}^{\varkappa},$$

from which

$$d\xi^{\varkappa} + \delta\left(-\frac{1}{c} \underset{1}{e}^{\varkappa}\right) = A^1 \underset{1}{e}^{\varkappa}, \tag{10.28}$$

which shows that

$$d\xi^{\varkappa} + \delta\left(-\frac{1}{c} \underset{1}{e}^{\varkappa}\right) = 0$$

along one of the hypersurfaces $f(\xi) =$ constant, and this means that the vector $\underset{1}{e}^{\varkappa}$ is concurrent[1] along the hypersurfaces. But because $\underset{1}{e}^{\varkappa}$ is in the direction of the normal, the hypersurfaces are umbilical and of constant mean curvature, their orthogonal trajectories being geodesic Ricci curves. Thus we have

THEOREM 10.5. *If a V_n, $n < 4$, $n \neq 8$ admits a group G_r of motions of order $\frac{1}{2}n(n-1) + 1$, then either (I) there exists a family of ∞^1 geodesic*

[1] YANO [6].

hypersurfaces whose orthogonal trajectories are geodesics, the hypersurfaces being S_{n-1}'s of the same constant curvature, or (II) there exists a family of ∞^1 umbilical hypersurfaces of constant mean curvature whose orthogonal trajectories are geodesic Ricci curves, the hypersurfaces being S_{n-1}'s of the same constant curvature. In both cases, the group leaves invariant the family of hypersurfaces and that of their orthogonal trajectories.

Of course geodesic hypersurfaces are special cases of umbilical hypersurfaces. But case I and case II in which hypersurfaces are umbilical but not geodesic are, as we shall see, essentially different. So we shall study these two cases separately.

§ 11. Case I.

Since the space admits a covariant constant vector field, according to a well-known theorem[1] there exists a coordinate system with respect to which the ds^2 of the space takes the form

$$ds^2 = (d\xi^1)^2 + g_{\eta\xi}(\xi^\zeta)d\xi^\eta d\xi^\xi; \quad \xi, \eta, \zeta, \varphi, \psi = 2, 3, \ldots, n, \tag{11.1}$$

the form $g_{\eta\xi}(\xi^\zeta)d\xi^\eta d\xi^\xi$ being the fundamental form of an S_{n-1}.

Conversely, if there exists a coordinate system with respect to which the fundamental form of the space takes the form (11.1), $g_{\eta\xi}(\xi^\zeta)d\xi^\eta d\xi^\xi$ being the fundamental form of an S_{n-1}, it is clear that we have case I and that the space admits a group G_r of motions of order $r = \frac{1}{2}n(n-1) + 1$:

$$'\xi^1 = \xi^1 + t, \quad '\xi^\zeta = f^\zeta(\xi^\varphi, a), \tag{11.2}$$

where $'\xi^\zeta = f^\zeta(\xi^\varphi, a)$ represents the complete group of motions of order $\frac{1}{2}n(n-1)$ in an S_{n-1}. Thus we have

THEOREM 11.1. *In order that case I in Theorem 10.8 occurs, it is necessary and sufficient that there exist a coordinate system with respect to which the ds^2 of the space takes the form (11.1), $g_{\eta\xi}(\xi^\zeta)d\xi^\eta d\xi^\xi$ being the fundamental form of an S_{n-1}.*

With respect to this coordinate system, the metric tensors have the form

$$g_{\lambda\varkappa} = \begin{pmatrix} 1 & 0 \\ 0 & g_{\eta\xi}(\xi^\zeta) \end{pmatrix}, \quad g^{\lambda\varkappa} = \begin{pmatrix} 1 & 0 \\ 0 & g^{\eta\xi}(\xi^\zeta) \end{pmatrix}.$$

[1] YANO [6].

Consequently, calculating the Christoffel symbol of the V_n, we find

$$(11.3)\qquad \{^{\xi}_{\zeta\eta}\} = '\{^{\xi}_{\zeta\eta}\};\ \{^{1}_{\mu\lambda}\} = 0;\ \{^{\varkappa}_{1\lambda}\} = 0,$$

where $'\{^{\xi}_{\zeta\eta}\}$ denotes the Christoffel symbol of an S_{n-1} with $g_{\eta\xi}(\xi^{\zeta})$ as metric tensor.

For the components $K_{\nu\mu\lambda}^{\cdot\cdot\cdot\varkappa}$ of the curvature tensor of V_n, we get

$$(11.4)\qquad K_{\varphi\zeta\eta}^{\cdot\cdot\cdot\xi} = 'K_{\varphi\zeta\eta}^{\cdot\cdot\cdot\xi};\ K_{\nu\mu\lambda}^{\cdot\cdot\cdot 1} = 0;\ K_{1\mu\lambda}^{\cdot\cdot\cdot\varkappa} = 0;\ K_{\nu\mu 1}^{\cdot\cdot\cdot\varkappa} = 0,$$

where $'K_{\varphi\zeta\eta}^{\cdot\cdot\cdot\xi}$ denote the components of the curvature tensor of S_{n-1} belonging to $'\{^{\xi}_{\zeta\eta}\}$.

But for an S_{n-1} we know that

$$(11.5)\qquad 'K_{\varphi\zeta\eta}^{\cdot\cdot\cdot\xi} = \frac{'K}{(n-1)(n-2)}(A_{\varphi}^{\xi} g_{\zeta\eta} - A_{\zeta}^{\xi} g_{\varphi\eta}),$$

$'K$ being constant. Consequently we have for the Ricci tensor

$$(11.6)\qquad K_{\zeta\eta} = \frac{'K}{n-1} g_{\zeta\eta};\ K_{1\eta} = 0,\ K_{11} = 0,$$

and for the scalar curvature

$$(11.7)\qquad K = 'K.$$

Thus if we put

$$(11.8)\qquad L_{\mu\lambda} \stackrel{\text{def}}{=} -K_{\mu\lambda} + \frac{1}{2(n-1)} K g_{\mu\lambda},$$

we find

$$(11.9)\qquad \begin{cases} L_{11} = \dfrac{'K}{2(n-1)};\quad L_{\zeta\eta} = -\dfrac{'K}{2(n-1)} g_{\zeta\eta};\ L_{1\eta} = 0, \\[2ex] L_{1}^{\cdot 1} = \dfrac{'K}{2(n-1)};\quad L_{\zeta}^{\cdot\xi} = -\dfrac{'K}{2(n-1)} A_{\zeta}^{\xi};\ L_{1}^{\cdot\xi} = 0;\ L_{\zeta}^{\cdot 1} = 0, \end{cases}$$

and for the conformal curvature tensor, we get

$$(11.10)\ C_{\nu\mu\lambda}^{\cdot\cdot\cdot\varkappa} \stackrel{\text{def}}{=} K_{\nu\mu\lambda}^{\cdot\cdot\cdot\varkappa} + \frac{1}{n-2}(A_{\nu}^{\varkappa} L_{\mu\lambda} - A_{\mu}^{\varkappa} L_{\nu\lambda} + L_{\nu}^{\cdot\varkappa} g_{\mu\lambda} - L_{\mu}^{\cdot\varkappa} g_{\nu\lambda}) = 0.$$

Consequently since we assumed $n > 4$, the V_n is a C_n.[1]

Conversely, if we assume that the space is conformally Euclidean and

[1] SCHOUTEN [8], p. 306. The C_n denotes a conformally Euclidean V_n.

admits a covariant constant vector field, then there exists a coordinate system with respect to which

$$ds^2 = (d\xi^1)^2 + g_{\eta\xi}(\xi^\zeta)d\xi^\eta d\xi^\xi$$

and

$$\{^{\,\xi}_{\zeta\eta}\} = '\{^{\,\xi}_{\zeta\eta}\},\quad K_{\varphi\zeta\eta}{}^{\xi} = 'K_{\varphi\zeta\eta}{}^{\xi},\quad K_{\zeta\eta} = 'K_{\zeta\eta},\quad K = 'K, \tag{11.11}$$

the other components of $\{^{\,\kappa}_{\mu\lambda}\}$, $K_{\nu\mu\lambda}{}^{\xi}$ and $K_{\mu\lambda}$ being zero.

From these equations we find

$$\left\{\begin{array}{ll} L_{11} = \dfrac{'K}{2(n-1)}, & L_{\zeta\eta} = -\,'K_{\zeta\eta} + \dfrac{'K}{2(n-1)}\,g_{\zeta\eta}, \\[2ex] L_1{}^{1} = \dfrac{'K}{2(n-1)}, & L_\zeta{}^{\xi} = -\,'K_\zeta{}^{\xi} + \dfrac{'K}{2(n-1)}\,A^\xi_\zeta, \end{array}\right. \tag{11.12}$$

the other components of $L_{\mu\lambda}$ and $L_\mu{}^{\kappa}$ being zero.

First, from

$$C_{1\zeta\eta}{}^{1} = \frac{1}{n-2}\,(L_{\zeta\eta} + L_1{}^{1} g_{\zeta\eta}) = 0,$$

we find

$$'K_{\zeta\eta} = \frac{'K}{n-1}\,g_{\zeta\eta},$$

and consequently

$$L_{\zeta\eta} = -\,\frac{'K}{2(n-1)}\,g_{\zeta\eta},\quad L_\zeta{}^{\xi} = -\,\frac{'K}{2(n-1)}\,A^\xi_\zeta.$$

Next from

$$C_{\varphi\zeta\eta}{}^{\xi} = K_{\varphi\zeta\eta}{}^{\xi} + \frac{1}{n-2}\,(A^\xi_\varphi L_{\zeta\eta} - A^\xi_\zeta L_{\varphi\eta} + L_\varphi{}^{\xi} g_{\zeta\eta} - L_\zeta{}^{\xi} g_{\varphi\eta}) = 0,$$

we find

$$'K_{\varphi\zeta\eta}{}^{\xi} = \frac{'K}{(n-1)(n-2)}\,(A^\xi_\varphi\, g_{\zeta\eta} - A^\xi_\zeta g_{\varphi\eta}),$$

which shows that the hypersurfaces $\xi' =$ constant are S_{n-1}'s with the same constant curvature. Thus we have

THEOREM 11.2. *In order that case I in Theorem* 10.8 *occur, it is necessary and sufficient that the space be conformally Euclidean and admit a covariant constant vector field.*

I. Adati and the present author [1] have shown that in order that a V_n be subprojective space of Kagan [2] it is necessary and sufficient that the space be conformally Euclidean and admit a concircular vector field. Thus according to Theorem 11.2, the space under consideration is a subprojective space of Kagan.

We can also give another geometrical characterization.

First there exists a covariant constant vector field $u^\varkappa$:

$$\nabla_\mu u_\lambda = 0. \tag{11.13}$$

It may be assumed that $u^\varkappa$ is a unit vector.

Since u_λ is a gradient, we can put $u_\lambda = \partial_\lambda f$.

From (11.13) we find

$$K_{\nu\mu\lambda\varkappa} u^\varkappa = 0. \tag{11.14}$$

The sectional curvature [3] determined by a plane containing $u^\varkappa$ and a unit vector $v^\varkappa$ orthogonal to $u^\varkappa$ at a point of the V_n is given by

$$-K_{\nu\mu\lambda\varkappa} u^\nu v^\mu u^\lambda v^\varkappa.$$

Since the V_n admits a transitive group of motions which transform the field $u^\varkappa$ into itself and every vector orthogonal to $u^\varkappa$ into a vector with the same property, the sectional curvature is a constant. But from (11.14) we get

$$-K_{\nu\mu\lambda\varkappa} u^\nu v^\mu u^\lambda v^\varkappa = 0 \tag{11.15}$$

for any $v^\varkappa$, which shows that this sectional curvature is always zero.

On the other hand, the hypersurfaces

$$f(\xi) = \text{constant} \tag{11.16}$$

are geodesic and of the same constant curvature. Consequently, representing one of them by its parametric equations

$$\xi^\varkappa = B^\varkappa(\eta^a); \quad B_b^\varkappa \overset{\text{def}}{=} \partial_b \xi^\varkappa; \tag{11.17}$$

$$a, b, c, d = 1, 2, \ldots, n-1,$$

---

[1] Yano and Adati [1].

[2] Rachevsky [1].

[3] Two vectors at a point determine a 2-plane. The Gaussian curvature at the point of the two-dimensional subspace described by the geodesics passing through the point and being tangent to the 2-plane is called the sectional curvature at the point determined by the 2-plane.

we have the equations of Gauss[1]

$$'K_{dcba} = B^{\nu\mu\lambda\varkappa}_{dcba} K_{\nu\mu\lambda\varkappa} \tag{11.18}$$

where

$$'K_{dcba} = 'k('g_{da}\,'g_{cb} - 'g_{ca}\,'g_{db}) \tag{11.19}$$

and

$$'g_{cb} \overset{\text{def}}{=} B^{\mu\lambda}_{cb} g_{\mu\lambda}. \tag{11.20}$$

The $'k$ in (11.19) could be different for each of the hypersurfaces (11.16). But, since $'k$ represents the sectional curvature determined by any 2-plane orthogonal to $u^\varkappa$, and because the space admits a transitive group of motions which leaves invariant the field $u^\varkappa$, $'k$ should be constant.

Now on putting

$$B^b_\lambda \overset{\text{def}}{=} 'g^{ba} g_{\lambda\varkappa} B^\varkappa_a,$$

we obtain

$$B^\varkappa_a B^a_\lambda = A^\varkappa_\lambda - u^\varkappa u_\lambda, \quad 'g_{ba} B^b_\lambda B^a_\varkappa = g_{\lambda\varkappa} - u_\lambda u_\varkappa. \tag{11.21}$$

On transvecting both members of (11.18) with $B^{dcba}_{\omega\tau\sigma\rho}$ we obtain

$$\begin{aligned} &'k('g_{da}\,'g_{cb} - 'g_{ca}\,'g_{db}) B^{dcba}_{\omega\tau\sigma\rho} \\ &\qquad = (A^\nu_\omega - u^\nu u_\omega)(A^\mu_\tau - u^\mu u_\tau)(A^\lambda_\sigma - u^\lambda u_\sigma)(A^\varkappa_\rho - u^\varkappa u_\rho) K_{\nu\mu\lambda\varkappa} \end{aligned}$$

from which

$$\begin{aligned} K_{\nu\mu\lambda\varkappa} = {}&'k[(g_{\varkappa\nu} g_{\mu\lambda} - g_{\varkappa\mu} g_{\nu\lambda}) \\ &- (u_\nu g_{\mu\lambda} - u_\mu g_{\nu\lambda}) u_\varkappa + (u_\nu g_{\mu\varkappa} - u_\mu g_{\nu\varkappa}) u_\lambda]. \end{aligned} \tag{11.22}$$

Conversely, suppose that the curvature tensor of V_n has the form (11.22) where $'k$ is now some constant and where $u_\lambda = \partial_\lambda f$ is some covariant constant vector field. Then the hypersurfaces $f(\xi) = \text{const.}$ are geodesic and their orthogonal trajectories are also geodesic.

Representing one of these hypersurfaces by (11.17), from (11.18) and (11.22), we find (11.19). (11.19) shows that all hypersurfaces are S_{n-1}'s with the same constant curvature. Hence

THEOREM 11.3. *In order that case I in Theorem* 10.8 *occur, it is necessary and sufficient that the curvature tensor of the space be of the form*

[1] SCHOUTEN [8], p. 242.

(11.22), *where* $'k$ *is constant and where* $u_\varkappa$ *is a covariant constant unit vector field.*

From the equation (11.22), we get

$$\nabla_\omega K_{\nu\mu\lambda}^{\cdot\cdot\cdot\varkappa} = 0, \tag{11.23}$$

which shows that the V_n under consideration is symmetric in the sense of E. Cartan. [1]

§ 12. Case II.

In this case, the normals to the hypersurfaces are Ricci directions. Thus on account of a well-known theorem [2] the space admits what we call a concircular transformation, and consequently there exists a coordinate system with respect to which

$$ds^2 = (d\xi^1)^2 + f(\xi^1) f_{\eta\xi}(\xi^\zeta) d\xi^\eta d\xi^\xi, \tag{12.1}$$

$g_{\eta\zeta} d\xi^\eta d\xi^\xi = f(\xi^1) f_{\eta\xi}(\xi^\zeta) d\xi^\eta d\xi^\xi$ being the fundamental form of an S_{n-1} with constant curvature. When the function $f(\xi^1)$ is a constant, the case reduces to case I, so we assume hereafter that $f(\xi^1)$ is not a constant.

On calculating the Christoffel symbols of V_n, we obtain, for the non-vanishing components of $\{^{\varkappa}_{\mu\lambda}\}$,

$$\{^{1}_{\zeta\eta}\} = -\tfrac{1}{2}\frac{f'}{f} g_{\zeta\eta}, \quad \{^{\xi}_{\zeta 1}\} = \{^{\xi}_{1\zeta}\} = +\tfrac{1}{2}\frac{f'}{f} A^{\xi}_{\zeta}, \quad \{^{\xi}_{\zeta\eta}\} = '\{^{\xi}_{\zeta\eta}\}, \tag{12.2}$$

where $f' = df/d\xi^1$ and $'\{^{\xi}_{\zeta\eta}\}$ are Christoffel symbols formed with $g_{\eta\xi} = f(\xi^1) f_{\eta\xi}(\xi^\zeta)$ or, what is the same here, with $f_{\eta\xi}(\xi^\zeta)$.

Now calculating the curvature tensor of V_n, we get, for the non-vanishing components of $K_{\nu\mu\lambda}^{\cdot\cdot\cdot\varkappa}$,

$$\left\{\begin{aligned} K_{\varphi 1\eta}^{\cdot\cdot\cdot 1} &= -K_{1\varphi\eta}^{\cdot\cdot\cdot 1} = +\left(\tfrac{1}{2}\frac{f''}{f} - \tfrac{1}{4}\frac{f'^2}{f^2}\right) g_{\varphi\eta}, \\ K_{\varphi 11}^{\cdot\cdot\cdot\xi} &= -K_{1\varphi 1}^{\cdot\cdot\cdot\xi} = -\left(\tfrac{1}{2}\frac{f''}{f} - \tfrac{1}{4}\frac{f'^2}{f^2}\right) A^{\xi}_{\varphi}, \\ K_{\varphi\zeta\eta}^{\cdot\cdot\cdot\xi} &= -K_{\zeta\varphi\eta}^{\cdot\cdot\cdot\xi} = 'K_{\varphi\zeta\eta}^{\cdot\cdot\cdot\xi} - \tfrac{1}{4}\frac{f'^2}{f^2}(A^{\xi}_{\varphi} g_{\zeta\eta} - A^{\xi}_{\zeta} g_{\varphi\eta}), \end{aligned}\right. \tag{12.3}$$

where $'K_{\varphi\zeta\eta}^{\cdot\cdot\cdot\xi}$ is the curvature tensor formed with $g_{\eta\xi}$.

---

[1] CARTAN [1, 2, 6, 8, 11], Cf. SCHOUTEN [8], p. 163, p. 370.
[2] YANO [4].

From (12.3), we get, for the non-vanishing components of $K_{\nu\mu\lambda\kappa}$,

$$(12.4)\qquad \begin{cases} K_{\varphi 1\eta 1} = \left(\frac{1}{2}\frac{f''}{f'} - \frac{1}{4}\frac{f'^2}{f^2}\right) g_{\varphi\eta}, \\ K_{\varphi\zeta\eta\xi} = {}'K_{\varphi\zeta\eta\xi} - \frac{1}{4}\frac{f'^2}{f^2}(g_{\varphi\xi}g_{\zeta\eta} - g_{\zeta\xi}g_{\varphi\eta}). \end{cases}$$

From the first equation of (12.4) we see that the sectional curvature determined by two mutually orthogonal unit vectors u^κ with $u^1 = 1$, $u^\xi = 0$ and v^κ with $v^1 = 0$, is

$$(12.5)\qquad -K_{\varphi 1\eta 1}v^\varphi v^\eta = -\left(\frac{1}{2}\frac{f''}{f'} - \frac{1}{4}\frac{f'^2}{f^2}\right),$$

and this does not depend on v^κ. Since there is always a motion of the transitive group which transforms the field u^κ into itself and every vector orthogonal to it into a vector with the same property, it follows that this sectional curvature is a constant.

From the second equation of (12.4) we see that the sectional curvature determined by two mutually orthogonal unit vectors v^κ with $v^1 = 0$ and w^κ with $w^1 = 0$ is

$$-K_{\varphi\zeta\eta\xi}v^\varphi w^\zeta v^\eta w^\xi = -\left({}'K_{\varphi\zeta\eta\xi}v^\varphi w^\zeta v^\eta w^\xi + \frac{1}{4}\frac{f'^2}{f^2}\right).$$

Since this must be independent of the choice of v^ξ and w^ξ, we must have

$$(12.6)\qquad {}'K_{\varphi\zeta\eta\xi} = {}'k(g_{\varphi\xi}g_{\zeta\eta} - g_{\zeta\xi}g_{\varphi\eta})$$

and consequently

$$(12.7)\qquad -K_{\varphi\zeta\eta\xi}v^\varphi w^\zeta v^\eta w^\xi = {}'k - \frac{1}{4}\frac{f'^2}{f^2}.$$

Since the group is transitive, this scalar must also be constant.

The equation (12.6) shows that the hypersurfaces $\xi^1 =$ constant are S_{n-1}'s. But we know that these must be all of the same constant curvature. Thus $'k$ is also constant. Hence $f'^2/4f^2$ is a constant

$$(12.8)\qquad \frac{1}{4}\frac{f'^2}{f^2} = k^2,$$

k being different from zero, from which by integration

$$(12.9)\qquad f = a^2 e^{2k\xi^1},$$

where a^2 is an arbitrary positive constant.

On the other hand, we have

$$g_{\zeta\eta} = f(\xi^1) f_{\zeta\eta}(\xi^\xi),$$
$$'\{^{\ \xi}_{\zeta\eta}\} = \tfrac{1}{2} f^{\xi\varphi}(\partial_\zeta f_{\eta\varphi} + \partial_\eta f_{\zeta\varphi} - \partial_\varphi f_{\zeta\eta}),$$

and consequently

$$'K_{\varphi\zeta\eta}^{\cdots\xi} = F_{\varphi\zeta\eta}^{\cdots\xi},$$

where $F_{\varphi\zeta\eta}^{\cdots\xi}$ is the curvature tensor formed from $f_{\zeta\eta}(\xi^\xi)$. Consequently we have

$$(12.10) \qquad 'K_{\varphi\zeta\eta\xi} = f(\xi^1) F_{\varphi\zeta\eta\xi}.$$

Thus from (12.6) and (12.7) we obtain

$$(12.11) \qquad F_{\varphi\zeta\eta\xi} = F(f_{\varphi\xi} f_{\zeta\eta} - f_{\zeta\xi} f_{\varphi\eta})$$

where

$$(12.12) \qquad F = f(\xi^1)'k$$

is a constant. Here F and $'k$ are constants and $f(\xi^1)$ is not a constant, consequently we must have

$$'k = 0, \quad F = 0$$

from which

$$(12.13) \qquad 'K_{\varphi\zeta\eta\xi} = 0, \quad F_{\varphi\zeta\eta\xi} = 0.$$

Substituting (12.9) and the first equation of (12.13) into (12.4) we obtain, for the non-vanishing components of $K_{\nu\mu\lambda\varkappa}$,

$$K_{\varphi 1\eta 1} = +k^2 g_{\varphi\eta}, \quad K_{\varphi\zeta\eta\xi} = -k^2(g_{\varphi\xi} g_{\zeta\eta} - g_{\zeta\xi} g_{\varphi\eta}).$$

Hence

$$(12.14) \qquad K_{\nu\mu\lambda\varkappa} = -k^2(g_{\nu\varkappa} g_{\mu\lambda} - g_{\mu\varkappa} g_{\nu\lambda}).$$

Thus the space is of negative constant curvature.

In fact, by (12.9) and the second equation of (12.13), the fundamental form of V_n can be written as

$$(12.15) \qquad ds^2 = (d\xi^1)^2 + a^2 e^{2k\xi^1}[(d\xi^2)^2 + \ldots + (d\xi^n)^2],$$

which is a well-known fundamental form of an S_n of negative constant curvature.

Conversely, if the V_n is an S_n of negative constant curvature, then there exists a coordinate system with respect to which the fundamental form takes the form (12.15). If we put

$$ae^{k\xi^1} = \frac{1}{ku}$$

then (12.15) becomes

$$ds^2 = \frac{(du)^2 + (d\xi^2)^2 + \ldots + (d\xi^n)^2}{k^2u^2}. \tag{12.16}$$

Thus we can see that the space admits a group of motions of order $\frac{1}{2}n(n-1)+1$ given by

$$'u = \alpha u, \quad '\xi^\eta = \alpha(a^\eta_\zeta \xi^\zeta + b^\eta) \tag{12.17}$$

where $'\xi^\eta = a^\eta_\zeta \xi^\zeta + b^\eta$ represents a general motion in a Euclidean R_{n-1}. Thus we have

THEOREM 12.1. *In order that case II in Theorem 10.8 occur it is necessary and sufficient that the space be of negative constant curvature.*

Since an S_n cannot admit a parallel vector field, case I is not a special case of II.

Gathering the results obtained in the last three sections, we can state

THEOREM 12.2. *In order that a V_n, $n < 4$, $n \neq 8$, admit a group of motions of order $\frac{1}{2}n(n-1)+1$, it is necessary and sufficient that the space be a product of a straight line and an S_{n-1} (this is equivalent to the fact that the space is a C_n and admits a parallel vector field) or that the space be an S_n of negative constant curvature.*

In this theorem, the cases $n = 3$, $n = 4$ and $n = 8$ are exceptional. E. Cartan[1] has studied the case $n = 3$ in detail and he obtained

THEOREM 12.3. *A simply connected complete V_3 admitting a G_4 of motions is homeomorphic to one of the following spaces.*

(1) *a Euclidean space.*
(2) *a product space of a straight line and a sphere.*
(3) *a spherical space.*

The case $n = 4$ was our exceptional case. But S. Ishihara[2] has studied this case and obtained

---

[1] CARTAN [6], p. 305.
[2] ISHIHARA [1].

THEOREM 12.4. *A simply connected complete V_4 admitting a transitive group of motions is homeomorphic to one of the following spaces:*

(1) *a Euclidean space of four dimensions.*

(2) *a sphere of four dimensions.*

(3) *a complex projective space of two complex dimensions.*

(4) *a product space of two spheres of two dimensions.*

(5) *a product space of a straight line and a sphere of three dimensions.*

(6) *a product space of a Euclidean plane and a sphere of two dimensions.*

THEOREM 12.5. *In a V_4 there exists no group of motions of order* 9. *If a V_4 admits a group of motions of order* 8, *then the group is transitive and the space is a Kählerian manifold whose holomorphic sectional curvature is constant.*

CHAPTER V

GROUPS OF AFFINE MOTIONS

§ 1. Groups of affine motions.

Consider an L_n with a linear connexion $\Gamma^{\varkappa}_{\mu\lambda}$. Since the linear connexion $\Gamma^{\varkappa}_{\mu\lambda}$ is a linear differential geometric object, Theorems 2.1 and 2.2 of Ch. III give

THEOREM 1.1. *If an L_n admits an infinitesimal affine motion, it admits also a G_1 of affine motions generated by the infinitesimal one.*

THEOREM 1.2. *In order that an L_n admit a G_1 of affine motions, it is necessary and sufficient that there exist a coordinate system with respect to which the components $\Gamma^{\varkappa}_{\mu\lambda}$ of the linear connexion are independent of one of the coordinates.*

To study the projective differential geometry inaugurated by O. Veblen [1], J. H. C. Whitehead [2] considered an A_{n+1} whose linear connection satisfies

$$\Gamma^{\varkappa}_{\mu 0} = A^{\varkappa}_{\mu}, \quad \partial_0 \Gamma^{\varkappa}_{\mu\lambda} = 0, \qquad (\varkappa, \lambda, \mu, \ldots = 0, 1, 2, \ldots, n). \tag{1.1}$$

According to Theorem 1.2, this A_n admits a G_1 of affine motions. In fact, if we put

$$v^{\varkappa} \overset{\text{def}}{=} \underset{0}{e}^{\varkappa}.$$

(1.1) can be written as

$$\nabla_{\mu} v^{\varkappa} = A^{\varkappa}_{\mu}, \quad \nabla_{\mu} \nabla_{\lambda} v^{\varkappa} + R_{\nu\mu\lambda}^{\cdots\varkappa} v^{\nu} = 0. \tag{1.2}$$

Thus the A_{n+1} considered by J. H. C. Whitehead can be characterized as an A_{n+1} which admits a concurrent vector field and an affine motion. The authors of the School of Princeton consider in such a space tensors

[1] VEBLEN [1, 2].
[2] WHITEHEAD [1].

or tensor densities whose components are of the form

$$\mathfrak{p}^{\varkappa\lambda}_{\cdot\cdot\mu} = e^{N\xi^0} f^{\varkappa\lambda}_{\cdot\cdot\mu}(\xi^1, \ldots, \xi^n),$$

from which

$$\underset{v}{\pounds}\, \mathfrak{p}^{\varkappa\lambda}_{\cdot\cdot\mu} = N \mathfrak{p}^{\varkappa\lambda}_{\cdot\cdot\mu}. \tag{1.3}$$

This means that they consider the quantities whose Lie derivatives are proportional to the quantities themselves.

If we choose a coordinate system with respect to which $v^\varkappa = \xi^\varkappa$, the equation $\underset{v}{\pounds}\Gamma^\varkappa_{\mu\lambda} = 0$ takes the form

$$\underset{v}{\pounds}\, \Gamma^\varkappa_{\mu\lambda} = \xi^\rho \partial_\rho \Gamma^\varkappa_{\mu\lambda} + \Gamma^\varkappa_{\mu\lambda} = 0, \tag{1.4}$$

hence

THEOREM 1.3. *In order that an L_n admit a G_1 of affine motions, it is necessary and sufficient that there exist a coordinate system with respect to which the $\Gamma^\varkappa_{\mu\lambda}(\xi)$ are homogeneous functions of degree -1 of the coordinates.*

To study projective differential geometry, D. van Dantzig considered [1] an L_{n+1} whose components of the linear connexion are homogeneous functions of degree -1 of the coordinates $\xi^\varkappa$; $\varkappa = 0, 1, \ldots, n$. According to Theorem 1.3, this L_{n+1} admits a G_1 of affine motions. [2] The authors of the School of Delft [3] consider in such a space the tensors or tensor densities whose components are homogeneous functions of degree r. Thus

$$\underset{v}{\pounds}\, \mathfrak{p}_{\lambda_1 \cdots \lambda_t}^{\cdot \; \cdot \; \varkappa_1 \cdots \varkappa_s} = [r - s + t + (n+1)w]\mathfrak{p}_{\lambda_1 \cdots \lambda_t}^{\cdot \; \cdot \; \varkappa_1 \cdots \varkappa_s} \tag{1.5}$$

for a tensor of weight w. This means they also consider those quantities whose Lie derivatives are proportional to the quantities themselves. The number $r - s + t + (n+1)w$ was called the *excess*.

Since a linear connexion is a linear differential geometric object, Theorems 2.3, 2.4, 2.5 and 2.6 of Ch. III hold for a group of affine motions.

§ 2. Groups of affine motions in a space with absolute parallelism. [4]

An L_n is said to possess *absolute parallelism* or *teleparallelism* if for

[1] VAN DANTZIG [1].

[2] VAN DANTZIG [2, 3].

[3] SCHOUTEN and HAANTJES [1].

[4] ROBERTSON [1].

any points P and Q of the space the parallel displacement of any quantity from P to Q along a curve joining P and Q gives a result at Q which does not depend on the choice of the curve. In such a space we fix a point $\underset{0}{\xi}{}^{\varkappa}$ and consider n linearly independent contravariant vectors $\underset{a}{e}{}^{\varkappa}(\underset{0}{\xi})$ at this point, $(a, b, c, \ldots = 1, 2, \ldots, n)$.

Since the parallel displacement does not depend on the curve, on displacing the vectors $\underset{a}{e}{}^{\varkappa}(\underset{0}{\xi})$ from $\underset{0}{\xi}{}^{\varkappa}$ to an arbitrary point $\xi^{\varkappa}$ of the space along any curve joining $\underset{0}{\xi}{}^{\varkappa}$ and $\xi^{\varkappa}$, we get fields of vectors $\underset{a}{e}{}^{\varkappa}(\xi)$, and for these fields holds

$$\nabla_{\mu}\underset{a}{e}{}^{\varkappa} \overset{\text{def}}{=} \partial_{\mu}\underset{a}{e}{}^{\varkappa} + \Gamma^{\varkappa}_{\mu\lambda}\underset{a}{e}{}^{\lambda} = 0, \tag{2.1}$$

from which

$$\Gamma^{\varkappa}_{\mu\lambda} = -\overset{a}{e}_{\lambda}\partial_{\mu}\underset{a}{e}{}^{\varkappa} = \underset{a}{e}{}^{\varkappa}\partial_{\mu}\overset{a}{e}_{\lambda}, \tag{2.2}$$

where the $\overset{a}{e}_{\lambda}$ are defined by $\overset{a}{e}_{\lambda}\underset{b}{e}{}^{\lambda} = \delta^{a}_{b}$.

It is well-known that if an L_n admits absolute parallelism, then $R_{\nu\mu\lambda}{}^{\varkappa} = 0$ and conversely, if $R_{\nu\mu\lambda}{}^{\varkappa} = 0$, then the L_n admits absolute parallelism, the $S_{\mu\lambda}{}^{\varkappa}$ being not necessarily zero.

If an L_n with absolute parallelism admits an infinitesimal affine motion $\xi^{\varkappa} \to \xi^{\varkappa} + v^{\varkappa}dt$, then from $\nabla_{\mu}\underset{a}{e}{}^{\varkappa} = 0$, $\underset{v}{\pounds}\Gamma^{\varkappa}_{\mu\lambda} = 0$ and

$$\underset{v}{\pounds}(\nabla_{\mu}\underset{a}{e}{}^{\varkappa}) - \nabla_{\mu}(\underset{v}{\pounds}\underset{a}{e}{}^{\varkappa}) = (\underset{v}{\pounds}\Gamma^{\varkappa}_{\mu\lambda})\underset{a}{e}{}^{\lambda},$$

we obtain

$$\nabla_{\mu}(\underset{v}{\pounds}\underset{a}{e}{}^{\varkappa}) = 0,$$

which shows that the vectors $\underset{v}{\pounds}\underset{a}{e}{}^{\varkappa}$ are also absolutely parallel and consequently that

$$\underset{v}{\pounds}\underset{b}{e}{}^{\varkappa} = c^{a}_{b}\underset{a}{e}{}^{\varkappa}, \quad c^{a}_{b} = \text{constants}. \tag{2.2}$$

Thus we have[1]

THEOREM 2.1. *In order that an L_n with absolute parallelism admit an infinitesimal affine motion, it is necessary and sufficient that the Lie*

[1] G. T., p. 18.

derivatives of n linearly independent absolutely parallel contravariant vectors be linear combinations of these vectors with constant coefficients.

If all the constants c_b^a are zero, that is, if

$$\underset{v}{\pounds}\underset{a}{e}^{\varkappa} = 0, \tag{2.3}$$

the affine motion is said to be *particular*.

If we choose a coordinate system with respect to which $v^\varkappa = \delta_1^\varkappa$, the conditions for a particular and a general affine motion become

$$\partial_1 \underset{a}{e}^{\varkappa} = 0 \text{ and } \partial_1 \underset{b}{e}^{\varkappa} = c_b^a \underset{a}{e}^{\varkappa}$$

respectively, from which

$$\underset{a}{e}^{\varkappa} = \underset{a}{f}^{\varkappa}(\xi^2, \ldots, \xi^n) \text{ and } \underset{b}{e}^{\varkappa} = u_b^a(\xi^1)\underset{a}{f}^{\varkappa}(\xi^2, \ldots, \xi^n) \tag{2.4}$$

respectively, where $u_b^a(\xi^1)$ are functions of ξ^1 satisfying

$$\partial_1 u_b^a(\xi^1) = c_b^c u_c^a(\xi^1) \tag{2.5}$$

and consequently also satisfying

$$u_b^a(\xi^1 + t) = u_c^a(\xi^1)u_b^c(t).\text{[1]} \tag{2.6}$$

Conversely when the $\underset{a}{e}^{\varkappa}$ have the property (2.4), the space evidently admits a one-parameter group of affine motions given by $\xi^\varkappa \rightarrow \xi^\varkappa + \underset{1}{e}^{\varkappa}dt$. Thus we have

THEOREM 2.2. *In order that an L_n with absolute parallelism admit a G_1 of particular affine motions, it is necessary and sufficient that there exist a coordinate system with respect to which all the components of the absolutely parallel contravariant vectors are independent of one of these variables.*

THEOREM 2.3. *In order that an L_n with absolute parallelism admit a G_1 of general affine motions, it is necessary and sufficient that there exist a coordinate system with respect to which the components of the absolutely parallel contravariant vectors have the form* $\underset{b}{e}^{\varkappa} = u_b^a(\xi^1)\underset{a}{f}^{\varkappa}(\xi^2, \ldots, \xi^n)$, *with* $u_b^a(\xi^1)$ *satisfying* (2.5) *or* (2.6).

An A_n with absolute parallelism is an E_n. In E_n we can take a rectilinear coordinate system $(\varkappa)$. Then $\underset{v}{\pounds}\underset{\sigma}{e}^{\varkappa} = 0$ gives $v^\varkappa =$ const, and conse-

[1] VON NEUMANN [1]; WEYL [3], p. 25.

quently this means that the particular affine motion is a translation. In E_n, $\underset{v}{\pounds} \underset{\mu}{e}^\varkappa = c_\mu^\lambda \underset{\lambda}{e}^\varkappa$ gives $v^\varkappa = c_\lambda^\varkappa \xi^\lambda + p^\varkappa$ and this means that the general affine motion is a general affine transformation.

§ 3. Infinitesimal transformations which carry affine conics into affine conics.

It is evident that in an A_n an infinitesimal affine motion carries an arbitrary *affine conic* [1]

$$\frac{\delta^3 \xi^\varkappa}{ds^3} + k \frac{d\xi^\varkappa}{ds} = 0 \quad (k = \text{constant}) \tag{3.1}$$

into an affine conic and transforms every affine parameter into an affine parameter.

Conversely we assume that an infinitesimal transformation $\xi^\varkappa \to \xi^\varkappa + v^\varkappa dt$ carries an arbitrary affine conic into an affine conic and the affine parameter on it into an affine parameter on the deformed affine conic.

First we get from $\pounds d\xi^\varkappa = 0$

$$\pounds \frac{d\xi^\varkappa}{ds} = - \frac{d\xi^\varkappa}{ds} \frac{\pounds ds}{ds}, \tag{3.2}$$

where we have dropped v from $\underset{v}{\pounds}$ for the sake of simplicity.

Secondly, from the formula

$$\pounds \delta u^\varkappa - \delta \pounds u^\varkappa = (\pounds \Gamma_{\mu\lambda}^\varkappa) d\xi^\mu u^\lambda,$$

we get

$$\pounds \frac{\delta u^\varkappa}{ds} = (\pounds \Gamma_{\mu\lambda}^\varkappa) \frac{d\xi^\mu}{ds} u^\lambda + \frac{\delta(\pounds u^\varkappa)}{ds} - \frac{\delta u^\varkappa}{ds} \frac{\pounds ds}{ds}. \tag{3.3}$$

Thus putting $u^\varkappa = d\xi^\varkappa / ds$ in (3.3), we get

$$\pounds \frac{\delta^2 \xi^\varkappa}{ds^2} = (\pounds \Gamma_{\mu\lambda}^\varkappa) \frac{d\xi^\mu}{ds} \frac{d\xi^\lambda}{ds} - 2 \frac{\delta^2 \xi^\varkappa}{ds^2} \frac{\pounds ds}{ds} - \frac{d\xi^\varkappa}{ds} \frac{d}{ds} \frac{\pounds ds}{ds}. \tag{3.4}$$

Putting then $u^\varkappa = \delta^2 \xi^\varkappa / ds^2$ in (3.3), we find

$$\begin{aligned} \pounds \frac{\delta^3 \xi^\varkappa}{ds^3} = {} & 3(\pounds \Gamma_{\mu\lambda}^\varkappa) \frac{d\xi^\mu}{ds} \frac{\delta^2 \xi^\lambda}{ds^2} + (\nabla_\nu \pounds \Gamma_{\mu\lambda}^\varkappa) \frac{d\xi^\nu}{ds} \frac{d\xi^\mu}{ds} \frac{d\xi^\lambda}{ds} \\ & - 3 \frac{\delta^3 \xi^\varkappa}{ds^3} \frac{\pounds ds}{ds} - 3 \frac{\delta^2 \xi^\varkappa}{ds^2} \frac{d}{ds} \frac{\pounds ds}{ds} - \frac{d\xi^\varkappa}{ds} \frac{d^2}{ds^2} \frac{\pounds ds}{ds}. \end{aligned} \tag{3.5}$$

---

[1] YANO and TAKANO [1]; cf. SCHOUTEN [8], p. 299.

But because the transformation transforms every affine parameter into an affine parameter, we must have

$$'s = (1 + adt)s + bdt; \quad a, b = \text{constants}. \tag{3.6}$$

From this it follows that

$$\frac{\pounds ds}{ds} = a \tag{3.7}$$

and that the equations (3.2) and (3.5) become

$$\pounds \frac{d\xi^\varkappa}{ds} = - a \frac{d\xi^\varkappa}{ds} \tag{3.8}$$

and

$$\pounds \frac{\delta^3 \xi^\varkappa}{ds^3} = 3(\pounds \Gamma^\varkappa_{\mu\lambda}) \frac{d\xi^\mu}{ds} \frac{\delta^2 \xi^\lambda}{ds^2} + (\nabla_\nu \pounds \Gamma^\varkappa_{\mu\lambda}) \frac{d\xi^\nu}{ds} \frac{d\xi^\mu}{ds} \frac{d\xi^\lambda}{ds} - 3a \frac{\delta^3 \xi^\varkappa}{ds^3}, \tag{3.9}$$

respectively.

On the other hand, under the transformation (3.6) of s, k is transformed as follows:

$$'k = \frac{1}{(1 + adt)^2} k = (1 - 2adt)k,$$

hence

$$\pounds k = - 2ak. \tag{3.10}$$

Thus, from (3.8), (3.9) and (3.10), we obtain

$$\pounds \left(\frac{\delta^3 \xi^\varkappa}{ds^3} + k \frac{d\xi^\varkappa}{ds} \right) = 3(\pounds \Gamma^\varkappa_{\mu\lambda}) \frac{d\xi^\mu}{ds} \frac{\delta^2 \xi^\lambda}{ds^2} + (\nabla_\nu \pounds \Gamma^\varkappa_{\mu\lambda}) \frac{d\xi^\nu}{ds} \frac{d\xi^\mu}{ds} \frac{d\xi^\lambda}{ds} - 3a \left(\frac{\delta^3 \xi^\varkappa}{ds^3} + k \frac{d\xi^\varkappa}{ds} \right). \tag{3.11}$$

Thus, in order that every affine conic be transformed into an affine conic and the affine parameter on it into an affine parameter on the deformed conic, the equation

$$3(\underset{}{\pounds}\Gamma^{\varkappa}_{\mu\lambda})\frac{d\xi^{\mu}}{ds}\frac{\delta^2\xi^{\lambda}}{ds^2} + (\nabla_{\nu}\pounds\Gamma^{\varkappa}_{\mu\lambda})\frac{d\xi^{\nu}}{ds}\frac{d\xi^{\mu}}{ds}\frac{d\xi^{\lambda}}{ds} = 0$$

must be satisfied for every value of $\dfrac{d\xi^{\varkappa}}{ds}$ and $\dfrac{\delta^2\xi^{\varkappa}}{ds^2}$, and consequently it is necessary that $\pounds\Gamma^{\varkappa}_{\mu\lambda} = 0$. Thus we have

THEOREM 3.1.[1] *In order that an infinitesimal transformation transform every affine conic of an A_n into an affine conic and the affine parameters on it into affine parameters on the deformed conic, it is necessary and sufficient that the transformation be an affine motion.*

§ 4. Some theorems on affine and projective motions.

We consider an A_n which admits a G_r of affine motions with the infinitesimal operators $\underset{a}{\pounds} f = \underset{a}{v}^{\mu}\partial_{\mu} f$ such that the rank of $\underset{a}{v}^{\varkappa}$ in a neighbourhood is $r \leq n$. Then we have

$$\underset{a}{\pounds}\Gamma^{\varkappa}_{\mu\lambda} = 0. \tag{4.1}$$

In order that an $'A_n$, projectively related to the A_n, admit the G_r as a group of affine motions, it is necessary and sufficient that there exist a covariant vector field p_{λ} such that

$$\underset{a}{\pounds}(\Gamma^{\varkappa}_{\mu\lambda} + p_{\mu}A^{\varkappa}_{\lambda} + p_{\lambda}A^{\varkappa}_{\mu}) = 0$$

or

$$\underset{a}{\pounds}p_{\lambda} = 0. \tag{4.1}$$

But according to Theorem 3.1 of Ch. III, this system of partial differential equations is completely integrable, hence

THEOREM 4.1.[2] *If an A_n admits a G_r of affine motions with the infinitesimal operators $\underset{a}{\pounds} f = \underset{a}{v}^{\mu}\partial_{\mu} f$ such that the rank of $\underset{a}{v}^{\varkappa}$ in a neighbourhood is $r \leq n$, there exists always an $'A_n$ which is (not trivially) projectively related to A_n and which admits the same G_r as a group of affine motions.*

---

[1] YANO and TAKANO [1]; G. T., p. 16.
[2] KNEBELMAN [3]; YANO and IMAI [1].

We next consider an A_n which admits a G_r of projective motions such that the rank of $\underset{a}{v}^\varkappa$ in a neighbourhood is $r \leq n$. Then we have

$$\underset{a}{\pounds}\Gamma^\varkappa_{\mu\lambda} = \underset{a}{p}_\mu A^\varkappa_\lambda + \underset{a}{p}_\lambda A^\varkappa_\mu. \tag{4.2}$$

In order that an $'A_n$, projectively related to the A_n, admit the same G_r as a group of affine motions, it is necessary and sufficient that there exist a covariant vector field p_λ such that

$$\underset{a}{\pounds}(\Gamma^\varkappa_{\mu\lambda} + p_\mu A^\varkappa_\lambda + p_\lambda A^\varkappa_\mu) = 0$$

or

$$\underset{a}{\pounds}p_\lambda = -\underset{a}{p}_\lambda. \tag{4.3}$$

On the other hand, substituting (4.2) in the identity $(\underset{c}{\pounds}\underset{b}{\pounds}\Gamma^\varkappa_{\mu\lambda}) = c^a_{cb}\underset{a}{\pounds}\Gamma^\varkappa_{\mu\lambda}$ we get

$$\underset{c}{\pounds}\underset{b}{p}_\lambda - \underset{b}{\pounds}\underset{c}{p}_\lambda = c^a_{cb}\underset{a}{p}_\lambda,$$

which shows that r covariant vectors $\underset{a}{p}_\lambda$ form a complete system with respect to G_r. Thus (4.3) is completely integrable and we have

THEOREM 4.2.[1] *When an A_n admits a G_r of projective motions such that the rank of $\underset{a}{v}^\varkappa$ is $r \leq n$, there exists an $'A_n$ which is (not trivially) projectively related to A_n and which admits the same G_r as a group of affine motions.*

From this we obtain

THEOREM 4.3.[2] *In order that a G_r in an X_n such that the rank of $\underset{a}{v}^\varkappa$ is $r \leq n$, can be regarded as a group of affine motions in a D_n, it is necessary and sufficient that the G_r be a subgroup of the ordinary projective group.*

The necessity is evident. Conversely, if the group G_r is a subgroup of the ordinary projective group, it is a group of projective motions in a D_n. Consequently according to Theorem 4.2, there exists an A_n which is projectively related to D_n, and is itself a D_n and admits G_r as a group of affine motions. Thus Theorem 4.3 is proved.

---

[1] KNEBELMAN [3].

[2] YANO and TASHIRO [1].

§ 5. Integrability conditions of $\underset{v}{\pounds}\Gamma^{\kappa}_{\mu\lambda} = 0$.

We consider the integrability conditions of $\underset{v}{\pounds}\Gamma^{\kappa}_{\mu\lambda} = 0$, which can be written as

$$(5.1) \qquad \begin{cases} \nabla_\lambda v^\kappa = v_\lambda^{\cdot\kappa} - 2S_{\mu\lambda}^{\cdot\cdot\kappa} v^\mu, [1] \\ \nabla_\mu v_\lambda^{\cdot\kappa} = - R_{\nu\mu\lambda}^{\cdot\cdot\cdot\kappa} v^\nu. \end{cases}$$

From (4.13) and (4.14) of Ch. I, we have

$$(5.2) \qquad \underset{v}{\pounds} S_{\mu\lambda}^{\cdot\cdot\kappa} = 0, \quad \underset{v}{\pounds} R_{\nu\mu\lambda}^{\cdot\cdot\cdot\kappa} = 0$$

respectively. Then applying the formula (4.9) of Chapter I to $S_{\mu\lambda}^{\cdot\cdot\kappa}$ and $R_{\nu\mu\lambda}^{\cdot\cdot\cdot\kappa}$, we obtain

$$(5.3) \qquad \underset{v}{\pounds} \nabla_\nu S_{\mu\lambda}^{\cdot\cdot\kappa} = 0, \quad \underset{v}{\pounds} \nabla_\omega R_{\nu\mu\lambda}^{\cdot\cdot\cdot\kappa} = 0$$

respectively. Repeating the same process, we have

$$(5.4) \qquad \begin{cases} \underset{v}{\pounds} \nabla_{\nu_2\nu_1} S_{\mu\lambda}^{\cdot\cdot\kappa} = 0, & \underset{v}{\pounds} \nabla_{\omega_2\omega_1} R_{\nu\mu\lambda}^{\cdot\cdot\cdot\alpha} = 0, \\ \underset{v}{\pounds} \nabla_{\nu_3\nu_2\nu_1} S_{\mu\lambda}^{\cdot\cdot\kappa} = 0, & \underset{v}{\pounds} \nabla_{\omega_3\omega_2\omega_1} R_{\nu\mu\lambda}^{\cdot\cdot\cdot\kappa} = 0, \\ \dots\dots\dots\dots & \end{cases}$$

Thus we have

THEOREM 5.1. [2] *In order that an L_n admit a group of affine motions, it is necessary and sufficient that there exist a positive integer N such that the first N sets of equations* (5.2), (5.3) *and* (5.4) *are compatible in v^κ and $v_\lambda^{\cdot\kappa}$ and that all their solutions satisfy the $(N+1)$st set of equations. If there exist $n^2 + n - r$ linearly independent equations in the first N sets, then the space admits an r-parameter complete group of affine motions.*

For an L_n with absolute parallelism, we have $R_{\nu\mu\lambda}^{\cdot\cdot\cdot\kappa} = 0$, if we replace (5.2), (5.3) and (5.4) by

$$(5.5) \qquad \underset{v}{\pounds} S_{\mu\lambda}^{\cdot\cdot\kappa} = 0, \quad \underset{v}{\pounds} \nabla_\nu S_{\mu\lambda}^{\cdot\cdot\kappa} = 0, \quad \underset{v}{\pounds} \nabla_{\nu_2\nu_1} S_{\mu\lambda}^{\cdot\cdot\kappa} = 0, \dots,$$

Theorem 5.1 holds.

We now consider an A_n for which $\nabla_\omega R_{\nu\mu\lambda}^{\cdot\cdot\cdot\kappa} = 0$, (Cartan's symmetric

---

[1] Cf. SCHOUTEN [8], p. 346.

[2] KNEBELMAN [2]; G. T., p. 19.

space). Then the Ricci identity

$$(5.6)\qquad 0 = \nabla_\pi \nabla_\omega R_{\nu\mu\lambda}^{\cdot\cdot\cdot\varkappa} - \nabla_\omega \nabla_\pi R_{\nu\mu\lambda}^{\cdot\cdot\cdot\varkappa}$$
$$= R_{\pi\omega\rho}^{\cdot\cdot\cdot\varkappa} R_{\nu\mu\lambda}^{\cdot\cdot\cdot\rho} - R_{\pi\omega\nu}^{\cdot\cdot\cdot\rho} R_{\rho\mu\lambda}^{\cdot\cdot\cdot\varkappa} - R_{\pi\omega\mu}^{\cdot\cdot\cdot\rho} R_{\nu\rho\lambda}^{\cdot\cdot\cdot\varkappa}$$
$$- R_{\pi\omega\lambda}^{\cdot\cdot\cdot\rho} R_{\nu\mu\rho}^{\cdot\cdot\cdot\varkappa}$$

shows that, if we take $v_\lambda^{\cdot\varkappa} = f^{\nu\mu} R_{\nu\mu\lambda}^{\cdot\cdot\cdot\varkappa}$, then $\underset{v}{\pounds} R_{\nu\mu\lambda}^{\cdot\cdot\cdot\varkappa} = 0$ is satisfied identically, $v^\varkappa$ being arbitrary. Thus we have

THEOREM 5.2.[1] *An A_n for which $\nabla_\omega R_{\nu\mu\lambda}^{\cdot\cdot\cdot\varkappa} = 0$ admits a transitive group of affine motions.*

On the other hand, A. Nijenhuis[2] proved that the generators of the holonomy group of an L_n span the $\overset{\varkappa}{\lambda}$-domain of the curvature tensor $R_{\nu\mu\lambda}^{\cdot\cdot\cdot\varkappa}$ and its covariant derivative. Thus the generators of the holonomy group of an A_n with $\nabla_\omega R_{\nu\mu\lambda}^{\cdot\cdot\cdot\varkappa} = 0$ span the $\overset{\varkappa}{\lambda}$-domain of the curvature tensor $R_{\nu\mu\lambda}^{\cdot\cdot\cdot\varkappa}$. Since the generators of the isotropy group span the $\overset{\varkappa}{\lambda}$-domain of the set $\underset{a}{v_\lambda^{\cdot\varkappa}} = \nabla_\lambda \underset{a}{v^\varkappa}$, we have

THEOREM 5.2.[3] *In an A_n for which $\nabla_\omega R_{\nu\mu\lambda}^{\cdot\cdot\cdot\varkappa} = 0$ the isotropy group contains the holonomy group.*

We now consider the conditions of complete integrability of $\underset{v}{\pounds}\Gamma_{\mu\lambda}^{\varkappa} = 0$, that is, those of (5.1). In order that we have complete integrability, the equations

$$\underset{v}{\pounds} S_{\mu\lambda}^{\cdot\cdot\varkappa} = v^\rho \nabla_\rho S_{\mu\lambda}^{\cdot\cdot\varkappa} - S_{\mu\lambda}^{\cdot\cdot\rho} v_\rho^{\cdot\varkappa} + S_{\rho\lambda}^{\cdot\cdot\varkappa} v_\mu^{\cdot\rho} + S_{\mu\rho}^{\cdot\cdot\varkappa} v_\lambda^{\cdot\rho} = 0$$

and

$$\underset{v}{\pounds} R_{\nu\mu\lambda}^{\cdot\cdot\cdot\varkappa} = v^\rho \nabla_\rho R_{\nu\mu\lambda}^{\cdot\cdot\cdot\varkappa} - R_{\nu\mu\lambda}^{\cdot\cdot\cdot\rho} v_\rho^{\cdot\varkappa} + R_{\rho\mu\lambda}^{\cdot\cdot\cdot\varkappa} v_\nu^{\cdot\rho} + R_{\nu\rho\lambda}^{\cdot\cdot\cdot\varkappa} v_\mu^{\cdot\rho} + R_{\nu\mu\rho}^{\cdot\cdot\cdot\varkappa} v_\lambda^{\cdot\rho} = 0$$

should be satisfied identically for any $v^\varkappa$ and $v_\lambda^{\cdot\varkappa}$, from which we can easily deduce $S_{\mu\lambda}^{\cdot\cdot\varkappa} = 0$ and $R_{\nu\mu\lambda}^{\cdot\cdot\cdot\varkappa} = 0$.

In this case, the group has its maximum order $n^2 + n$. Thus we have

THEOREM 5.3.[4] *In order that an L_n admit a G_r of affine motions of the maximum order $n^2 + n$, it is necessary and sufficient that the L_n be an E_n,[5] the group being a general affine group.*

---

[1] CARTAN [6, 8, 11].
[2] NIJENHUIS [2], Ch. II, § 13.
[3] SCHOUTEN [8], p. 363.
[4] EISENHART [4], p. 234; G. T., p. 20.
[5] E_n stands for an ordinary affine space.

In an L_n with absolute parallelism, a particular affine motion $\xi^\varkappa \to \xi^\varkappa + v^\varkappa dt$ satisfies the equation

$$\underset{v}{\pounds}\underset{a}{e}^\varkappa = v^\mu \nabla_\mu \underset{a}{e}^\varkappa - \underset{a}{e}^\mu v_\mu^{\cdot\varkappa} = 0,$$

from which, because of $\nabla_\mu \underset{a}{e}^\varkappa = 0$,

$$v_\mu^{\cdot\varkappa} = 0. \tag{5.7}$$

Thus the conditions of complete integrability of the equations $\underset{v}{\pounds}\underset{a}{e}^\varkappa = 0$ or of $v_\lambda^{\cdot\varkappa} = 0$ are that

$$\underset{v}{\pounds} S_{\mu\lambda}^{\cdot\cdot\varkappa} = v^\nu \nabla_\nu S_{\mu\lambda}^{\cdot\cdot\varkappa} = 0$$

are identically satisfied for any v^ν, from which it follows that $\nabla_\nu S_{\mu\lambda}^{\cdot\cdot\varkappa} = 0$. Thus we have

THEOREM 5.4.[1] *In order that an L_n with absolute parallelism admit a group of particular affine motions of the maximum order n, it is necessary and sufficient that the covariant derivative of the torsion tensor vanish.*

In this case, the differential equations $v_\lambda^{\cdot\varkappa} = 0$ admit solutions $v^\varkappa$ whose initial values can be arbitrarily assigned. Thus when an L_n with absolute parallelism admits a group of particular affine motions of the maximum order n, the group is simply transitive.

§ 6. An L_n with absolute parallelism which admits a simply transitive group of particular affine motions.[2]

We consider an L_n with absolute parallelism and denote n linearly independent absolutely parallel vectors by $\underset{b}{e}^\varkappa$ and the components of the linear connexion of the space by $\overset{+}{\Gamma}{}_{\mu\lambda}^\varkappa$. Then we have $\overset{+}{\nabla}_\mu \underset{b}{e}^\varkappa = 0$, from which

$$\overset{+}{\Gamma}{}_{\mu\lambda}^\varkappa = -\overset{b}{e}_\lambda \partial_\mu \underset{b}{e}^\varkappa = +\underset{b}{e}^\varkappa \partial_\mu \overset{b}{e}_\lambda, \tag{6.1}$$

$$\overset{+}{S}{}_{\mu\lambda}^{\cdot\cdot\varkappa} \overset{\text{def}}{=} \overset{+}{\Gamma}{}_{[\mu\lambda]}^\varkappa = +\underset{b}{e}^\varkappa \partial_{[\mu} \overset{b}{e}_{\lambda]}. \tag{6.2}$$

$$\overset{+}{R}{}_{\nu\mu\lambda}^{\cdot\cdot\cdot\varkappa} = 0. \tag{6.3}$$

---

[1] G. T., p. 20.

[2] SCHOUTEN [8], p. 185.

We assume that the space admits a simply transitive group of particular affine motions, then according to Theorem 5.4, we have

$$\overset{+}{\nabla}_\nu \overset{+}{S}_{\mu\lambda}^{\cdot\cdot\kappa} = 0. \tag{6.4}$$

Thus if we put

$$c_{cb}^a \overset{\text{def}}{=} -2\overset{+}{S}_{\mu\lambda}^{\cdot\cdot\kappa} \underset{c}{e^\mu} \underset{b}{e^\lambda} \overset{a}{e_\kappa}; \quad \overset{+}{S}_{\mu\lambda}^{\cdot\cdot\kappa} = -\tfrac{1}{2} c_{cb}^a \overset{c}{e_\mu} \overset{b}{e_\lambda} \underset{a}{e^\kappa}, \tag{6.5}$$

the c_{bc}^a are scalars and $\overset{+}{\nabla}_\nu c_{bc}^a = 0$, which shows that the c_{bc}^a are constants. Thus from (6.2) and (6.5), we get

$$\partial_{[\mu} \overset{a}{e_{\lambda]}} = -\tfrac{1}{2} c_{cb}^a \overset{c}{e_\mu} \overset{b}{e_\lambda}, \tag{6.6}$$

and

$$\underset{c}{e^\mu} \partial_\mu \underset{b}{e^\kappa} - \underset{b}{e^\mu} \partial_\mu \underset{c}{e^\kappa} = c_{cb}^a \underset{a}{e^\kappa}, \tag{6.7}$$

which shows that n vectors $\underset{b}{e^\kappa}$ generate a simply transitive group.

Conversely, if n vectors $\underset{b}{e^\kappa}$ generate a simply transitive group, then we have (6.6) and (6.7), and we can easily see that (6.4) holds. Thus according to Theorem 5.4, the space admits a simply transitive group of particular affine motions. Thus we have

THEOREM 6.1. *In order that an L_n with absolute parallelism admit a simply transitive group of particular affine motions, it is necessary and sufficient that n absolutely parallel vectors generate a simply transitive group.*

We assume again that the space admits a simply transitive group of particular affine motions and denote by $\underset{B}{e^\kappa}$ $(A, B, C = \dot{1}, \dot{2}, \ldots, \dot{n})$ n vectors which generate the group. Then we have

$$\underset{B}{\pounds} \underset{b}{e^\kappa} \overset{\text{def}}{=} \underset{B}{e^\mu} \partial_\mu \underset{b}{e^\kappa} - \underset{b}{e^\mu} \partial_\mu \underset{B}{e^\kappa} = 0, \tag{6.8}$$

which can also be written as

$$\underset{B}{e^\kappa} \partial_\mu \overset{B}{e_\lambda} = \underset{b}{e^\kappa} \partial_\lambda \overset{b}{e_\mu}. \tag{6.9}$$

This shows that with respect to the linear connexion

$$\overset{-}{\Gamma}_{\mu\lambda}^\kappa \overset{\text{def}}{=} \overset{+}{\Gamma}_{\lambda\mu}^\kappa, \tag{6.10}$$

the vectors $\underset{B}{e}{}^{\varkappa}$ are absolutely parallel, from which

$$\bar{R}_{\nu\mu\lambda}{}^{\cdot\cdot\cdot\varkappa} = 0. \tag{6.11}$$

Conversely, we assume that the linear connexion $\bar{\Gamma}^{\varkappa}_{\mu\lambda}$ defined by (6.10) from a linear connexion $\overset{+}{\Gamma}{}^{\varkappa}_{\mu\lambda}$ of an L_n with absolute parallelism, is of zero curvature. Then denoting by $\underset{b}{e}{}^{\varkappa}$ and $\underset{B}{e}{}^{\varkappa}$ n linearly independent vector fields absolutely parallel with respect to $\overset{+}{\Gamma}{}^{\varkappa}_{\mu\lambda}$ and $\bar{\Gamma}^{\varkappa}_{\mu\lambda}$ respectively, we have

$$\overset{+}{\Gamma}{}^{\varkappa}_{\mu\lambda} = \underset{b}{e}{}^{\varkappa}\partial_\mu \overset{b}{e}_\lambda, \quad \bar{\Gamma}^{\varkappa}_{\mu\lambda} = \underset{B}{e}{}^{\varkappa}\partial_\mu \overset{B}{e}_\lambda \tag{6.12}$$

and consequently

$$\underset{b}{e}{}^{\varkappa}\partial_\mu \overset{b}{e}_\lambda = \underset{B}{e}{}^{\varkappa}\partial_\lambda \overset{B}{e}_\mu,$$

from which

$$\underset{B}{\pounds}\,\underset{b}{e}{}^{\varkappa} = \underset{B}{e}{}^{\mu}\partial_\mu \underset{b}{e}{}^{\varkappa} - \underset{b}{e}{}^{\mu}\partial_\mu \underset{B}{e}{}^{\varkappa} = 0. \tag{6.13}$$

This equation shows that the space with affine connexion $\overset{+}{\Gamma}{}^{\varkappa}_{\mu\lambda}$ admits a simply transitive group of particular affine motions. Thus we have

THEOREM 6.2. *In order that an L_n with absolute parallelism $\overset{+}{\Gamma}{}^{\varkappa}_{\mu\lambda}$ admit a simply transitive group of particular affine motions, it is necessary and sufficient that the linear connexion $\bar{\Gamma}^{\varkappa}_{\mu\lambda} = \overset{+}{\Gamma}{}^{\varkappa}_{\lambda\mu}$ be of zero curvature.*

Again we suppose that an L_n with absolute parallelism admits a simply transitive group of particular affine motions, then we have the second equation of (6.5). Denoting by $\underset{B}{e}{}^{\varkappa}$ the vectors generating the simply transitive group of particular affine motions, we have

$$\underset{C}{e}{}^{\mu}\partial_\mu \underset{B}{e}{}^{\varkappa} - \underset{B}{e}{}^{\mu}\partial_\mu \underset{C}{e}{}^{\varkappa} = c^{A}_{CB}\underset{A}{e}{}^{\varkappa}, \tag{6.14}$$

from which

$$\bar{S}_{\mu\lambda}{}^{\cdot\cdot\varkappa} \overset{\text{def}}{=} \bar{\Gamma}^{\varkappa}_{[\mu\lambda]} = -\tfrac{1}{2}c^{A}_{CB}\overset{C}{e}_\mu \overset{B}{e}_\lambda \underset{A}{e}{}^{\varkappa}. \tag{6.15}$$

Thus from the second equation of (6.5), (6.15) and

$$\overset{+}{S}_{\mu\lambda}{}^{\cdot\cdot\varkappa} = -\bar{S}_{\mu\lambda}{}^{\cdot\cdot\varkappa}, \tag{6.16}$$

we obtain

$$c^{a}_{cb}\, \overset{c}{e}_{\mu}\, \overset{b}{e}_{\lambda}\, \underset{a}{e}^{\kappa} = - c^{A}_{CB}\, \overset{C}{e}_{\mu}\, \overset{B}{e}_{\lambda}\, \underset{A}{e}^{\kappa}. \tag{6.17}$$

Now the linear connexion $\overset{+}{\Gamma}{}^{\kappa}_{\mu\lambda}(\overset{-}{\Gamma}{}^{\kappa}_{\mu\lambda})$ defines an absolute parallelism in the L_n and consequently, if we give n vectors $\underset{b}{e}^{\kappa}(\underset{0}{\xi})$ $(\underset{B}{e}^{\kappa}(\underset{0}{\xi}))$ at a fixed point $\underset{0}{\xi}$ of the space, then the vectors $\underset{b}{e}^{\kappa}(\xi)$ $(\underset{B}{e}^{\kappa}(\xi))$ at every point of the space are automatically determined. Now for convenience we choose the vectors $\underset{b}{e}^{\kappa}$ and $\underset{B}{e}^{\kappa}$ in such a way that $\underset{b}{e}^{\kappa}(\underset{0}{\xi}) = \underset{B}{e}^{\kappa}(\underset{0}{\xi})$. Then from (6.17) we have at the point $(\underset{0}{\xi})$

$$c^{a}_{cb} \overset{*}{=} - \delta^{c}_{C}\, \delta^{b}_{B}\, \delta^{A}_{a}\, c^{A}_{CB}, \tag{6.18}$$

which also holds at all points of the space.

The space discussed in this paragraph is exactly a group space.[1] The group generated by $\underset{b}{e}^{\kappa}(\underset{B}{e}^{\kappa})$ is called the first (the second) parameter group.

From

$$\underset{B}{\pounds}\, \underset{b}{e}^{\kappa} = 0, \quad \underset{b}{\pounds}\, \underset{B}{e}^{\kappa} = 0, \tag{6.19}$$

we have

THEOREM 6.3. *The vectors defining the first (the second) parameter group are transformed into themselves by the second (the first) parameter group.*

§ 7. Semi-simple group space.

We consider a group space and adopt the notations used in the preceding paragraph. If we put

$$\Gamma^{\kappa}_{\mu\lambda} = \overset{+}{\Gamma}{}^{\kappa}_{(\mu\lambda)}, \tag{7.1}$$

we get

$$\overset{+}{\Gamma}{}^{\kappa}_{\mu\lambda} = \Gamma^{\kappa}_{\mu\lambda} + \overset{+}{S}{}_{\mu\lambda}^{\cdot\cdot\kappa}, \quad \overset{-}{\Gamma}{}^{\kappa}_{\mu\lambda} = \Gamma^{\kappa}_{\mu\lambda} - \overset{+}{S}{}_{\mu\lambda}^{\cdot\cdot\kappa}. \tag{7.2}$$

The linear connexions given by $\overset{+}{\Gamma}{}^{\kappa}_{\mu\lambda}$, $\Gamma^{\kappa}_{\mu\lambda}$ and $\overset{-}{\Gamma}{}^{\kappa}_{\mu\lambda}$ are called respectively (+)-*connexion*, (0)-*connexion* and (−)-*connexion* of the space.[2]

---

[1] EISENHART [4], p. 198; SCHOUTEN [8], IV.

[2] CARTAN [3]; CARTAN and SCHOUTEN [1]; EISENHART [4].

On substituting the first equation of (7.2) in $\overset{+}{R}_{\nu\mu\lambda}^{\cdot\cdot\cdot\varkappa} = 0$, we obtain

$$(7.3) \qquad R_{\nu\mu\lambda}^{\cdot\cdot\cdot\varkappa} = 2\overset{+}{S}_{\nu\mu}^{\cdot\cdot\rho}\overset{+}{S}_{\lambda\rho}^{\cdot\cdot\varkappa} + \overset{+}{S}_{\mu\lambda}^{\cdot\cdot\rho}\overset{+}{S}_{\nu\rho}^{\cdot\cdot\varkappa} + \overset{+}{S}_{\lambda\nu}^{\cdot\cdot\rho}\overset{+}{S}_{\mu\rho}^{\cdot\cdot\varkappa},$$

where $R_{\nu\mu\lambda}^{\cdot\cdot\cdot\varkappa}$ is the curvature tensor of $\Gamma_{\mu\lambda}^{\varkappa}$. From $R_{[\nu\mu\lambda]}^{\cdot\cdot\cdot\varkappa} = 0$ and (7.3) we obtain

$$(7.4) \qquad \overset{+}{S}_{[\nu\mu}^{\cdot\cdot\rho}\overset{+}{S}_{\lambda]\rho}^{\cdot\cdot\varkappa} = 0.$$

This equation can also be obtained from the Jacobi identity satisfied by the structural constants:

$$c_{[dc}^{e}c_{b]e}^{a} = 0.$$

From (7.3) and (7.4), we get

$$(7.5) \qquad R_{\nu\mu\lambda}^{\cdot\cdot\cdot\varkappa} = \overset{+}{S}_{\nu\mu}^{\cdot\cdot\rho}\overset{+}{S}_{\lambda\rho}^{\cdot\cdot\varkappa}.$$

For the covariant derivative of the torsion tensor $\overset{+}{S}_{\mu\lambda}^{\cdot\cdot\varkappa}$ with respect to $\Gamma_{\mu\lambda}^{\varkappa}$, we have

$$\nabla_\nu \overset{+}{S}_{\mu\lambda}^{\cdot\cdot\varkappa} = \overset{+}{\nabla}_\nu \overset{+}{S}_{\mu\lambda}^{\cdot\cdot\varkappa} - \overset{+}{S}_{\nu\mu}^{\cdot\cdot\rho}\overset{+}{S}_{\lambda\rho}^{\cdot\cdot\varkappa} - \overset{+}{S}_{\mu\lambda}^{\cdot\cdot\rho}\overset{+}{S}_{\nu\rho}^{\cdot\cdot\varkappa} - \overset{+}{S}_{\lambda\nu}^{\cdot\cdot\rho}\overset{+}{S}_{\mu\rho}^{\cdot\cdot\varkappa},$$

from which, because of (7.4),

$$(7.6) \qquad \nabla_\nu \overset{+}{S}_{\mu\lambda}^{\cdot\cdot\varkappa} = 0.$$

Thus from (7.5), we get

$$(7.7) \qquad \nabla_\omega R_{\nu\mu\lambda}^{\cdot\cdot\cdot\varkappa} = 0.$$

because of (7.6). Thus we have

THEOREM 7.1.[1] *Every group space is a symmetric A_n with respect to its* (0)-*connexion.*

We now suppose that the space is a semi-simple group space. According to E. Cartan,[2] in order that the group be semi-simple, it is necessary and sufficient that the rank of the matrix

$$(7.8) \qquad g_{cb} = c_{ac}^{d}c_{bd}^{a}$$

be n. Thus putting

$$(7.9) \qquad g_{\mu\lambda} \overset{\text{def}}{=} g_{cb}\overset{c}{e}_{\mu}\overset{b}{e}_{\lambda}$$

[1] SCHOUTEN [8], p. 191.

[2] CARTAN [3].

we can give a Riemannian metric

$$ds^2 = g_{\mu\lambda}\, d\xi^\mu\, d\xi^\lambda \tag{7.10}$$

to the space. The equation (7.8) can also be written as

$$g_{\mu\lambda} = 4\overset{+}{S}{}_{\varkappa\mu}^{\cdot\cdot\rho}\,\overset{+}{S}{}_{\lambda\rho}^{\cdot\cdot\varkappa}. \tag{7.11}$$

Now from (7.5) we obtain by contraction

$$R_{\mu\lambda} = \overset{+}{S}{}_{\varkappa\mu}^{\cdot\cdot\rho}\,\overset{+}{S}{}_{\lambda\rho}^{\cdot\cdot\varkappa}, \tag{7.12}$$

from which, because of (7.11),

$$R_{\mu\lambda} = \tfrac{1}{4} g_{\mu\lambda}. \tag{7.13}$$

Moreover from (7.7) and (7.13), we have

$$\nabla_\nu R_{\mu\lambda} = \tfrac{1}{4}\nabla_\nu g_{\mu\lambda} = 0, \tag{7.14}$$

which shows that the $\Gamma_{\mu\lambda}^{\varkappa}$ are the Christoffel symbols $\{{}^{\varkappa}_{\mu\lambda}\}$ formed with $g_{\mu\lambda}$. Thus we have

THEOREM 7.2.[1] *For a semi-simple group space, the* (0)*-connexion is Riemannian and the space is Einsteinian.*

From

$$\overset{-}{S}{}_{\mu\lambda}^{\cdot\cdot\varkappa} = -\tfrac{1}{2} c_{CB}^{A}\, \overset{C}{e}_\mu \overset{B}{e}_\lambda \underset{A}{e}^\varkappa, \quad \underset{b}{\pounds}\, \underset{B}{e}^\varkappa = \underset{b}{\pounds}\, \overset{A}{e}_\lambda = 0,$$

we obtain

$$\underset{b}{\pounds}\, \overset{-}{S}{}_{\mu\lambda}^{\cdot\cdot\varkappa} = 0 \text{ and consequently } \underset{b}{\pounds}\, \overset{+}{S}{}_{\mu\lambda}^{\cdot\cdot\varkappa} = 0. \tag{7.15}$$

From (7.11) and (7.15), we obtain

$$\underset{b}{\pounds}\, g_{\mu\lambda} = 0. \tag{7.16}$$

On the other hand

$$g_{\mu\lambda}\, \underset{c}{e}^\mu \underset{b}{e}^\lambda = g_{cb} \tag{7.17}$$

are constants. The equations (7.16) and (7.17) show that the infinitesimal transformations of the first parameter group are translations. Since a similar result holds for the second parameter group, we have

[1] CARTAN and SCHOUTEN [1]; EISENHART [4], p. 206.

THEOREM 7.3.[1] *The infinitesimal transformations of the first and of the second parameter group of a semi-simple group space are translations.*

§ 8. A group as group of affine motions.

We apply now Theorems 3.1, 3.2 and 3.3 of Ch. III to the case of the group of affine motions.

We consider a G_r in an X_n and we suppose first that the rank of $\underset{b}{v}^{\varkappa}$ in the neighbourhood under consideration is $r \leq n$. In this case we choose a coordinate system with respect to which we have (3.6) of Ch. III. Then the equations $\underset{a}{\pounds}\Gamma^{\varkappa}_{\mu\lambda} = 0$ become

$$(8.1)\quad \underset{a}{\pounds}\Gamma^{\varkappa}_{\mu\lambda} = \partial_\mu \partial_\lambda \underset{a}{v}^{\varkappa} + \underset{a}{v}^{\rho} \partial_\rho \Gamma^{\varkappa}_{\mu\lambda} - \Gamma^{\rho}_{\mu\lambda} \partial_\rho \underset{a}{v}^{\varkappa} + \Gamma^{\varkappa}_{\rho\lambda} \partial_\mu \underset{a}{v}^{\rho} + \Gamma^{\varkappa}_{\mu\rho} \partial_\lambda \underset{a}{v}^{\rho} = 0$$

and consequently, defining the functions $\Theta_{\alpha}{}^{\varkappa}_{\mu\lambda}(\Gamma, \xi)$ by

$$(8.2)\quad \underset{a}{v}^{\alpha} \Theta_{\alpha}{}^{\varkappa}_{\mu\lambda}(\Gamma, \xi) \overset{\text{def}}{=} -\partial_\mu \partial_\lambda \underset{a}{v}^{\varkappa} + \Gamma^{\rho}_{\mu\lambda} \partial_\rho \underset{a}{v}^{\varkappa} - \Gamma^{\varkappa}_{\rho\lambda} \partial_\mu \underset{a}{v}^{\rho} - \Gamma^{\varkappa}_{\mu\rho} \partial_\lambda \underset{a}{v}^{\rho},$$

we obtain

$$(8.3)\quad \underset{a}{\pounds}\Gamma^{\varkappa}_{\mu\lambda} = \underset{a}{v}^{\alpha}[\partial_\alpha \Gamma^{\varkappa}_{\mu\lambda} - \Theta_{\alpha}{}^{\varkappa}_{\mu\lambda}(\Gamma, \xi)],$$

from which

$$(8.4)\quad \partial_\alpha \Gamma^{\varkappa}_{\mu\lambda} = \Theta_{\alpha}{}^{\varkappa}_{\mu\lambda}(\Gamma, \xi).$$

As was shown in § 3 of Ch. III, we can prove that the system of partial differential equations (8.4) is completely integrable and that the solutions $\Gamma^{\varkappa}_{\mu\lambda}(\xi)$ are determined by the initial values of $\Gamma^{\varkappa}_{\mu\lambda}$ at a point $(\overset{0}{\xi})$, which in the case $r < n$, can be arbitrary functions of the variables ξ^{ξ} and, in the case $r = n$, arbitrary constants. Thus we have

THEOREM 8.1.[2] *A G_r in an X_n such that the rank of $\underset{b}{v}^{\varkappa}$ in the neighbourhood under consideration is $r \leq n$ can be regarded as a group of affine motions in an L_n whose components of the affine connexion contain n^3 arbitrary functions or constants.*

We next consider a G_r in an X_n such that the rank of $\underset{b}{v}^{\varkappa}$ in the neighbourhood under consideration is $q < r, n$. In this case we choose a coordinate system with respect to which we have (3.15) of Ch. III.

[1] CARTAN and SCHOUTEN [1]; EISENHART [4], p. 213.

[2] G. T., p. 26.

Then the equations $\underset{a}{\pounds}\Gamma^{\varkappa}_{\mu\lambda} = 0$ become

$$\text{(8.5)}\quad \begin{cases} \underset{j}{\pounds}\Gamma^{\varkappa}_{\mu\lambda} = \partial_\mu\partial_\lambda \underset{j}{v}^\varkappa + \underset{j}{v}^\alpha\partial_\alpha\Gamma^{\varkappa}_{\mu\lambda} - \Gamma^{\rho}_{\mu\lambda}\partial_\rho \underset{j}{v}^\varkappa + \Gamma^{\varkappa}_{\alpha\lambda}\partial_\mu \underset{j}{v}^\alpha + \Gamma^{\varkappa}_{\mu\alpha}\partial_\lambda \underset{j}{v}^\alpha = 0, \\ \underset{u}{\pounds}\Gamma^{\varkappa}_{\mu\lambda} = \varphi^j_u \underset{j}{\pounds}\Gamma^{\varkappa}_{\mu\lambda} + (\partial_\mu\partial_\lambda\varphi^i_u)\underset{i}{v}^\varkappa + (\partial_\mu\varphi^i_u)(\partial_\lambda \underset{i}{v}^\varkappa) + (\partial_\lambda\varphi^i_u)(\partial_\mu \underset{i}{v}^\varkappa) \\ \qquad - \Gamma^{\rho}_{\mu\lambda}(\partial_\rho\varphi^i_u)\underset{i}{v}^\varkappa + \Gamma^{\varkappa}_{\alpha\lambda}(\partial_\mu\varphi^i_u)\underset{i}{v}^\alpha + \Gamma^{\varkappa}_{\mu\alpha}(\partial_\lambda\varphi^i_u)\underset{i}{v}^\alpha = 0. \end{cases}$$

Thus, if we define the functions $\Theta_{\alpha}{}^{\varkappa}_{\mu\lambda}(\Gamma, \xi)$ and $\Xi_{u}{}^{\varkappa}_{\mu\lambda}(\Gamma, \xi)$ by

$$\text{(8.6)}\quad \underset{j}{v}^\alpha\Theta_{\alpha}{}^{\varkappa}_{\mu\lambda}(\Gamma, \xi) \overset{\text{def}}{=} -\partial_\mu\partial_\lambda \underset{j}{v}^\varkappa + \Gamma^{\rho}_{\mu\lambda}\partial_\rho \underset{j}{v}^\varkappa - \Gamma^{\varkappa}_{\alpha\lambda}\partial_\mu \underset{j}{v}^\alpha - \Gamma^{\varkappa}_{\mu\alpha}\partial_\lambda \underset{j}{v}^\alpha$$

and

$$\text{(8.7)}\quad \begin{aligned} \Xi_{u}{}^{\varkappa}_{\mu\lambda}(\Gamma, \xi) = {} & (\partial_\mu\partial_\lambda\varphi^i_u)\underset{i}{v}^\varkappa + (\partial_\mu\varphi^i_u)(\partial_\lambda \underset{i}{v}^\varkappa) + (\partial_\lambda\varphi^i_u)(\partial_\mu \underset{i}{v}^\varkappa) \\ & - \Gamma^{\rho}_{\mu\lambda}(\partial_\rho\varphi^i_u)\underset{i}{v}^\varkappa + \Gamma^{\varkappa}_{\alpha\lambda}(\partial_\mu\varphi^i_u)\underset{i}{v}^\alpha + \Gamma^{\varkappa}_{\mu\alpha}(\partial_\lambda\varphi^i_u)\underset{i}{v}^\alpha, \end{aligned}$$

we obtain

$$\text{(8.8)}\quad \begin{cases} \underset{j}{\pounds}\Gamma^{\varkappa}_{\mu\lambda} = \underset{j}{v}^\alpha[\partial_\alpha\Gamma^{\varkappa}_{\mu\lambda} - \Theta_{\alpha}{}^{\varkappa}_{\mu\lambda}(\Gamma, \xi)] = 0, \\ \underset{u}{\pounds}\Gamma^{\varkappa}_{\mu\lambda} = \varphi^i_u \underset{i}{\pounds}\Gamma^{\varkappa}_{\mu\lambda} + \Xi_{u}{}^{\varkappa}_{\mu\lambda}(\Gamma, \xi) = 0, \end{cases}$$

from which

$$\text{(8.9)}\quad \partial_\alpha\Gamma^{\varkappa}_{\mu\lambda} = \Theta_{\alpha}{}^{\varkappa}_{\mu\lambda}(\Gamma, \xi), \quad \Xi_{u}{}^{\varkappa}_{\mu\lambda}(\Gamma, \xi) = 0.$$

Using the method of § 3 of Ch. III, we can prove that if $\Xi_{u}{}^{\varkappa}_{\mu\lambda}(\Gamma, \xi) = 0$ is compatible at some point of the space, then the mixed system (8.9) is completely integrable and the solutions $\Gamma^{\varkappa}_{\mu\lambda}(\xi)$ are determined by their initial values at this point which satisfy $\Xi_{u}{}^{\varkappa}_{\mu\lambda}(\Gamma, \xi) = 0$. Thus we have

THEOREM 8.2.[1] *Consider a G_r in an X_n such that the rank of $\underset{b}{v}^\varkappa$ in the neighbourhood under consideration is $q < r, n$. If, in a coordinate system with respect to which* (3.15) *of Ch.* III *is valid, the equations $\Xi_{u}{}^{\varkappa}_{\mu\lambda}(\Gamma, \xi) = 0$ are compatible in $\Gamma^{\varkappa}_{\mu\lambda}$ at some point of the space, then the group can be regarded as a group of affine motions in an L_n.*

We finally consider a multiply transitive G_r in an X_n. Then the rank

[1] G. T., p. 27.

of $\underset{b}{v}^{\varkappa}$ in a neighbourhood under consideration is $n < r$. Thus, if we put

$$\text{(8.10)} \qquad \det(\underset{b}{v}^{\varkappa}) \neq 0, \quad \underset{u}{v}^{\varkappa} = \varphi_u^a \underset{a}{v}^{\varkappa}; \qquad \begin{array}{l} a, b, c = 1, 2, \ldots, \mathrm{n}, \\ u, v, w = \mathrm{n}+1, \ldots, \mathrm{r}, \end{array}$$

we can state

THEOREM 8.3. *If, for a multiply transitive G_r in X_n, the equations* $\Xi_{u\,\mu\lambda}^{\;\;\varkappa}(\Gamma, \xi) = 0$ *are compatible at a fixed point of the space, the group can be regarded as a group of affine motions in an L_n.*

We now consider analogous problems for groups of particular affine motions in a space with absolute parallelism.

We first consider a G_r in X_n such that the rank of $\underset{b}{v}^{\varkappa}$ in a neighbourhood is $r \leq n$. In this case, if we choose a coordinate system such that (3.6) of Ch. III is valid, the equations $\underset{a}{\pounds}\underset{A}{e}^{\varkappa} = 0$ $(A, B, C, \ldots = 1, 2, \ldots, n)$ become

$$\text{(8.11)} \qquad \underset{a}{\pounds}\underset{A}{e}^{\varkappa} = \underset{a}{v}^{\alpha}\partial_{\alpha}\underset{A}{e}^{\varkappa} - \underset{A}{e}^{\mu}\partial_{\mu}\underset{a}{v}^{\varkappa} = 0.$$

Consequently, defining the functions $\Theta_{\alpha\,A}^{\;\;\varkappa}(e, \xi)$ by

$$\text{(8.12)} \qquad \underset{a}{v}^{\alpha}\Theta_{\alpha\,A}^{\;\;\varkappa}(e, \xi) = \underset{A}{e}^{\mu}\partial_{\mu}\underset{a}{v}^{\varkappa},$$

we have

$$\text{(8.13)} \qquad \underset{a}{\pounds}\underset{A}{e}^{\varkappa} = \underset{a}{v}^{\alpha}(\partial_{\alpha}\underset{A}{e}^{\varkappa} - \Theta_{\alpha\,A}^{\;\;\varkappa}(e, \xi)),$$

from which

$$\text{(8.14)} \qquad \partial_{\alpha}\underset{A}{e}^{\varkappa} = \Theta_{\alpha\,A}^{\;\;\varkappa}(e, \xi).$$

As is shown in § 3 of Ch. III, we can prove that (8.14) is completely integrable, and that the solutions $\underset{A}{e}^{\varkappa}(\xi)$ are determined by the initial values at a point $(\underset{0}{\xi})$, which, in the case $r < n$, can be arbitrary functions of the variables ξ^{ξ} and in the case $r = n$, arbitrary constants. Thus we have

THEOREM 8.4. *A G_r in an X_n such that the rank of $\underset{a}{v}^{\varkappa}$ in a neighbourhood is $r \leq n$ can be regarded as a group of particular affine motions in an L_n with absolute parallelism and the components of the absolutely parallel vectors can contain n^2 arbitrary functions when $r < n$, and n^2 arbitrary constants when $r = n$.*

We consider next an intransitive G_r in an X_n and suppose that the rank of $\underset{a}{v}^\varkappa$ in a neighbourhood is $q < r, n$. In this case we can choose a coordinate system such that (3.15) of Ch. III holds. Then the equations $\underset{a}{\pounds}\underset{A}{e}^\varkappa = 0$ become

$$(8.15)\qquad \begin{cases} \underset{i}{\pounds}\underset{A}{e}^\varkappa = \underset{i}{v}^\alpha \partial_\alpha \underset{A}{e}^\varkappa - \underset{A}{e}^\mu \partial_\mu \underset{i}{v}^\varkappa = 0, \\ \underset{u}{\pounds}\underset{A}{e}^\varkappa = \varphi^i_u \underset{i}{\pounds}\underset{A}{e}^\varkappa - \underset{A}{e}^\mu (\partial_\mu \varphi^i_u)\underset{i}{v}^\varkappa = 0. \end{cases}$$

Thus, if we define the functions $\Theta_{\alpha}{}^{\varkappa}_{A}(e, \xi)$ and $\Xi_u{}^\varkappa_A(e, \xi)$ by

$$(8.16)\qquad \underset{i}{v}^\alpha \Theta_\alpha{}^\varkappa_A(e, \xi) = \underset{A}{e}^\mu \partial_\mu \underset{i}{v}^\varkappa$$

and

$$(8.17)\qquad \Xi_u{}^\varkappa_A(e, \xi) = -\underset{A}{e}^\mu(\partial_\mu \varphi^i_u)\underset{i}{v}^\varkappa,$$

respectively, we have

$$(8.18)\qquad \underset{i}{\pounds}\underset{A}{e}^\varkappa = \underset{i}{v}^\alpha[\partial_\alpha \underset{A}{e}^\varkappa - \Theta_\alpha{}^\varkappa_A(e, \xi)] = 0,$$

and

$$(8.19)\qquad \underset{u}{\pounds}\underset{A}{e}^\varkappa = \varphi^i_u \underset{i}{\pounds}\underset{A}{e}^\varkappa + \Xi_u{}^\varkappa_A(e, \xi) = 0,$$

from which

$$(8.20)\qquad \partial_\alpha \underset{A}{e}^\varkappa = \Theta_\alpha{}^\varkappa_A(e, \xi), \quad \Xi_u{}^\varkappa_A(e, \xi) = 0.$$

Using the method of § 3 of Ch. III, we can prove that if the equations $\Xi_u{}^\varkappa_A(e, \xi) = 0$ are compatible at a point of the space, then (8.20) is completely integrable. Thus we have

THEOREM 8.5.[1] *Consider an intransitive G_r in an X_n such that the rank of $\underset{a}{v}^\varkappa$ in a neighbourhood is $q < r, n$. If, in a coordinate system with respect to which* (3.15) *of Ch.* III *holds, the equations $\Xi_u{}^\varkappa_A(e, \xi) = 0$ are compatible at a point, then the group can be regarded as a group of particular affine motions in an L_n with absolute parallelism.*

We finally consider a multiply transitive G_r in an X_n. Then the rank of $\underset{a}{v}^\varkappa$ is $n < r$. Thus (8.10) is valid, and we have

[1] G. T., p. 29.

THEOREM 8.6. *If, for a multiply transitive G_r in an X_n, the equations $\Xi_{uA}^{\ \varkappa}(e, \xi) = 0$ are compatible in $\underset{A}{e^\varkappa}$ at a point of the space, the group can be regarded as a group of particular affine motions in an L_n with absolute parallelism.*

§ 9. Groups of affine motions in an L_n or an A_n.

Let an L_n with a linear connexion $\Gamma^\varkappa_{\mu\lambda}$ admit an infinitesimal affine motion $\xi^\varkappa \to \xi^\varkappa + v^\varkappa dt$. Then $\underset{v}{\pounds}\Gamma^\varkappa_{\mu\lambda} = 0$. But if we put

$$\overset{0}{\Gamma}{}^\varkappa_{\mu\lambda} = \Gamma^\varkappa_{(\mu\lambda)}, \qquad S_{\mu\lambda}^{\cdot\cdot\varkappa} = \Gamma^\varkappa_{[\mu\lambda]}, \tag{9.1}$$

then the equations $\underset{v}{\pounds}\Gamma^\varkappa_{\mu\lambda} = 0$ are equivalent to

$$\underset{v}{\pounds}\overset{0}{\Gamma}{}^\varkappa_{\mu\lambda} = 0, \quad \underset{v}{\pounds}S_{\mu\lambda}^{\cdot\cdot\varkappa} = 0. \tag{9.2}$$

Thus the integrability conditions of $\underset{v}{\pounds}\Gamma^\varkappa_{\mu\lambda} = 0$ are

$$\left\{\begin{array}{ll} \underset{v}{\pounds}S_{\mu\lambda}^{\cdot\cdot\varkappa} = 0, & \underset{v}{\pounds}\overset{0}{R}{}_{\nu\mu\lambda}^{\cdot\cdot\cdot\varkappa} = 0, \\ \underset{v}{\pounds}\overset{0}{\nabla}_\nu S_{\mu\lambda}^{\cdot\cdot\varkappa} = 0, & \underset{v}{\pounds}\overset{0}{\nabla}_\omega \overset{0}{R}{}_{\nu\mu\lambda}^{\cdot\cdot\cdot\varkappa} = 0, \\ \underset{v}{\pounds}\overset{0}{\nabla}_{\nu_2\nu_1} S_{\mu\lambda}^{\cdot\cdot\varkappa} = 0, & \underset{v}{\pounds}\overset{0}{\nabla}_{\omega_2\omega_1} \overset{0}{R}{}_{\nu\mu\lambda}^{\cdot\cdot\cdot\varkappa} = 0, \\ \cdot\ \cdot\ \cdot\ \cdot\ \cdot & \cdot\ \cdot\ \cdot\ \cdot\ \cdot \end{array}\right. \tag{9.3}$$

where $\overset{0}{\nabla}_\nu$ denotes the covariant differentiation with respect to $\overset{0}{\Gamma}{}^\varkappa_{\mu\lambda}$ and $\overset{0}{R}{}_{\nu\mu\lambda}^{\cdot\cdot\cdot\varkappa}$ the curvature tensor belonging to $\overset{0}{\Gamma}{}^\varkappa_{\mu\lambda}$.

As we know, the space admits a complete G_r of affine motions, if and only if there exists an integer N such that the first N sets of equations (9.3) are compatible in $v^\varkappa$ and $\overset{0}{\nabla}_\lambda v^\varkappa = v_\lambda^{\cdot\varkappa}$ and are equivalent to a set of $n^2 + n - r$ linearly independent equations and that all $v^\varkappa$ and $v_\lambda^{\cdot\varkappa}$ satisfying the first N sets satisfy also the $(N + 1)$st set of equations.

In this case, the rank of the matrix formed by the coefficients of $v^\varkappa$ and $v_\lambda^{\cdot\varkappa}$ in the first N sets is equal to $n^2 + n - r$.

We now consider the equations

$$\underset{v}{\pounds}S_{\mu\lambda}^{\cdot\cdot\varkappa} = v^\sigma \overset{0}{\nabla}_\sigma S_{\mu\lambda}^{\cdot\cdot\varkappa} - S_{\mu\lambda}^{\cdot\cdot\rho} v_\rho^{\cdot\varkappa} + S_{\sigma\lambda}^{\cdot\cdot\varkappa} v_\mu^{\cdot\sigma} + S_{\mu\sigma}^{\cdot\cdot\varkappa} v_\lambda^{\cdot\sigma} = 0.$$

In these equations, the coefficients of v^σ are $\overset{0}{\nabla}_\sigma S_{\mu\lambda}^{\cdot\cdot\varkappa}$ and those of $v_\rho^{\cdot\sigma}$ are

$$(9.4) \qquad T_{\mu\lambda\cdot\sigma}^{\cdot\cdot\varkappa\cdot\rho} \overset{\text{def}}{=} A_\mu^\rho S_{\sigma\lambda}^{\cdot\cdot\varkappa} + A_\lambda^\rho S_{\mu\sigma}^{\cdot\cdot\varkappa} - A_\sigma^\varkappa S_{\mu\lambda}^{\cdot\cdot\rho}.$$

We consider the $\overset{\rho}{\sigma}$-rank of $T_{\mu\lambda\cdot\sigma\cdot}^{\cdot\cdot\varkappa\cdot\rho}$.

First we shall prove

LEMMA 1. *If the $\overset{\rho}{\sigma}$-rank of T is less than n, then the components of the torsion tensor of the form $S_{\alpha_3\alpha_2}^{\cdot\cdot\alpha_1}$ ($\alpha_j \neq \alpha_i$ if $j \neq i$; $i, j, k, \ldots = 1, 2, \ldots, n$) are all zero.*

In fact, if we consider the n-rowed determinant formed by the components of T (upper diagram), then its value is $(-S_{\alpha_3\alpha_2}^{\cdot\cdot\alpha_1})^n$. The $\overset{\rho}{\sigma}$-rank of T is less than n and consequently we have

$\varkappa$, $\mu\lambda$ \ ρ, σ	α_1, α_j
α_i, $\alpha_3\alpha_2$	$-\delta_j^i S_{\alpha_3\alpha_2}^{\cdot\cdot\alpha_1}$
	$(i, j = 1, 2, \ldots, n)$

$$(9.5) \qquad S_{\alpha_3\alpha_2}^{\cdot\cdot\alpha_1} = 0.$$

LEMMA 2. *If the $\overset{\rho}{\sigma}$-rank of T is less than n, then the components of the torsion tensor of the form $S_{\alpha_2\alpha_1}^{\cdot\cdot\alpha_1}$ are all zero.*

In fact, taking account of Lemma 1, we consider an n-rowed determinant formed by the components of T (lower diagram):

$\varkappa$, $\mu\lambda$ \ ρ, σ	α_1, α_2	α_r, α_2
α_2, $\alpha_1\alpha_2$	$S_{\alpha_2\alpha_1}^{\cdot\cdot\alpha_1}$	*
α_1, $\alpha_s\alpha_1$	0	$\delta_s^r S_{\alpha_2\alpha_1}^{\cdot\cdot\alpha_1}$
	$(r, s = 2, \ldots, n)$	

Since the $\overset{\rho}{\sigma}$-rank of T is less than n, we have

$$(9.6) \qquad S_{\alpha_2\alpha_1}^{\cdot\cdot\alpha_1} = 0.$$

From the lemmas 1 and 2, we have

LEMMA 3. *If the $\overset{\rho}{\sigma}$-rank of T is less than n, then the torsion tensor vanishes identically.*

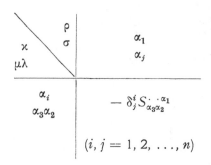

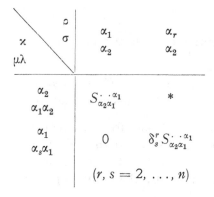

We now suppose that the torsion tensor does not vanish identically and we denote by r the order of the complete group of affine motions admitted by the space. Then the $\overset{\rho}{\sigma}$-rank of T is $n^2 + n - r$. Thus from Lemma 3, we have

$$n^2 + n - r \geqq n$$

that is

$$r \leqq n^2. \tag{9.7}$$

On the other hand, we can easily verify that a space with an asymmetric linear connexion $\Gamma^{\varkappa}_{\mu\lambda}$ for which

$$\begin{cases} \Gamma^1_{1n} = \Gamma^2_{2n} = \ldots = \Gamma^{n-1}_{n-1n} = a - b \\ \Gamma^1_{n1} = \Gamma^2_{n2} = \ldots = \Gamma^{n-1}_{nn-1} = a + b \\ \Gamma^n_{nn} = 2a, \end{cases} \tag{9.8}$$

with a and b non zero constants, the other Γ's being zero, admits a group of affine motions whose n^2 infinitesimal operators are

$$\partial_\lambda f, \quad \xi^\varkappa \partial_\alpha f. \qquad (\alpha = 1, 2, \ldots, n-1) \tag{9.9}$$

For the curvature tensor $R_{\nu\mu\lambda}{}^{\cdot\cdot\cdot\varkappa}$ holds in this case

$$R_{1nn}{}^{\cdot\cdot\cdot 1} = R_{2nn}{}^{\cdot\cdot\cdot 2} = \ldots = R_{n-1nn}{}^{\cdot\cdot\cdot n-1} = (a-b)^2, \tag{9.10}$$

and the other components of $R_{\nu\mu\lambda}{}^{\cdot\cdot\cdot\varkappa}$ not related to these are zero. Thus the curvature tensor of the space is zero or not zero according as $a = b$ or $a \neq b$. This proves a theorem of Egorov.[1]

THEOREM 9.1. *The maximum order of a complete group of affine motions in an* L_n $(\neq A_n)$ *is equal to* n^2.

When the space is an A_n, that is, when the torsion tensor vanishes, we cannot apply the above arguments.

If an A_n admits an infinitesimal affine motion $\xi^\varkappa \to \xi^\varkappa + v^\varkappa dt$, we have $\underset{v}{\pounds}\Gamma^{\varkappa}_{\mu\lambda} = 0$ and the integrability conditions are given by

$$\underset{v}{\pounds} R_{\nu\mu\lambda}{}^{\cdot\cdot\cdot\varkappa} = 0, \quad \underset{v}{\pounds} \nabla_\omega R_{\nu\mu\lambda}{}^{\cdot\cdot\cdot\varkappa} = 0, \quad \ldots. \tag{9.11}$$

In order that the space admit a complete G_r of affine motions, it is necessary and sufficient that there exist a positive integer N such that the first N sets of the equations (9.11) are algebraically compatible in $v^\varkappa$ and $\nabla_\lambda v^\varkappa$ and equivalent to a set of $n^2 + n - r$ linearly independent equations and that all $v^\varkappa$ and $\nabla_\lambda v^\varkappa$ satisfying the first N sets satisfy

[1] EGOROV [3].

the $(N+1)$st set. We shall consider the equations

$$\underset{v}{\pounds} R_{\nu\mu\lambda}{}^{\cdot\cdot\cdot\varkappa} = v^\sigma \nabla_\sigma R_{\nu\mu\lambda}{}^{\cdot\cdot\cdot\varkappa} - R_{\nu\mu\lambda}{}^{\cdot\cdot\cdot\rho} \nabla_\rho v^\varkappa + R_{\sigma\mu\lambda}{}^{\cdot\cdot\cdot\varkappa} \nabla_\nu v^\sigma + R_{\nu\sigma\lambda}{}^{\cdot\cdot\cdot\varkappa} \nabla_\mu v^\sigma + R_{\nu\mu\sigma}{}^{\cdot\cdot\cdot\varkappa} \nabla_\lambda v^\sigma = 0.$$

In these equations, the coefficients of v^σ are $\nabla_\sigma R_{\nu\mu\lambda}{}^{\cdot\cdot\cdot\varkappa}$ and those of $\nabla_\rho v^\sigma$ are

$$(9.12) \quad T_{\nu\mu\lambda\cdot\sigma}^{\cdot\cdot\cdot\varkappa\cdot\rho} = A^\rho_\nu R_{\sigma\mu\lambda}{}^{\cdot\cdot\cdot\varkappa} + A^\rho_\mu R_{\nu\sigma\lambda}{}^{\cdot\cdot\cdot\varkappa} + A^\rho_\lambda R_{\nu\mu\sigma}{}^{\cdot\cdot\cdot\varkappa} - A^\varkappa_\sigma R_{\nu\mu\lambda}{}^{\cdot\cdot\cdot\rho}.$$

First we prove

LEMMA 1. *If the* $^\rho_\sigma$*-rank of T is less than n, then*

$$(9.13) \qquad R_{\alpha_4\alpha_3\alpha_2}{}^{\cdot\cdot\cdot\alpha_1} = 0, \qquad (\alpha_1 \neq \alpha_2, \alpha_3, \alpha_4)$$

$\varkappa$ $\nu\mu\lambda$ \ ρ σ	α_1 α_j
α_i $\alpha_4\alpha_3\alpha_2$	$-\delta^i_j R_{\alpha_4\alpha_3\alpha_2}{}^{\cdot\cdot\cdot\alpha_1}$
	$(i, j = 1, 2, \ldots, n)$

$\varkappa$ $\nu\mu\lambda$ \ ρ σ	α_1 α_1	α_r α_1
α_2 $\alpha_1\alpha_2\alpha_1$	$2R_{\alpha_1\alpha_2\alpha_1}{}^{\cdot\cdot\cdot\alpha_2}$	0
α_2 $\alpha_1\alpha_2\alpha_s$	*	$\delta^r_s R_{\alpha_1\alpha_2\alpha_1}{}^{\cdot\cdot\cdot\alpha_2}$
	$(r, s = 2, 3, \ldots, n)$	

In fact, we consider the n-rowed determinant formed by the components of T (upper diagram).

Since the $^\rho_\sigma$-rank of T is less than n, we have (9.13). It should be noticed that the indices $\alpha_4, \alpha_3, \alpha_2$ different from α_1 may be equal.

LEMMA 2. *If the* $^\rho_\sigma$*-rank of T is less than n, then*

$$(9.14) \qquad R_{\alpha_1\alpha_2\alpha_1}{}^{\cdot\cdot\cdot\alpha_2} = 0.$$

In fact, taking account of Lemma 1, we consider the n-rowed determinant formed by the components of T (lower diagram).

Since the $^\rho_\sigma$-rank of T is less than n, we have (9.14).

LEMMA 3. *If the* $^\rho_\sigma$*-rank of T is less than n, then*

$$(9.15) \qquad R_{\alpha_3\alpha_2\alpha_3}{}^{\cdot\cdot\cdot\alpha_3}(R_{\alpha_3\alpha_2\alpha_3}{}^{\cdot\cdot\cdot\alpha_3} - R_{\alpha_3\alpha_2\alpha_1}{}^{\cdot\cdot\cdot\alpha_1}) = 0.$$

In fact, taking account of the lemmas 1 and 2, we consider the n-rowed determinant formed with the following components of T:

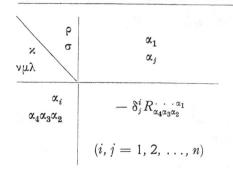

$\kappa \backslash \rho$ $\nu\mu\lambda \backslash \sigma$	α_1 α_3	α_3 α_3	α_1 α_2	α_r α_2
α_3 $\alpha_3\alpha_2\alpha_1$	$R_{\alpha_3\alpha_2\alpha_3}^{\cdot\cdot\cdot\alpha_3} - R_{\alpha_3\alpha_2\alpha_1}^{\cdot\cdot\cdot\alpha_1}$	0	0	0
α_3 $\alpha_3\alpha_2\alpha_3$	*	$R_{\alpha_3\alpha_2\alpha_3}^{\cdot\cdot\cdot\alpha_3}$	0	0
α_3 $\alpha_3\alpha_1\alpha_3$	*	*	$R_{\alpha_3\alpha_2\alpha_3}^{\cdot\cdot\cdot\alpha_3}$	0
α_3 $\alpha_3\alpha_s\alpha_3$	*	*	*	$\delta_s^r R_{\alpha_3\alpha_2\alpha_3}^{\cdot\cdot\cdot\alpha_3}$

$(r, s = 4, 5, \ldots, n).$

Since the rank of T is less than n, we have (9.15)

LEMMA 4. *If the* $\substack{\rho\\ \sigma}$*-rank of* T *is less than* n, *then*

$$R_{\alpha_2\alpha_1\alpha_3}^{\cdot\cdot\cdot\alpha_1}(R_{\alpha_1\alpha_2\alpha_3}^{\cdot\cdot\cdot\alpha_1} + R_{\alpha_3\alpha_2\alpha_1}^{\cdot\cdot\cdot\alpha_1} - R_{\alpha_3\alpha_2\alpha_3}^{\cdot\cdot\cdot\alpha_3}) = 0. \tag{9.16}$$

In fact, taking account of the lemmas 1 and 2, we consider the n-rowed determinant formed by the following components of T:

$\kappa \backslash \rho$ $\nu\mu\lambda \backslash \sigma$	α_1 α_3	α_3 α_3	α_1 α_2	α_r α_3
α_1 $\alpha_3\alpha_2\alpha_3$	$R_{\alpha_1\alpha_2\alpha_3}^{\cdot\cdot\cdot\alpha_1} + R_{\alpha_3\alpha_2\alpha_1}^{\cdot\cdot\cdot\alpha_1} - R_{\alpha_3\alpha_2\alpha_3}^{\cdot\cdot\cdot\alpha_3}$	0	0	0
α_1 $\alpha_2\alpha_1\alpha_3$	*	$R_{\alpha_2\alpha_1\alpha_3}^{\cdot\cdot\cdot\alpha_1}$	0	0
α_2 $\alpha_1\alpha_2\alpha_3$	*	*	$-R_{\alpha_2\alpha_1\alpha_3}^{\cdot\cdot\cdot\alpha_1}$	0
α_1 $\alpha_2\alpha_1\alpha_s$	*	*	*	$\delta_s^r R_{\alpha_2\alpha_1\alpha_3}^{\cdot\cdot\cdot\alpha_1}$

$(r, s = 4, 5, \ldots, n).$

Since the ${}^{\rho}_{\sigma}$-rank of T is less than n, we have (9.16).

We now assume that the ${}^{\rho}_{\sigma}$-rank of T is less than n. Then multiplying (9.16) by $R^{\cdot\,\cdot\,\cdot\alpha_3}_{\alpha_3\alpha_2\alpha_3}$ and using (9.15), we get

$$(R^{\cdot\,\cdot\,\cdot\alpha_1}_{\alpha_2\alpha_1\alpha_3})^2 R^{\cdot\,\cdot\,\cdot\alpha_3}_{\alpha_3\alpha_2\alpha_3} = 0,$$

that is

$$R^{\cdot\,\cdot\,\cdot\alpha_1}_{\alpha_2\alpha_1\alpha_3} R^{\cdot\,\cdot\,\cdot\alpha_3}_{\alpha_3\alpha_2\alpha_3} = 0.$$

Substituting this in (9.16) we find

$$R^{\cdot\,\cdot\,\cdot\alpha_1}_{\alpha_2\alpha_1\alpha_3}(R^{\cdot\,\cdot\,\cdot\alpha_1}_{\alpha_1\alpha_2\alpha_3} + R^{\cdot\,\cdot\,\cdot\alpha_1}_{\alpha_3\alpha_2\alpha_1}) = 0,$$

from which, interchanging α_2 and α_3,

$$R^{\cdot\,\cdot\,\cdot\alpha_1}_{\alpha_3\alpha_1\alpha_2}(R^{\cdot\,\cdot\,\cdot\alpha_1}_{\alpha_1\alpha_3\alpha_2} + R^{\cdot\,\cdot\,\cdot\alpha_1}_{\alpha_2\alpha_3\alpha_1}) = 0.$$

Adding these two equations and using $R^{\cdot\,\cdot\,\cdot\alpha_1}_{[\alpha_3\alpha_2\alpha_1]} = 0$, we find

$$-(R^{\cdot\,\cdot\,\cdot\alpha_1}_{\alpha_2\alpha_1\alpha_3})^2 - (R^{\cdot\,\cdot\,\cdot\alpha_2}_{\alpha_3\alpha_1\alpha_2})^2 - (R^{\cdot\,\cdot\,\cdot\alpha_1}_{\alpha_3\alpha_2\alpha_1})^2 = 0,$$

from which it follows that

$$R^{\cdot\,\cdot\,\cdot\alpha_1}_{\alpha_2\alpha_1\alpha_3} = R^{\cdot\,\cdot\,\cdot\alpha_2}_{\alpha_3\alpha_1\alpha_2} = R^{\cdot\,\cdot\,\cdot\alpha_1}_{\alpha_3\alpha_2\alpha_1} = 0. \tag{9.17}$$

Consequently, from (9.15) we find

$$R^{\cdot\,\cdot\,\cdot\alpha_3}_{\alpha_3\alpha_2\alpha_3} = 0. \tag{9.18}$$

Hence

LEMMA 5. *If the rank of T is less than n, all the components of the curvature tensor vanish.*

We now assume that the curvature tensor does not vanish identically and we denote by r the order of the complete group of affine motions admitted by the space. Then the ${}^{\rho}_{\sigma}$-rank of T is $n^2 + n - r$. Thus from Lemma 5, we have

$$n^2 + n - r \geq n$$

that is

$$r \leq n^2. \tag{9.19}$$

On the other hand, we can easily verify that a space with a symmetric linear connexion $\Gamma^{\varkappa}_{\mu\lambda}$ for which

$$\Gamma^1_{1n} = \Gamma^2_{2n} = \ldots = \Gamma^{n-1}_{n-1\,n} = \tfrac{1}{2}\Gamma^n_{nn} = a; \quad a = \text{constant} \neq 0, \tag{9.20}$$

the other Γ's not related to these being zero, admits an n^2-parameter group of affine motions generated by

(9.21) $$p_\lambda,\ \xi^\varkappa p_\alpha \quad (\alpha = 1, 2, \ldots, n-1).$$

For the curvature tensor holds in this case

(9.22) $$R_{1nn}^{\cdot\cdot\cdot 1} = R_{2nn}^{\cdot\cdot\cdot 2} = \ldots = R_{n-1nn}^{\cdot\cdot\cdot n-1} = a^2,$$

and the other components of $R_{\nu\mu\lambda}^{\cdot\cdot\cdot\varkappa}$ not related to these are zero.

Thus we have a theorem of Egorov.[1]

THEOREM 9.2. *The maximum order of a complete group of affine motions in an A_n, $n \geq 4$, with non zero curvature is n^2.*

§ 10. L_n's admitting an n^2-parameter complete group of motions.

We consider in this paragraph an L_n (with non-vanishing torsion) which admits the group G_{n^2} of affine motions. In this case $T_{\mu\lambda\cdot\sigma}^{\cdot\cdot\varkappa\cdot\rho}$ defined by (9.4) should have the $^\rho_\sigma$-rank n. We shall prove two lemmas:

LEMMA 1. *Under the above-mentioned assumptions, we have*

(10.1) $$S_{\alpha_3\alpha_2}^{\cdot\cdot\alpha_1} = 0.$$

In fact, we consider the $(n+1)$-rowed determinant formed by the following components of T:

$^{\varkappa}_{\mu\lambda}$ \ $^{\rho}_{\sigma}$	$^{\alpha_1}_{\alpha_1}$	$^{\alpha_4}_{\alpha_2}$	$^{\alpha_1}_{\alpha_2}$	$^{\alpha_1}_{\alpha_s}$
$^{\alpha_1}_{\alpha_3\alpha_2}$	$-S_{\alpha_3\alpha_2}^{\cdot\cdot\alpha_1}$	0	0	0
$^{\alpha_1}_{\alpha_4\alpha_3}$	*	$-S_{\alpha_3\alpha_2}^{\cdot\cdot\alpha_1}$	0	0
$^{\alpha_2}_{\alpha_2\alpha_3}$	*	*	$-S_{\alpha_3\alpha_2}^{\cdot\cdot\alpha_1}$	0
$^{\alpha_r}_{\alpha_3\alpha_2}$	*	*	*	$-\delta_s^r S_{\alpha_3\alpha_2}^{\cdot\cdot\alpha_1}$

$(r, s = 3, 4, \ldots, n)$

Since the rank of T is n, we have (10.1).

[1] EGOROV [1].

LEMMA 2. *Under the same assumptions as in Lemma* 1, *we have*

$$S_{\alpha_3\alpha_1}^{\cdot\cdot\,\alpha_1} = S_{\alpha_3\alpha_2}^{\cdot\cdot\,\alpha_2}. \tag{10.2}$$

In fact, taking account of Lemma 1, we consider the $(n+1)$-rowed determinant formed with the following components of T:

$\varkappa$ \ ρ $\mu\lambda$ \ σ	α_2 α_2	α_1 α_3	α_3 α_1	α_r α_2
α_1 $\alpha_2\alpha_1$	$S_{\alpha_2\alpha_1}^{\cdot\cdot\,\alpha_1}$	0	0	0
α_3 $\alpha_2\alpha_1$	*	$S_{\alpha_2\alpha_3}^{\cdot\cdot\,\alpha_3} - S_{\alpha_2\alpha_1}^{\cdot\cdot\,\alpha_1}$	0	0
α_1 $\alpha_2\alpha_3$	*	*	$S_{\alpha_2\alpha_1}^{\cdot\cdot\,\alpha_1} - S_{\alpha_2\alpha_3}^{\cdot\cdot\,\alpha_3}$	0
α_1 $\alpha_s\alpha_1$	*	*	*	$\delta_s^r S_{\alpha_2\alpha_1}^{\cdot\cdot\,\alpha_1}$

$$(r, s = 3, 4, \ldots, n).$$

Since the rank of T is n, we have

$$S_{\alpha_2\alpha_1}^{\cdot\cdot\,\alpha_1}(S_{\alpha_2\alpha_1}^{\cdot\cdot\,\alpha_1} - S_{\alpha_2\alpha_3}^{\cdot\cdot\,\alpha_3}) = 0,$$

from which, interchanging α_1 and α_3,

$$S_{\alpha_2\alpha_3}^{\cdot\cdot\,\alpha_3}(S_{\alpha_2\alpha_3}^{\cdot\cdot\,\alpha_3} - S_{\alpha_2\alpha_1}^{\cdot\cdot\,\alpha_1}) = 0.$$

Adding these two equations, we obtain

$$(S_{\alpha_2\alpha_1}^{\cdot\cdot\,\alpha_1} - S_{\alpha_2\alpha_3}^{\cdot\cdot\,\alpha_3})^2 = 0,$$

or

$$S_{\alpha_2\alpha_1}^{\cdot\cdot\,\alpha_1} = S_{\alpha_2\alpha_3}^{\cdot\cdot\,\alpha_3},$$

which proves Lemma 2. Lemmas 1 and 2 show that the torsion tensor has the form

$$S_{\mu\lambda}^{\cdot\cdot\,\varkappa} = S_{[\mu} A_{\lambda]}^{\varkappa}, \tag{10.3}$$

and this means that the linear connexion is semi-symmetric.[1] Thus we have a theorem of Egorov.[2]

[1] FRIEDMANN and SCHOUTEN [1]; SCHOUTEN [8], p. 126.

[2] EGOROV [5].

THEOREM 10.1. *If an L_n, $n \geq 4$, with non-vanishing torsion admits a complete G_{n^2} of affine motions, the connexion is semi-symmetric.* ($n \geq 4$)

REMARK. The theorem does not hold for $n = 3$. As an example take L_3 with

$$\Gamma^1_{32} = - \Gamma^1_{23} = 1, \tag{10.3}$$

the other Γ's being zero. This L_3 admits the 3^2-parameter group of affine motions generated by

$$p_1, p_2, p_3, \xi^1 p_1 + \xi^2 p_2, \xi^1 p_1 + \xi^3 p_3, \xi^2 p_1, \xi^3 p_2, \xi^3 p_1, \xi^2 p_3, \tag{10.4}$$

but its connexion is not semi-symmetric.

The theorem does not hold for $n = 2$ either. This has been examined by J. Levine. [1] As an example we take L_2 with

$$\Gamma^1_{21} = - \Gamma^1_{12} = 1, \tag{10.5}$$

the other Γ's being zero. This L_2 admits the 2^2-parameter group of affine motions generated by

$$p_1, p_2, \xi^1 p_2, \xi^2 p_1, \tag{10.6}$$

but its connexion is not semi-symmetric.

§ 11. A_n's which admit a group of affine motions leaving invariant a symmetric covariant tensor of valence 2.

We prove the following theorem of Egorov. [2]

THEOREM 11.1. *Let an A_n admit a group of affine motions which leaves invariant a symmetric covariant tensor $H_{\mu\lambda}$ of valence 2 and of rank m. Then we have:*

1°. *The order r of the group satisfies the inequality*

$$r \leq n^2 + n - nm + \tfrac{1}{2}m(m-1). \tag{11.1}$$

2°. *If the equality in* (11.1) *holds, the group is transitive.*

Proof of 1°. If $\xi^\varkappa \to \xi^\varkappa + v^\varkappa dt$ is an affine motion, we have

$$\underset{v}{\pounds} H_{\mu\lambda} = v^\sigma \nabla_\sigma H_{\mu\lambda} + H_{\sigma\lambda} \nabla_\mu v^\sigma + H_{\mu\sigma} \nabla_\lambda v^\sigma = 0. \tag{11.2}$$

The coefficients of $\nabla_\rho v^\sigma$ in (11.2) are

$$T_{\mu\lambda\sigma}^{\cdot\cdot\cdot\rho} = A^\rho_\mu H_{\sigma\lambda} + A^\rho_\lambda H_{\mu\sigma}. \tag{11.3}$$

Since $H_{\mu\lambda}$ is symmetric and of rank m, we can choose, at an arbitrary

[1] LEVINE [4].
[2] EGOROV [7].

point of the space, a coordinate system with respect to which, at this point,

(11.4) $H_{\mu\lambda} \overset{*}{=} \delta_{\mu\lambda} H_\lambda$ (not summed for λ)

$$H_\alpha \neq 0 \text{ for } \alpha = 1, 2, \ldots, m; \; H_\eta = 0 \text{ for } \eta = m+1, \ldots, n.$$

Then we have at this point

(11.5) $T_{\mu\lambda\sigma}^{\cdot\cdot\cdot\rho} \overset{*}{=} (\delta_\mu^\rho \delta_{\sigma\lambda} + \delta_\lambda^\rho \delta_{\mu\sigma}) H_\sigma$ (not summed for σ).

Thus to find the ${}^\rho_\sigma$-rank of T we have only to consider the matrix $'T$ for which $\mu\lambda$ indicates rows and ${}^\rho_\sigma$ columns and for which

$$'T : T_{\mu\lambda\sigma}^{\cdot\cdot\cdot\rho}; \; \mu \geq \lambda, \; \sigma \leq m, \; \lambda \leq m.$$

Now the only non zero elements of $'T$ in the columns ${}^\rho_\sigma$ are those in the rows $\rho\sigma$ or $\sigma\rho$. Thus the rank of $'T$ and consequently the ${}^\rho_\sigma$-rank of T is equal to the number of the rows

(11.6) $$\rho\sigma; \; \rho \geq \sigma, \; \sigma \leq m$$

and

(11.7) $$\sigma\rho; \; \sigma \geq \rho, \; \sigma \leq m.$$

The number of (11.6) is $(n - m)m$ and that of (11.7) is $\frac{1}{2}m(m + 1)$. Consequently, the rank of T is

$$(n - m)m + \tfrac{1}{2}m(m + 1) = nm - \tfrac{1}{2}m(m - 1),$$

and we have

$$r \leq n^2 + n - [nm - \tfrac{1}{2}m(m - 1)] = n^2 + n - mn + \tfrac{1}{2}m(m - 1).$$

Proof of 2°. If the equality in (11.1) holds, then the integrability conditions of $\underset{v}{\pounds}\Gamma_{\mu\lambda}^\varkappa$ are equivalent to

(11.8) $$T_{\mu\lambda\sigma}^{\cdot\cdot\cdot\rho} \nabla_\rho v^\sigma = 0.$$

Thus the initial values of $v^\varkappa$ can be chosen arbitrarily and consequently the group is transitive.

§ 12. A_n's which admit a group of affine motions leaving invariant an alternating covariant tensor of valence 2.

We prove the following theorem of Egorov.[1]

THEOREM 12.1. *Let an A_n admit a group of affine motions which leaves invariant an alternating covariant tensor $S_{\mu\lambda}$ of valence 2 and of rank 2k. Then we have:*

[1] EGOROV [7].

1°. *The order* r *of the group satisfies the inequality*

$$r \leq n^2 - (n - k)(2k - 1) + 2k. \tag{12.1}$$

2°. *If the space is projectively Euclidean and* $S_{\mu\lambda} = R_{[\mu\lambda]}$, *then*

$$r \leq n^2 - (n - k)(2k - 1).$$

3°. *If the equality holds in* 1° *or in* 2°, *the group is transitive.*

PROOF OF 1°. If $\xi^\varkappa \rightarrow \xi^\varkappa + v^\varkappa dt$ is an affine motion, we have

$$\underset{v}{\pounds} S_{\mu\lambda} = v^\sigma \nabla_\sigma S_{\mu\lambda} + S_{\sigma\lambda} \nabla_\mu v^\sigma + S_{\mu\sigma} \nabla_\lambda v^\sigma = 0. \tag{12.2}$$

The coefficients of $\nabla_\rho v^\sigma$ in these equations are

$$U_{\mu\lambda\sigma}^{\cdot\cdot\cdot\rho} = A_\mu^\rho S_{\sigma\lambda} + A_\lambda^\rho S_{\mu\sigma}. \tag{12.3}$$

Since the $S_{\mu\lambda}$ is alternating and of rank $2k$, we can choose, at an arbitrary point of the space, a coordinate system with respect to which, at that point,

$$S_{2a,2a-1} = - S_{2a-1,2a} \quad (a = 1, 2, \ldots, k) \tag{12.4}$$

the other S's being zero.

We now consider the following matrix formed by components of U:

$\mu\lambda$ \ ρ σ	i 2	k 1	p 4	r 3	$\ldots$ $\ldots$	u $2k$	x $2k-1$
$j1$	$\delta_j^i S_{21}$	0	0	0	$\ldots$	0	0
$l2$	0	$\delta_l^k S_{12}$	0	0	$\ldots$	0	0
$q3$	0	0	$\delta_q^p S_{43}$	0	$\ldots$	0	0
$s4$	0	0	0	$\delta_s^r S_{34}$	$\ldots$	0	0
$\vdots$	$\vdots$	$\vdots$	$\vdots$	$\vdots$	$\ldots$	$\vdots$	$\vdots$
$v, 2k-1$	0	0	0	0	$\ldots$	$\delta_v^u S_{2k,2k-1}$	0
$y, 2k$	0	0	0	0	$\ldots$	0	$\delta_y^x S_{2k-1,2k}$

$$2 \leq i, j \leq n;\ 3 \leq k, l \leq n;\ 4 \leq p, q \leq n;\ 5 \leq r, s \leq n;$$

$$\cdots\cdots\cdot$$

$$2k \leq u, v \leq n;\ 2k + 1 \leq x, y \leq n.$$

The rank of this matrix is

$$(n-1)+(n-2)+\ldots+(n-2k) = 2kn - k(2k+1)$$
$$= n + (n-k)(2k-1) - 2k,$$

and consequently

$$r \leq n^2 + n - [n + (n-k)(2k-1) - 2k] = n^2 - (n-k)(2k-1) + 2k.$$

PROOF OF 2°. Since the space is projectively Euclidean, we can choose a coordinate system with respect to which

$$\Gamma^{\varkappa}_{\mu\lambda} = p_\mu A^{\varkappa}_{\lambda} + p_\lambda A^{\varkappa}_{\mu}. \tag{12.5}$$

We then fix a point $\overset{}{\underset{0}{(\xi)}}$ in the space and effect a linear transformation in such a way that the tensor $S_{\mu\lambda} \overset{\text{def}}{=} R_{[\mu\lambda]}$ takes the form (12.4) at the point $\underset{0}{(\xi)}$. Since a linear transformation of coordinates does not change the form (12.5), we have, in this special coordinate system, at $\underset{0}{(\xi)}$, (12.5) and

$$R_{[2a,2a-1]} = -R_{[2a-1,2a]} \neq 0; \quad a = 1, 2, \ldots, k, \tag{12.6}$$

the other $R_{[\mu\lambda]}$ being all zero.

Since (12.6) holds at $\underset{0}{(\xi)}$, it holds also in a neighbourhood containing the point $\underset{0}{(\xi)}$. From this fact we shall prove that at least one of the expressions $R_{2a,2a}$, $R_{(2a,2a-1)}$ and $R_{2a-1,2a-1}$ is different from zero.

From (12.5), we have

$$R_{\mu\lambda} = -n\partial_\mu p_\lambda + \partial_\lambda p_\mu + (n-1)p_\mu p_\lambda, \tag{12.7}$$

from which it follows

$$R_{(\mu\lambda)} = -(n-1)[\partial_{(\mu} p_{\lambda)} - p_\mu p_\lambda], \tag{12.8}$$

$$R_{[\mu\lambda]} = -(n+1)\partial_{[\mu} p_{\lambda]}. \tag{12.9}$$

Now if all expressions $R_{2a,2a}$, $R_{(2a,2a-1)}$ and $R_{2a-1,2a-1}$ were zero, we should have

$$\partial_{2a} p_{2a} = p_{2a} p_{2a}, \tag{12.10}$$

$$R_{2a,2a-1} = R_{[2a,2a-1]} = -(n+1)\partial_{[2a} p_{2a-1]}, \tag{12.11}$$

$$\partial_{(2a} p_{2a-1)} = p_{2a} p_{2a-1}, \tag{12.12}$$

$$\partial_{2a-1} p_{2a-1} = p_{2a-1} p_{2a-1}. \tag{12.13}$$

From (12.11) and (12.12), we find

$$\partial_{2a} p_{2a-1} = p_{2a} p_{2a-1} - \frac{1}{n+1} R_{2a,2a-1}, \tag{12.14}$$

$$\partial_{2a-1} p_{2a} = p_{2a} p_{2a-1} + \frac{1}{n+1} R_{2a,2a-1}. \tag{12.15}$$

Substituting(12.13) and (12.14) in

$$\partial_{[2a} \partial_{2a-1]} p_{2a-1} = 0$$

and taking account of the other equations, we find

$$p_{2a-1} = \tfrac{1}{3} \partial_{2a-1} (\log |R_{2a,2a-1}|). \tag{12.16}$$

Similarly

$$p_{2a} = \tfrac{1}{3} \partial_{2a} (\log |R_{2a-1,2a}|). \tag{12.17}$$

From (12.16) and (12.17), we have

$$\partial_{[2a} p_{2a-1]} = 0.$$

Substituting this in (12.11), we obtain

$$R_{[2a,2a-1]} = 0,$$

which is in contradiction to (12.6). Hence it is proved that at least one of the expressions $R_{2a,2a}$, $R_{(2a,2a-1)}$ and $R_{2a-1,2a-1}$ is different from zero.

Now if $\xi^\varkappa \to \xi^\varkappa + v^\varkappa dt$ is an affine motion of the space, we have $\underset{v}{\pounds} R_{\mu\lambda} = 0$, from which

$$\underset{v}{\pounds} R_{(\mu\lambda)} = 0, \quad \underset{v}{\pounds} R_{[\mu\lambda]} = 0. \tag{12.18}$$

We have seen that if the rank of $R_{[\mu\lambda]}$ is $m \neq 0$, then $R_{(\mu\lambda)}$ and $R_{[\mu\lambda]}$ are both not identically zero. We put

$$\underset{v}{\pounds} R_{(\mu\lambda)} = v^\sigma \nabla_\sigma R_{(\mu\lambda)} + T_{\mu\lambda\sigma}^{\cdot\cdot\cdot\rho} \nabla_\rho v^\sigma, \tag{12.19}$$

$$\underset{v}{\pounds} R_{[\mu\lambda]} = v^\sigma \nabla_\sigma R_{[\mu\lambda]} + U_{\mu\lambda\sigma}^{\cdot\cdot\cdot\rho} \nabla_\rho v^\sigma, \tag{12.20}$$

where

$$T_{\mu\lambda\sigma}^{\cdot\cdot\cdot\rho} = A_\mu^\rho R_{(\sigma\lambda)} + A_\lambda^\rho R_{(\mu\sigma)}, \tag{12.21}$$

$$U_{\mu\lambda\sigma}^{\cdot\cdot\cdot\rho} = A_\mu^\rho R_{[\sigma\lambda]} + A_\lambda^\rho R_{[\mu\sigma]}. \tag{12.22}$$

Putting $R_{[\mu\lambda]} = S_{\mu\lambda}$, we consider a matrix constructed in the following way. If $R_{2a-1,2a-1} \neq 0$, we add, to the matrix considered in the proof of 1°, other rows and columns containing the components $T_{2a-1,2a-1,2a-1}{}^{2a-1}$ and $T_{2a,2a-1,2a-1}{}^{2a}$ of T. If $R_{(2a,2a-1)} \neq 0$, we add rows and columns containing the components $T_{2a-1,2a-1,2a}{}^{2a-1}$ and $T_{2a,2a-1,2a-1}{}^{2a-1}$. Finally, if $R_{2a,2a} \neq 0$, we add rows and columns containing the components $T_{2a,2a,2a}{}^{2\alpha}$ and $T_{2a,2a-1,2a}{}^{2a-1}$.

The rank of the matrix thus formed is $n + (n - k)(2k - 1)$ and consequently we have

$$r \leq n^2 + n - [n + (n - k)(2k - 1)] = n^2 - (n - k)(2k - 1),$$

which proves the second part of the theorem.

The third part can be proved by the same argument as was used in the proof of the second part of Theorem 11.1.

§ 13. Groups of affine motions in an A_n of order greater than $n^2 - n + 5$.

Let H_n be the linear homogeneous group in n variables $x^\varkappa$:

$$'x^\varkappa = a^\varkappa_\lambda x^\lambda, \quad \text{Det}\,(a^\varkappa_\lambda) \neq 0. \tag{13.1}$$

Then each element of H_n can be regarded as a non-singular real matrix $(a^\varkappa_\lambda)$. To denote various subgroups of H_n, we shall use the following notations throughout:

$$
\begin{array}{lll}
H^+_n & (a^\varkappa_\lambda): & \text{Det}\,(a^\varkappa_\lambda) > 0, \\
P_n & (a^\varkappa_\lambda): & \text{Det}\,(a^\varkappa_\lambda) = 1, \\
K & (a^\varkappa_\lambda): & a^\varkappa_\lambda = a\delta^\varkappa_\lambda,\ a\text{: positive number}, \\
L & (a^\varkappa_\lambda): & a^1_1 = 1,\ a^\xi_1 = 0,\ \text{Det}\,(a^\xi_\eta) = 1, \\
L' & (a^\varkappa_\lambda): & a^1_1 = 1,\ a^1_\eta = 0,\ \text{Det}\,(a^\xi_\eta) = 1, \\
M & (a^\varkappa_\lambda): & a^1_1 > 0,\ a^\xi_1 = 0,\ \text{Det}\,(a^\xi_\eta) = 1, \\
M' & (a^\varkappa_\lambda): & a^1_1 > 0,\ a^1_\eta = 0,\ \text{Det}\,(a^\xi_\eta) = 1, \\
I(b) & & \begin{pmatrix} e^{(1+b)t} & 0 & 0 & \dots & 0 \\ 0 & e^{bt} & 0 & \dots & 0 \\ 0 & 0 & e^{bt} & \dots & 0 \\ \dots & \dots & \dots & \dots & \dots \\ 0 & 0 & 0 & \dots & e^{bt} \end{pmatrix}
\end{array} \tag{13.2}
$$

$$\xi, \eta = 2, 3, \dots, n,$$

where b is a real constant and t runs over all real numbers. The groups L and M leave the direction $x^\varkappa = \delta_1^\varkappa$ invariant and the groups L' and M' leave invariant the hyperplane $x^1 = 0$. The orders of these groups are given by

(13.3)

group	H_n	H_n^+	P_n	K	L	L'	M	M'	$I(b)$
order	n^2	n^2	n^2-1	1	n^2-n-1	n^2-n-1	n^2-n	n^2-n	1

H. C. Wang and the present author [1] proved

THEOREM 13.1. *Each closed and connected subgroup of H_n of order greater than or equal to $n^2 - 2n + 5$ is, but for a coordinate transformation, one of the groups: H_n^+, P_n, $K \times L$, $K \times L'$, $K \times M$, $K \times M'$, $I(b) \times L$, $I(b) \times L'$, L, L'.*

If an A_n admits a group of affine motions G of order r and if we take a point in A_n and consider all the transformations of the group which leave this point invariant, then such transformations form a subgroup $G(P)$, called the isotropic subgroup at P. This subgroup consists of the transformations

$$T_\zeta: \ '\xi^\varkappa = h^\varkappa(\xi, \zeta) \tag{13.4}$$

such that

$$\underset{0}{\xi}^\varkappa = h^\varkappa(\underset{0}{\xi}, \zeta), \tag{13.5}$$

where $\underset{0}{\xi}^\varkappa$ are coordinates of the point P and ζ denotes the parameters.

To each transformation T_ζ in $G(P)$, there corresponds a linear transformation

$$\tau(T_\zeta): \quad d'\xi^\varkappa = \partial_\lambda h^\varkappa(\underset{0}{\xi}, \zeta) d\xi^\lambda \tag{13.6}$$

of the tangent space at the point P. By a method analogous to that used in § 8 of Ch. IV, we can prove that this linear representation τ of $G(P)$ is an isomorphism in the sense of topological groups.

Now consider the matrix $\underset{a}{e}^\varkappa$ $(a, b, c = 1, 2, \ldots, n)$ of the components of a basis of the infinitesimal group of G and denote by q the maximum rank of this matrix.

---

[1] WANG and YANO [1]

A point is called an ordinary point if, at this point, the matrix assumes the maximum rank q, and is called a singular point if otherwise.

Let an A_n admit a group G of affine motions of order greater than or equal to $n^2 - n + 5$. We confine ourselves to an open domain containing only ordinary points. Let $G(P)$ denote the isotropic subgroup at P. Then evidently the order of $G(P) =$ the order of $\tau(G(P)) \geq n^2 - 2n + 5$. Thus by Theorem 13.1, the connected component of the identity $\tilde{G}(P)$ of $\tau(G(P))$ must, but for a coordinate transformation, be one of the groups: H_n^+, P_n, $K \times L$, $K \times L'$, $K \times M$, $K \times M'$, $I(b) \times L$, $I(b) \times L'$, L, L'.

1°. *The case in which* $\tilde{G}(P)$ *is conjugate*[1] *to* H_n^+ *or* P_n.

In these two cases, the group G is transitive. Because, if G is not transitive, there would be an invariant subspace passing through P, and consequently $\tilde{G}(P)$ would leave invariant a proper linear subspace of the tangent space at the point P, which is impossible.

1) Case $\tilde{G}(P) = H_n^+$. In this case, G is of order $n^2 + n$. Thus by Theorem 5.3, in order that $\tilde{G}(P) = H_n^+$, it is necessary and sufficient that the space be locally an E_n.

2) Case $\tilde{G}(P) = P_n$. In this case, since the group $\tilde{G}(P)$ is of order $n^2 - 1$ and the group G is transitive, we know that the order of G is $n^2 + n - 1$. Since $\tilde{G}(P) = P_n$, $\nabla_\lambda v^\varkappa$ must satisfy

$$\nabla_\lambda v^\lambda = 0 \tag{13.7}$$

and the integrability conditions of $\underset{v}{\pounds}\Gamma^\varkappa_{\mu\lambda} = 0$:

$$\underset{v}{\pounds} R_{\nu\mu\lambda}^{\cdots\varkappa} = 0, \quad \underset{v}{\pounds}\nabla_\omega R_{\nu\mu\lambda}^{\cdots\varkappa} = 0, \quad \ldots. \tag{13.8}$$

must be satisfied identically by any $v^\varkappa$ and $\nabla_\lambda v^\varkappa$ satisfying (13.7). Thus comparing $\underset{v}{\pounds} R_{\nu\mu\lambda}^{\cdots\varkappa} = 0$ with (13.7), we see that there must exist functions $F_{\nu\mu\lambda}^{\cdots\varkappa}$ such that

$$v^\sigma \nabla_\sigma R_{\nu\mu\lambda}^{\cdots\varkappa} - R_{\nu\mu\lambda}^{\cdots\rho}\nabla_\rho v^\varkappa + R_{\sigma\mu\lambda}^{\cdots\varkappa}\nabla_\nu v^\sigma + R_{\nu\sigma\lambda}^{\cdots\varkappa}\nabla_\mu v^\sigma + R_{\nu\mu\sigma}^{\cdots\varkappa}\nabla_\lambda v^\sigma = -F_{\nu\mu\lambda}^{\cdots\varkappa}\nabla_\sigma v^\sigma$$

become identities in $v^\varkappa$ and $\nabla_\lambda v^\varkappa$. Thus we must have

$$\nabla_\sigma R_{\nu\mu\lambda}^{\cdots\varkappa} = 0, \tag{13.9}$$

$$R_{\nu\mu\lambda}^{\cdots\rho} A_\sigma^\varkappa - R_{\sigma\mu\lambda}^{\cdots\varkappa} A_\nu^\rho - R_{\nu\sigma\lambda}^{\cdots\varkappa} A_\mu^\rho - R_{\nu\mu\sigma}^{\cdots\varkappa} A_\lambda^\rho = F_{\nu\mu\lambda}^{\cdots\varkappa} A_\sigma^\rho. \tag{13.10}$$

[1] This means "equal to but for a coordinate transformation".

By contraction with respect to ρ and σ, we find from (13.10)

$$F_{\nu\mu\lambda}^{\cdot\cdot\cdot\varkappa} = -\frac{2}{n} R_{\nu\mu\lambda}^{\cdot\cdot\cdot\varkappa},$$

and by contraction with respect to $\varkappa$ and σ, we get

$$nR_{\varkappa\mu\lambda}^{\cdot\cdot\cdot\rho} - R_{\mu\lambda}A_\nu^\rho + R_{\nu\lambda}A_\mu^\rho + (R_{\nu\mu} - R_{\mu\nu})A_\lambda^\rho = -\frac{2}{n} R_{\nu\mu\lambda}^{\cdot\cdot\cdot\rho} \tag{13.11}$$

where $R_{\mu\lambda} = R_{\varkappa\mu\lambda}^{\cdot\cdot\cdot\varkappa}$. Contracting again in (13.11) with respect to ρ and ν, we find $R_{\mu\lambda} = 0$ for $n > 2$. Thus we have, from (13.11), $R_{\nu\mu\lambda}^{\cdot\cdot\cdot\varkappa} = 0$.

2°. *The case in which $\tilde{G}(P)$ is conjugate to $K \times L, K \times L', K \times M$ or $K \times M'$.* In these cases, the group G is transitive. We shall prove this by the method of contradiction.

We first suppose that $\tilde{G}(P)$ is conjugate to $K \times L$ or $K \times M$ and that the group G is intransitive. Then the invariant subspace passing through P should be one-dimensional, because the linear space tangent to this subspace at P is left invariant by $K \times L$ or $K \times M$ which fixes one and only one direction. Thus the rank of $\nabla_\lambda v^\varkappa$ is equal to 1 at P, and consequently, is equal to 1 at every point of the domain under consideration. It follows that, through every point of this domain, there passes one and only one invariant curve.

Now take an invariant curve passing through a point Q which is not on the invariant curve passing through P and which is in the domain under consideration, and consider all the geodesics joining P to the points on the invariant curve passing through Q. These geodesics constitute a two-dimensional surface.

This surface is left invariant by the isotropic subgroup $G(P)$. Consequently, the corresponding linear group $\tilde{G}(P)$ must leave invariant the two-dimensional plane tangent to this surface at P which contradicts our assumption.

We next suppose that $\tilde{G}(P) = K \times L'$ or $K \times M'$ and that the group G is intransitive. The invariant subspace passing through P should be $(n - 1)$-dimensional, because the linear space tangent to this subspace at P is left invariant by $K \times L'$ or $K \times M'$ which fixes one and only one hyperplane.

Thus the rank of the matrix $\nabla_\lambda v^\varkappa$ is equal to $n - 1$ at P, and consequently, is equal to $n - 1$ at every point under consideration. It follows that, through every point of the domain, there passes one and only one invariant hypersurface.

Now consider a geodesic through P which intersects these invariant hypersurfaces, then the points of the intersection can be transformed by an affine motion corresponding to an element of K into one another (except, of course, the point P), which is a contradiction.

Thus, in these cases, the group G is transitive, and consequently, two isotropic groups at any two ordinary points in the domain under consideration are conjugate to each other.

The groups $K\times L, K\times L', K\times M, K\times M'$ are respectively of order n^2-n, n^2-n, n^2-n+1, n^2-n+1 and the group G is transitive. Hence the group G is respectively of order n^2, n^2, n^2+1, n^2+1.

Now, at the point P of the domain, we choose the normal coordinates [1] $\xi^\varkappa$ whose origin is P, then, since the linear isotropy group $\tilde{G}(P)$ contains the K as a subgroup, the space admits a one-parameter group of affine motions

$$'\xi^\varkappa = e^t \xi^\varkappa. \tag{13.12}$$

In this coordinate system, the vector $v^\varkappa$ defining an infinitesimal transformation of this one-parameter group is given by $v^\varkappa = \xi^\varkappa$. Thus the integrability condition $\underset{v}{\pounds} R_{\nu\mu\lambda}^{\cdot\cdot\cdot\varkappa} = 0$ becomes

$$\xi^\omega \partial_\omega R_{\nu\mu\lambda}^{\cdot\cdot\cdot\varkappa} + 2R_{\nu\mu\lambda}^{\cdot\cdot\cdot\varkappa} = 0, \tag{13.13}$$

which shows that the $R_{\nu\mu\lambda}^{\cdot\cdot\cdot\varkappa}$ are homogeneous functions of degree -2 of the $\xi^\varkappa$.

But we know that the components of the curvature tensor are well defined at the origin of the normal coordinate system. Thus the components of the curvature tensor must vanish at P and consequently at any point of the domain.

Thus, in these cases, the space is locally affinely Euclidean.

3°. *The case in which* $\tilde{G}(P)$ *is conjugate to* $I(b)\times L$ or L.

In these cases, the group G is transitive. This can be proved by the same argument as the one used at the beginning of 2°.

Since the group is transitive, the isotropic groups at any two points of the domain under consideration are conjugate to each other. On the other hand, the isotropic group $G(Q)$ at an arbitrary point Q leaves invariant one and only one direction, which we denote by $u(Q)$. Thus, at every point Q of the domain under consideration, there is associated a direction $u(Q)$.

---

[1] SCHOUTEN [8], p. 155.

Consider a geodesic which passes through a point Q and is tangent to $U(Q)$, then since the isotropic group $G(Q)$ is an affine motion, it leaves this geodesic invariant. We take a point R different from Q on this geodesic and consider the transformations of $G(Q)$ which leaves invariant this point R. The linear representations of these transformations form the group L.

Now, we consider an affine frame at Q whose first axis is in the direction $u(Q)$ and we transport it parallelly along the geodesic to the point R. Then we have at R an affine frame whose first axis is tangent to the geodesic. The parallelism of vectors along a curve is preserved by an affine motion and hence the transformation of $G(Q)$ fixing the point R gives the same effect on the affine frame at R as on that at Q. This shows that the subgroup of $G(Q)$ leaving invariant R coincides with the subgroup of $G(R)$ leaving Q invariant. The subgroup of $G(R)$ fixing Q fixes the tangent to the geodesic and $u(R)$, and consequently, the tangent must coincide with $u(R)$, which shows that the geodesic is a streamline of the field of directions u.

Now, since the isotropic groups $I(b) \times L$ and L are respectively of order $n^2 - n$ and $n^2 - n - 1$ and the group is transitive, the group G is respectively of order n^2 and $n^2 - 1$.

Now since the group G of affine motions is transitive, we denote by T a transformation of G which carries a point Q into a point R. Then by the same method as in § 10 of Ch. IV, we can prove

$$Tu(Q) = u(R)$$

and that $u(Q)$ is a parallel field of directions.

If we denote this field of directions by $u^\varkappa(\xi)$, then we have

$$\underset{v}{\pounds} u^\varkappa = a u^\varkappa \tag{13.14}$$

and

$$\nabla_\lambda u^\varkappa = p_\lambda u^\varkappa, \tag{13.15}$$

where a is a certain scalar and p_λ a certain covariant vector field. From (13.15), we find

$$R_{\nu\mu\lambda}{}^{\cdot\cdot\cdot\varkappa} u^\lambda = p_{\nu\mu} u^\varkappa, \tag{13.16}$$

where

$$p_{\nu\mu} = 2\partial_{[\nu} p_{\mu]}. \tag{13.17}$$

We first suppose that $\tilde{G}(P) = I(b) \times L$. Then the equations $\underset{v}{\pounds} R_{\nu\mu\lambda}^{\cdot\cdot\cdot\varkappa} = 0$ must be satisfied by any $v^\varkappa$ and $\nabla_\lambda v^\varkappa$ satisfying

$$(1 + nb)\underset{v}{\pounds} u^\varkappa = (1 + b) u^\varkappa \nabla_\rho v^\rho. \tag{13.18}$$

We see that the conditions

$$\underset{v}{\pounds} u^\varkappa = v^\mu \nabla_\mu u^\varkappa - u^\mu \nabla_\mu v^\varkappa = 0 \tag{13.19}$$

and

$$\nabla_\rho v^\rho = 0 \tag{13.20}$$

taken together are stronger than (13.18). Hence any $v^\varkappa$ and $\nabla_\lambda v^\varkappa$ satisfying (13.19) and (13.20) must satisfy (13.18) and hence satisfy $\underset{v}{\pounds} R_{\nu\mu\lambda}^{\cdot\cdot\cdot\varkappa} = 0$.

Since the group is that of affine motions, the covariant differentiation and the Lie derivation are commutative and consequently, from (13.15) and (13.19), we find $\underset{v}{\pounds} p_\lambda = 0$. But the group $\tilde{G}(P)$ does not leave invariant a hyperplane and consequently we must have $p_\lambda = 0$. Consequently we have

$$\nabla_\lambda u^\varkappa = 0 \text{ and } R_{\nu\mu\lambda}^{\cdot\cdot\cdot\varkappa} u^\lambda = 0. \tag{13.21}$$

Thus the integrability condition $\underset{v}{\pounds} R_{\nu\mu\lambda}^{\cdot\cdot\cdot\varkappa} = 0$ must be satisfied by any $v^\varkappa$ and $\nabla_\lambda v^\varkappa$ satisfying

$$u^\lambda \nabla_\lambda v^\varkappa = 0 \text{ and } \nabla_\rho v^\rho = 0, \tag{13.22}$$

and consequently there must exist functions $F_{\nu\mu\lambda}^{\cdot\cdot\cdot\varkappa}$ and $G_{\nu\mu\lambda\cdot\sigma}^{\cdot\cdot\cdot\varkappa}$ such that

$$\nabla_\sigma R_{\nu\mu\lambda}^{\cdot\cdot\cdot\varkappa} = 0 \tag{13.23}$$

and

$$R_{\nu\mu\lambda}^{\cdot\cdot\cdot\rho} A_\sigma^\varkappa - R_{\sigma\mu\lambda}^{\cdot\cdot\cdot\varkappa} A_\nu^\rho - R_{\nu\sigma\lambda}^{\cdot\cdot\cdot\varkappa} A_\mu^\rho - R_{\nu\mu\sigma}^{\cdot\cdot\cdot\varkappa} A_\lambda^\rho = F_{\nu\mu\lambda}^{\cdot\cdot\cdot\varkappa} A_\sigma^\rho + G_{\nu\mu\lambda\cdot\sigma}^{\cdot\cdot\cdot\varkappa} u^\rho. \tag{13.24}$$

After some calculation, we can deduce from (13.24) $R_{\nu\mu\lambda}^{\cdot\cdot\cdot\varkappa} = 0$.

The case $\tilde{G}(P) = L$ is characterized by (13.19) and (13.20) and consequently the above discussion shows that when $\tilde{G}(P) = L$, the space is also locally affinely Euclidean.

4°. *The case in which $\tilde{G}(P)$ is conjugate to $I(b) \times L'$ or L' and the G is transitive.*

Since the group G is transitive, two isotropic groups at any two ordinary points in the domain under consideration are conjugate to one another.

On the other hand, the isotropic group $G(Q)$ at an ordinary point Q fixes one and only one hyperplane which we denote by $w(Q)$. Thus with every point Q of the domain, there is associated a hyperplane $w(Q)$.

The isotropic groups $I(b) \times L'$ and L' are respectively of order $n^2 - n$ and $n^2 - n - 1$ and the group G is transitive, hence the group G is respectively of order n^2 and $n^2 - 1$.

By exactly the same method as in § 10 of Ch. IV, we can prove that

$$Tw(Q) = w(R),$$

where T is an arbitrary affine motion carrying a point Q into a point R.

Furthermore, if we represent this hyperplane by a covariant vector $w_\lambda(\xi)$, then we can prove that

$$\underset{v}{\pounds} w_\lambda = a w_\lambda, \tag{13.25}$$

$$\nabla_\mu w_\lambda = p w_\mu w_\lambda, \tag{13.26}$$

where a and p are scalars. From (13.26) we find

$$-R_{\nu\mu\lambda}^{\cdot\cdot\cdot\varkappa} w_\varkappa = p_{\nu\mu} w_\lambda, \tag{13.27}$$

where

$$p_{\nu\mu} = 2\partial_{[\nu} p_{\mu]}. \tag{13.28}$$

We first suppose that $\tilde{G}(P) = I(b) \times L'$. Then the equations $\underset{v}{\pounds} R_{\nu\mu\lambda}^{\cdot\cdot\cdot\varkappa} = 0$ must be satisfied by any $v^\varkappa$ and $\nabla_\lambda v^\varkappa$ satisfying

$$(1 + nb) \underset{v}{\pounds} w_\lambda = (1 + b)\, w_\lambda \nabla_\rho v^\rho \tag{13.29}$$

We see that the conditions

$$\underset{v}{\pounds} w_\lambda = v^\sigma p w_\sigma w_\lambda - w_\mu \nabla_\lambda v^\mu = 0 \text{ and } \nabla_\rho v^\rho = 0 \tag{13.30}$$

taken together are stronger than (13.29). Hence any $v^\varkappa$ and $\nabla_\lambda v^\varkappa$ satisfying (13.30) must satisfy also (13.29) and hence satisfy $\underset{v}{\pounds} R_{\nu\mu\lambda}^{\cdot\cdot\cdot\varkappa} = 0$.

Since the group is that of affine motions, the covariant differentiation and the Lie derivation are commutative and consequently, from (13.26) and the first equation of (13.30), we find $\underset{v}{\pounds} p = 0$, which shows, since the group G is transitive, that the p is a constant.

Thus the integrability condition $\underset{v}{\pounds} R_{\nu\mu\lambda}^{\cdot\cdot\cdot\kappa} = 0$ must be satisfied by any v^κ and $\nabla_\lambda v^\kappa$ satisfying (11.30) and consequently there must exist functions $F_{\nu\mu\lambda}^{\cdot\cdot\cdot\kappa}$ and $G_{\nu\mu\lambda}^{\cdot\cdot\cdot\kappa\rho}$ such that

$$\nabla_\omega R_{\nu\mu\lambda}^{\cdot\cdot\cdot\kappa} = - p w_\omega G_{\nu\mu\lambda}^{\cdot\cdot\cdot\rho} w_\rho, \tag{13.31}$$

$$R_{\nu\mu\lambda}^{\cdot\cdot\cdot\rho} A_\sigma^\kappa - R_{\sigma\mu\lambda}^{\cdot\cdot\cdot\kappa} A_\nu^\rho - R_{\nu\sigma\lambda}^{\cdot\cdot\cdot\kappa} A_\mu^\rho - R_{\nu\mu\sigma}^{\cdot\cdot\cdot\kappa} A_\lambda^\rho = F_{\nu\mu\lambda}^{\cdot\cdot\cdot\kappa} A_\sigma^\rho + G_{\nu\mu\lambda}^{\cdot\cdot\cdot\kappa\rho} w_\sigma. \tag{13.32}$$

From (13.32) we can conclude, after some calculation, that

$$R_{\nu\mu\lambda}^{\cdot\cdot\cdot\kappa} = k(w_\nu A_\mu^\kappa - w_\mu A_\nu^\kappa) w_\lambda \tag{13.33}$$

where k is a constant.

Thus $\underset{v}{\pounds} R_{\nu\mu\lambda}^{\cdot\cdot\cdot\kappa} = 0$ becomes

$$\underset{v}{\pounds} R_{\nu\mu\lambda}^{\cdot\cdot\cdot\kappa} = 2a R_{\nu\mu\lambda}^{\cdot\cdot\cdot\kappa} = 0 \tag{13.34}$$

where a is given by $\underset{v}{\pounds} w_\lambda = a w_\lambda$.

When $1 + b \neq 0$ there exists an operator $\underset{v}{\pounds}$ such that $a \neq 0$, and thus we have $R_{\nu\mu\lambda}^{\cdot\cdot\cdot\kappa} = 0$. When $1 + b = 0$ then $\underset{v}{\pounds} w_\lambda = 0$ and thus $\underset{v}{\pounds} R_{\nu\mu\lambda}^{\cdot\cdot\cdot\kappa} = 0$ is satisfied by all the infinitesimal transformations $\underset{v}{\pounds}$ of the group G.

5°. *The case in which* $\tilde{G}(P)$ *is conjugate to* $I(b) \times L'$ *or* L' *and* G *is intransitive.*

Let us consider the invariant subspace through P. All the points in this invariant subspace are equivalent under the group and consequently isotropic groups at points of this invariant subspace are conjugate to each other. Thus the invariant subgroup must be $(n - 1)$-dimensional, because the plane tangent to this invariant subspace at a point must be left invariant by the linear isotropic group $I(b) \times L'$ or L' at this point which fixes one and only one hyperplane.

Take a point Q not in this invariant subspace. If the isotropic group at Q is one of the groups hitherto examined except $I(b) \times L'$ and L', then the group G must be transitive. Thus the isotropic group at Q must be also $I(b) \times L'$ or L'.

Consequently, passing through every ordinary point on the domain under consideration, there exists an $(n - 1)$-dimensional invariant subspace whose tangent hyperplane is left invariant by the isotropic group at the point of contact. We denote this hyperplane at Q by $w(Q)$.

The isotropic groups $I(b) \times L'$ and L' are respectively of order $n^2 - n$ and $n^2 - n - 1$, and the invariant subspaces are $(n - 1)$-dimensional. Hence the group G is of order $n^2 - 1$ and $n^2 - 2$ respectively.

Thus, if we denote by $f(x) =$ constant the family of invariant subspaces and if we put

$$sw_\lambda = \partial_\lambda f, \tag{13.35}$$

then, using the so-called adapted frames, we can prove that

$$\nabla_\mu (sw_\lambda) = 2p_{(\mu}(sw_{\lambda)}), \tag{13.36}$$

where p_λ is a certain covariant vector.

On the other hand, we know that

$$\underset{v}{\pounds} f = 0, \quad \underset{v}{\pounds}(sw_\lambda) = 0, \quad \underset{v}{\pounds}\nabla_\mu(sw_\lambda) = 0$$

and consequently, from (13.36), we find $\underset{v}{\pounds} p_\lambda = 0$. But the hyperplane represented by w_λ is the only one left invariant by the isotropic group and consequently we must have $p_\lambda = \frac{1}{2}hw_\lambda$, where h is a certain function of f.

Thus substituting this in (13.36), we get

$$\nabla_\mu (sw_\lambda) = h(sw_\mu)(sw_\lambda), \tag{13.37}$$

from which

$$R_{\nu\mu\lambda}^{\cdot\cdot\cdot\varkappa} w_\varkappa = 0. \tag{13.38}$$

We first suppose that $\tilde{G}(P) = I(b) \times L'$. Then the equations $\underset{v}{\pounds} R_{\nu\mu\lambda}^{\cdot\cdot\cdot\varkappa} = 0$ must be satisfied by any $v^\varkappa$ and $\nabla_\lambda v^\varkappa$ satisfying

$$(1 + nb)\underset{v}{\pounds} w_\lambda = (1 + b)w_\lambda \nabla_\rho v^\rho. \tag{13.39}$$

We see that the conditions

$$\underset{v}{\pounds} f = v^\sigma sw_\sigma = 0, \ \underset{v}{\pounds} w_\lambda = v^\mu \nabla_\mu w_\lambda + w_\mu \nabla_\lambda v^\mu = 0, \ \nabla_\rho v^\rho = 0 \tag{13.40}$$

taken together are stronger than (13.39). Hence any $v^\varkappa$ and $\nabla_\lambda v^\varkappa$ satisfying (13.40) must satisfy (13.39) and consequently also satisfy $\underset{v}{\pounds} R_{\nu\mu\lambda}^{\cdot\cdot\cdot\varkappa} = 0$.

The equations $\underset{v}{\pounds}(sw_\lambda) = 0$ and $\underset{v}{\pounds} w_\lambda = 0$ show that $\underset{v}{\pounds} s = 0$ and consequently that s is a function of f. Thus, from (13.35), we see that we can suppose $s = 1$. Thus the equation $\underset{v}{\pounds} w_\lambda = 0$ can be written as

$$\underset{v}{\pounds} w_\lambda = w_\mu \nabla_\lambda v^\mu = 0 \tag{13.41}$$

by virtue of (13.37) and the first equation of (13.40).

Thus the integrability condition $\underset{v}{\pounds} R_{\nu\mu\lambda}^{\cdot\cdot\cdot\varkappa} = 0$ must be satisfied by any $v^\varkappa$ and $\nabla_\lambda v^\varkappa$ satisfying

$$\underset{v}{\pounds} f = v^\sigma w_\sigma = 0, \; \underset{v}{\pounds} w_\lambda = w_\mu \nabla_\lambda v^\mu = 0, \; \nabla_\rho v^\rho = 0, \tag{13.42}$$

and consequently there must exist functions $E_{\nu\mu\lambda}^{\cdot\cdot\cdot\varkappa}$, $F_{\nu\mu\lambda}^{\cdot\cdot\cdot\varkappa}$ and $G_{\nu\mu\lambda}^{\cdot\cdot\cdot\varkappa\rho}$ such that

$$\nabla_\omega R_{\nu\mu\lambda}^{\cdot\cdot\cdot\varkappa} = w_\omega E_{\nu\mu\lambda}^{\cdot\cdot\cdot\varkappa}, \tag{13.43}$$

$$R_{\nu\mu\lambda}^{\cdot\cdot\cdot\rho} A_\sigma^\varkappa - R_{\sigma\mu\lambda}^{\cdot\cdot\cdot\varkappa} A_\nu^\rho - R_{\nu\sigma\lambda}^{\cdot\cdot\cdot\varkappa} A_\mu^\rho - R_{\nu\mu\sigma}^{\cdot\cdot\cdot\varkappa} A_\lambda^\rho = F_{\nu\mu\lambda}^{\cdot\cdot\cdot\varkappa} A_\sigma^\rho + G_{\nu\mu\lambda}^{\cdot\cdot\cdot\varkappa\rho} w_\sigma. \tag{13.44}$$

From (13.44) we can conclude that the curvature tensor $R_{\nu\mu\lambda}^{\cdot\cdot\cdot\varkappa}$ must be of the form

$$R_{\nu\mu\lambda}^{\cdot\cdot\cdot\varkappa} = k(w_\nu A_\mu^\varkappa - w_\mu A_\nu^\varkappa) w_\lambda. \tag{13.45}$$

But since we have $\underset{v}{\pounds} R_{\nu\mu\lambda}^{\cdot\cdot\cdot\varkappa} = 0$ and $\underset{v}{\pounds} w_\lambda = 0$, we find from this $\underset{v}{\pounds} k = 0$, which shows that k is a certain function of f. Thus the equations $\underset{v}{\pounds} R_{\nu\mu\lambda}^{\cdot\cdot\cdot\varkappa} = 0$, become

$$\underset{v}{\pounds} R_{\nu\mu\lambda}^{\cdot\cdot\cdot\varkappa} = 2a R_{\nu\mu\lambda}^{\cdot\cdot\cdot\varkappa} = 0,$$

where a is given by $\underset{v}{\pounds} w_\lambda = a w_\lambda$. When $1 + b \neq 0$, there exists an $\underset{v}{\pounds}$ such that $a \neq 0$ and thus $R_{\nu\mu\lambda}^{\cdot\cdot\cdot\varkappa} = 0$. When $1 + b = 0$, then $\underset{v}{\pounds} w_\lambda = 0$ and thus $\underset{v}{\pounds} R_{\nu\mu\lambda}^{\cdot\cdot\cdot\varkappa} = 0$ is satisfied by all the infinitesimal transformations $\underset{v}{\pounds}$ of the group G.

The case $\tilde{G}(P) = L'$ is characterized by (13.40) and consequently the above discussion shows that if $\tilde{G}(P) = L'$, the group has also the curvature tensor of the form (13.45).

Gathering all results, we obtain

THEOREM 13.2. *If an A_n admits a group G of affine motions of order greater than $n^2 - n + 5$, then we have for the linear isotropic group $\tilde{G}(P)$ at a point P, the order of $\tilde{G}(P)$, the group of affine motions, the order of G,* *and the structure of A_n, only the following*:

isotropic group $\tilde{G}(P)$	dimension of $\tilde{G}(P)$	group of affine motions G	order of G	structure of A_n
H_n	n^2	transitive	n^2+n	$R_{\nu\mu\lambda}{}^{\kappa} = 0$
H_n^+	,,	,,	,,	,,
P_n	n^2-1	,,	n^2+n-1	,,
$K \times L$	n^2-n	,,	n^2	,,
$K \times L'$	n^2-n	,,	,,	,,
$K \times M$	n^2-n+1	,,	n^2+1	,,
$K \times M'$	n^2-n+1	,,	,,	,,
$I(b) \times L$	n^2-n	,,	n^2	,,
L	n^2-n-1	,,	n^2-1	,,
$I(b) \times L'$	n^2-n	transitive	n^2	(i) $1+b \neq 0$, $R_{\nu\mu\lambda}{}^{\kappa}=0$, (ii) $1+b=0$, $\nabla_\mu w_\lambda = p w_\mu w_\lambda$, $R_{\nu\mu\lambda}{}^{\kappa} = k(w_\nu A_\mu^\kappa - w_\mu A_\nu^\kappa) w_\lambda$, p, k: constants
		intransitive	n^2-1	(i) $1+b \neq 0$, $R_{\nu\mu\lambda}{}^{\kappa}=0$, (ii) $1+b=0$, $\nabla_\mu w_\lambda = p w_\mu w_\lambda$ $(w_\lambda = \partial_\lambda f)$ $R_{\nu\mu\lambda}{}^{\kappa} = k(w_\nu A_\mu^\kappa - w_\mu A_\nu^\kappa) w_\lambda$, p, k: functions of f.
L'	n^2-n-1	transitive	n^2-1	$\nabla_\mu w_\lambda = p w_\mu w_\lambda$, $R_{\nu\mu\lambda}{}^{\kappa} = k(w_\nu A_\mu^\kappa - w_\mu A_\nu^\kappa) w_\lambda$, p, k: constants.
		intransitive	n^2-2	$\nabla_\mu w_\lambda = p w_\mu w_\lambda$ $(w_\lambda = \partial_\lambda f)$ $R_{\nu\mu\kappa}{}^{\kappa} = k(w_\nu A_\mu^\kappa - w_\mu A_\nu^\kappa) w_\lambda$, p, k: functions of f.

Chapter VI

GROUPS OF PROJECTIVE MOTIONS

§ 1. Groups of projective motions.

An infinitesimal projective motion $\xi^\varkappa \to \xi^\varkappa + v^\varkappa dt$ in an A_n is characterized by

$$\underset{v}{\pounds}\Gamma^\varkappa_{\mu\lambda} = 2p_{(\mu}A^\varkappa_{\lambda)} \tag{1.1}$$

or

$$\underset{v}{\pounds}\overset{p}{\Gamma}{}^\varkappa_{\mu\lambda} = 0; \quad \overset{p}{\Gamma}{}^\varkappa_{\mu\lambda} \overset{\text{def}}{=} \Gamma^\varkappa_{\mu\lambda} - \frac{2}{n+1} A^\varkappa_{(\mu}\Gamma^\rho_{\lambda)\rho}. \tag{1.2}$$

Since the projective connexion $\overset{p}{\Gamma}{}^\varkappa_{\mu\lambda}$ is a linear differential geometric object, according to Theorems 2.1 and 2.2 of Ch. III, we have

THEOREM 1.1. *If an A_n admits an infinitesimal projective motion, it admits also a one-parameter group of projective motions generated by this infinitesimal one*

THEOREM 1.2. *In order that an A_n admit a one-parameter group of projective motions, it is necessary and sufficient that there exist a coordinate system with respect to which the components $\overset{p}{\Gamma}{}^\varkappa_{\mu\lambda}$ of the projective connexion are independent of one of the coordinates.*

When the components $\overset{p}{\Gamma}{}^\varkappa_{\mu\lambda}$ are independent of ξ^1, the components $\Gamma^\varkappa_{\mu\lambda}$ have the form

$$\Gamma^\varkappa_{\mu\lambda} = f^\varkappa_{\mu\lambda}(\xi^2, \ldots, \xi^n) + 2A^\varkappa_{(\mu}p_{\lambda)}, \tag{1.3}$$

where the p_λ are functions of $\xi^1, \ldots, \xi^n$.

Conversely, if the components $\Gamma^\varkappa_{\mu\lambda}$ of the linear connexion have the form (1.3), then the components $\overset{p}{\Gamma}{}^{\varkappa}_{\mu\lambda}$ of the projective connexion are independent of the variable ξ^1. Thus

THEOREM 1.3.[1] *In order that an A_n admit a one-parameter group of projective motions, it is necessary and sufficient that there exist a coordinate system with respect to which the components $\Gamma^{\varkappa}_{\mu\lambda}$ of the linear connexion have the form* (1.3).

When we choose a coordinate system with respect to which $v^{\varkappa} = \xi^{\varkappa}$, then the equations $\underset{v}{\pounds}\overset{p}{\Gamma}{}^{\varkappa}_{\mu\lambda} = 0$ give

$$\underset{v}{\pounds}\overset{p}{\Gamma}{}^{\varkappa}_{\mu\lambda} = \xi^{\nu}\partial_{\nu}\overset{p}{\Gamma}{}^{\varkappa}_{\mu\lambda} + \overset{p}{\Gamma}{}^{\varkappa}_{\mu\lambda} = 0,$$

from which we have

THEOREM 1.4.[2] *In order that an A_n admit a one-parameter group of projective motions, it is necessary and sufficient that there exist a coordinate system with respect to which the components of the projective connexion are homogeneous functions of degree* -1 *of the coordinates.*

Furthermore, since the projective connexion $\overset{p}{\Gamma}{}^{\varkappa}_{\mu\lambda}$ is a linear differential geometric object, Theorems 2.3, 2.4, 2.5 and 2.6 of Ch. III hold for projective motions.

§ 2. Transformations carrying projective conics into projective conics.

We now ask for the condition that an infinitesimal transformation $\xi^{\varkappa} \to \xi^{\varkappa} + v^{\varkappa}dt$ transforms any projective conic[3] into a projective conic and projective parameters on them into projective parameters.

A projective conic and *a projective parameter t* on it are defined by the differential equations

$$(2.1)\qquad \begin{cases} \dfrac{d}{ds}\{t, s\} + \dfrac{da}{ds} + P_{\mu\lambda}\dfrac{\delta^2\xi^{\mu}}{ds^2}\dfrac{d\xi^{\lambda}}{ds} = 0, \\[2ex] \dfrac{\delta^3\xi^{\varkappa}}{ds^3} + [2\{t, s\} + a]\dfrac{d\xi^{\varkappa}}{ds} = 0, \end{cases}$$

[1] YANO and TOMONAGA [1]; G. T., p. 64.
[2] G. T., p. 65.
[3] YANO and TAKANO [1].

where

$$a \overset{\text{def}}{=} P_{\mu\lambda} \frac{d\xi^\mu}{ds} \frac{d\xi^\lambda}{ds}, \tag{2.2}$$

$$P_{\mu\lambda} \overset{\text{def}}{=} -\frac{1}{n^2-1}(nR_{\mu\lambda} + R_{\lambda\mu}) \tag{2.3}$$

and where $\{t, s\}$ denotes the Schwarzian derivative of t with respect to s.

We calculate first of all the Lie derivative of the left-hand members of (2.1). After some calculation, we get

$$\mathcal{L}\left[\frac{d}{ds}\{t, s\} + \frac{da}{ds} + P_{\mu\lambda}\frac{\delta^2\xi^\mu}{ds^2}\frac{d\xi^\lambda}{ds}\right]^1 \tag{2.4}$$

$$= -3\left[\frac{d}{ds}\{t, s\} + \frac{da}{ds} + P_{\mu\lambda}\frac{\delta^2\xi^\mu}{ds^2}\frac{d\xi^\lambda}{ds}\right]\frac{\mathcal{L}ds}{ds} - [2\{t, s\} + 3a]\frac{d}{ds}\frac{\mathcal{L}ds}{ds}$$

$$-\frac{d^3}{ds^3}\frac{\mathcal{L}ds}{ds} + \frac{d}{ds}\left[(\mathcal{L}P_{\mu\lambda})\frac{d\xi^\mu}{ds}\frac{d\xi^\lambda}{ds}\right] + (\mathcal{L}P_{\mu\lambda})\frac{\delta^2\xi^\mu}{ds^2}\frac{d\xi^\lambda}{ds}$$

$$+ (\mathcal{L}\Gamma^\sigma_{\nu\mu})P_{\sigma\lambda}\frac{d\xi^\nu}{ds}\frac{d\xi^\mu}{ds}\frac{d\xi^\lambda}{ds}.$$

$$\mathcal{L}\left[\frac{\delta^3\xi^\varkappa}{ds^3} + (2\{t, s\} + a)\frac{d\xi^\varkappa}{ds}\right] \tag{2.5}$$

$$= -3\left[\frac{\delta^3\xi^\varkappa}{ds^3} + (2\{t, s\} + a)\frac{d\xi^\varkappa}{ds}\right]\frac{\mathcal{L}ds}{ds} + 3(\mathcal{L}\Gamma^\varkappa_{\mu\lambda})\frac{\delta^2\xi^\mu}{ds^2}\frac{d\xi^\lambda}{ds}$$

$$+ (\nabla_\nu\mathcal{L}\Gamma^\varkappa_{\mu\lambda})\frac{d\xi^\nu}{ds}\frac{d\xi^\mu}{ds}\frac{d\xi^\lambda}{ds} - 3\frac{\delta^3\xi^\varkappa}{ds^3}\frac{d}{ds}\frac{\mathcal{L}ds}{ds} - 3\frac{d\xi^\varkappa}{ds}\frac{d^2}{ds^2}\frac{\mathcal{L}ds}{ds}$$

$$+ \frac{d\xi^\varkappa}{ds}(\mathcal{L}\Gamma_{\mu\lambda})\frac{d\xi^\mu}{ds}\frac{d\xi^\lambda}{ds}.$$

Hence, in order that the infinitesimal transformation $\xi^\varkappa \to \xi^\varkappa + v^\varkappa dt$ transform every projective conic into a projective conic and a projective parameter on it into a projective parameter on the deformed conic, it is necessary that the equations

$$-[2\{t, s\} + a]\frac{d}{ds}\frac{\mathcal{L}ds}{ds} - \frac{d^3}{ds^3}\frac{\mathcal{L}ds}{ds} + \frac{d}{ds}\left[(\mathcal{L}P_{\mu\lambda})\frac{d\xi^\mu}{ds}\frac{d\xi^\lambda}{ds}\right] \tag{2.6}$$

$$+ (\mathcal{L}P_{\mu\lambda})\frac{\delta^2\xi^\mu}{ds^2}\frac{d\xi^\lambda}{ds} + (\mathcal{L}\Gamma^\sigma_{\nu\mu})P_{\sigma\lambda}\frac{d\xi^\nu}{ds}\frac{d\xi^\mu}{ds}\frac{d\xi^\lambda}{ds} = 0$$

[1] We dropped v in $\underset{v}{\mathcal{L}}$.

and

$$(2.7)\qquad 3(\pounds\Gamma_{\mu\lambda})\frac{\delta^2\xi^\mu}{ds^2}\frac{d\xi^\lambda}{ds} + (\nabla_\nu\pounds\Gamma^{\varkappa}_{\mu\lambda})\frac{d\xi^\nu}{ds}\frac{d\xi^\mu}{ds}\frac{d\xi^\lambda}{ds}$$

$$-3\frac{\delta^2\xi^\varkappa}{ds^2}\frac{d}{ds}\frac{\pounds ds}{ds} - 3\frac{d\xi^\varkappa}{ds}\frac{d^2}{ds^2}\frac{\pounds ds}{ds} + \frac{d\xi^\varkappa}{ds}(\pounds P_{\mu\lambda})\frac{d\xi^\mu}{ds}\frac{d\xi^\lambda}{ds} = 0$$

be identically satisfied for any vector $\frac{\delta^2\xi^\varkappa}{ds^2}$ and $\frac{d\xi^\varkappa}{ds}$. Thus, from (2.7), we must have

$$(2.8)\qquad \pounds\Gamma^{\varkappa}_{\mu\lambda} = p_\mu A^{\varkappa}_{\lambda} + p_\lambda A^{\varkappa}_{\mu},$$

that is, $\xi^\varkappa \to \xi^\varkappa + v^\varkappa dt$ is an infinitesimal projective motion.

Since the converse is evident, we have

THEOREM 2.1. *In order that an infinitesimal transformation carry every projective conic into a projective conic and a projective parameter on it into a projective parameter on its deform, it is necessary and sufficient that the transformation be a projective motion.*

§ 3. Integrability conditions of $\underset{v}{\pounds}\Gamma^{\varkappa}_{\mu\lambda} = 2p_{(\mu}A^{\varkappa}_{\lambda)}$.

We consider the integrability conditions of $\underset{v}{\pounds}\Gamma^{\varkappa}_{\mu\lambda} = 2p_{(\mu}A^{\varkappa}_{\lambda)}$. Substituting this in

$$(3.1)\qquad \underset{v}{\pounds}R_{\nu\mu\lambda}^{\cdot\cdot\cdot\varkappa} = 2\nabla_{[\nu}\underset{v}{\pounds}\Gamma^{\varkappa}_{\mu]\lambda},$$

we find

$$(3.2)\qquad \underset{v}{\pounds}R_{\nu\mu\lambda}^{\cdot\cdot\cdot\varkappa} = -2A^{\varkappa}_{[\nu}\nabla_{\mu]}p_\lambda + 2\nabla_{[\nu}p_{\mu]}A^{\varkappa}_{\lambda},$$

from which, by contraction,

$$(3.3)\qquad \underset{v}{\pounds}P_{\mu\lambda} = \nabla_\mu p_\lambda.$$

Thus we are led to consider a system of partial differential equations

$$(3.4)\qquad \begin{cases}\nabla_\lambda v^\varkappa = v_\lambda^{\cdot\varkappa},\ \nabla_\mu v_\lambda^{\cdot\varkappa} = -R_{\nu\mu\lambda}^{\cdot\cdot\cdot\varkappa}v^\nu + 2p_{(\mu}A^{\varkappa}_{\lambda)},\\ \nabla_\mu p_\lambda = v^\sigma\nabla_\sigma P_{\mu\lambda} + P_{\sigma\lambda}v_\mu^{\cdot\sigma} + P_{\mu\sigma}v_\lambda^{\cdot\sigma}\end{cases}$$

with $n^2 + 2n$ unknown functions $v^\varkappa$, $v_\lambda^{\cdot\varkappa}$ and p_λ.

First, substituting (3.3) in (3.2), we find

(3.5) $$\underset{v}{\pounds} P_{\nu\mu\lambda}^{\cdot\cdot\cdot\varkappa} = 0,$$

where

(3.6) $$P_{\nu\mu\lambda}^{\cdot\cdot\cdot\varkappa} \overset{\text{def}}{=} R_{\nu\mu\lambda}^{\cdot\cdot\cdot\varkappa} + 2A_{[\nu}^{\varkappa} P_{\mu]\lambda} - 2P_{[\nu\mu]} A_{\lambda}^{\varkappa}$$

is Weyl's projective curvature tensor.[1]

Next we substitute $\underset{v}{\pounds}\Gamma_{\mu\lambda}^{\varkappa} = 2p_{(\mu} A_{\lambda)}^{\varkappa}$ in the equation

$$\underset{v}{\pounds}\nabla_\nu P_{\mu\lambda} - \nabla_\nu \underset{v}{\pounds} P_{\mu\lambda} = -(\underset{v}{\pounds}\Gamma_{\nu\mu}^{\sigma})P_{\sigma\lambda} - (\underset{v}{\pounds}\Gamma_{\nu\lambda}^{\sigma})P_{\mu\sigma}$$

which is obtained by applying the formula (4.9) of Ch. I to $P_{\mu\lambda}$. Then we obtain

$$\underset{v}{\pounds}\nabla_\nu P_{\mu\lambda} = \nabla_\nu \nabla_\mu p_\lambda - 2p_\nu P_{\mu\lambda} - p_\mu P_{\nu\lambda} - p_\lambda P_{\mu\nu},$$

from which

(3.7) $$\underset{v}{\pounds} P_{\nu\mu\lambda} = -P_{\nu\mu\lambda}^{\cdot\cdot\cdot\varkappa} p_\varkappa,$$

where

(3.8) $$P_{\nu\mu\lambda} \overset{\text{def}}{=} 2\nabla_{[\nu} P_{\mu]\lambda}.$$

We substitute $\underset{v}{\pounds}\Gamma_{\mu\lambda}^{\varkappa} = 2p_{(\mu} A_{\lambda)}^{\varkappa}$ and (3.5) in the equation

$$\begin{aligned}\underset{v}{\pounds}\nabla_\omega P_{\nu\mu\lambda}^{\cdot\cdot\cdot\varkappa} &- \nabla_\omega \underset{v}{\pounds} P_{\nu\mu\lambda}^{\cdot\cdot\cdot\varkappa} \\ &= (\underset{v}{\pounds}\Gamma_{\omega\rho}^{\varkappa})P_{\nu\mu\lambda}^{\cdot\cdot\cdot\rho} - (\underset{v}{\pounds}\Gamma_{\omega\nu}^{\rho})P_{\rho\mu\lambda}^{\cdot\cdot\cdot\varkappa} - (\underset{v}{\pounds}\Gamma_{\omega\mu}^{\rho})P_{\nu\rho\lambda}^{\cdot\cdot\cdot\varkappa} - (\underset{v}{\pounds}\Gamma_{\omega\lambda}^{\rho})P_{\nu\mu\rho}^{\cdot\cdot\cdot\varkappa}\end{aligned}$$

which is obtained by applying the formula (4.9) of Ch. I to $P_{\nu\mu\lambda}^{\cdot\cdot\cdot\varkappa}$. Then we get

(3.9) $$\begin{aligned}\underset{v}{\pounds}\nabla_\omega P_{\nu\mu\lambda}^{\cdot\cdot\cdot\varkappa} = &-2p_\omega P_{\nu\mu\lambda}^{\cdot\cdot\cdot\varkappa} \\ &+ A_\omega^\varkappa P_{\nu\mu\lambda}^{\cdot\cdot\cdot\rho} p_\rho - P_{\omega\mu\lambda}^{\cdot\cdot\cdot\varkappa} p_\nu - P_{\nu\omega\lambda}^{\cdot\cdot\cdot\varkappa} p_\mu - P_{\nu\mu\omega}^{\cdot\cdot\cdot\varkappa} p_\lambda.\end{aligned}$$

We next substitute $\underset{v}{\pounds}\Gamma_{\mu\lambda}^{\varkappa} = 2p_{(\mu} A_{\lambda)}^{\varkappa}$ and (3.7) in the equations

$$\underset{v}{\pounds}\nabla_\omega P_{\nu\mu\lambda} - \nabla_\omega \underset{v}{\pounds} P_{\nu\mu\lambda} = -(\underset{v}{\pounds}\Gamma_{\omega\nu}^{\rho})P_{\rho\mu\lambda} - (\underset{v}{\pounds}\Gamma_{\omega\mu}^{\rho})P_{\nu\rho\lambda} - (\underset{v}{\pounds}\Gamma_{\omega\lambda}^{\rho})P_{\nu\mu\rho}$$

---

[1] WEYL [1], EISENHART [3], T. Y. THOMAS [3], SCHOUTEN [8], p. 289.

which are also obtained by applying the formula (4.9) of Ch. I to the tensor $P_{\nu\mu\lambda}$. Then we obtain

$$(3.10)\qquad \underset{v}{\pounds}\nabla_\omega P_{\nu\mu\lambda} = -(\underset{v}{\pounds}P_{\omega\rho})P_{\nu\mu\lambda}^{\cdots\rho} - (\nabla_\omega P_{\nu\mu\lambda}^{\cdots\rho})p_\rho$$
$$- 3p_\omega P_{\nu\mu\lambda} - P_{\omega\mu\lambda}p_\nu - P_{\nu\omega\lambda}p_\mu - P_{\nu\mu\omega}p_\lambda.$$

This procedure can be continued as far as we wish. Thus we have

THEOREM 3.1. *In order that A_n admit a group of projective motions, it is necessary and sufficient that the equations* (3.5), (3.7), (3.9), (3.10) *and all equations of this kind obtained by further differentiations be algebraically compatible with respect to $v^\varkappa$, $\nabla_\lambda v^\varkappa$ and p_λ. If there are exactly $n^2 + 2n - r$ linearly independent equations among them, then the space admits an r-parameter complete group of projective motions.*

In order that the equations (3.4) be completely integrable, it is necessary and sufficient that the equations

$$(3.11)\qquad \underset{v}{\pounds}P_{\nu\mu\lambda}^{\cdots\varkappa} = v^\sigma\nabla_\sigma P_{\nu\mu\lambda}^{\cdots\varkappa} - P_{\nu\mu\lambda}^{\cdots\rho}\nabla_\rho v^\varkappa + P_{\sigma\mu\lambda}^{\cdots\varkappa}\nabla_\nu v^\sigma + P_{\nu\sigma\lambda}^{\cdots\varkappa}\nabla_\mu v^\sigma$$
$$+ P_{\nu\mu\sigma}^{\cdots\varkappa}\nabla_\lambda v^\sigma = 0$$

and

$$(3.12)\qquad \underset{v}{\pounds}P_{\nu\mu\lambda} = v^\sigma\nabla_\sigma P_{\nu\mu\lambda} + P_{\sigma\mu\lambda}\nabla_\nu v^\sigma + P_{\nu\sigma\lambda}\nabla_\mu v^\sigma + P_{\nu\mu\sigma}\nabla_\lambda v^\sigma$$
$$= - P_{\nu\mu\lambda}^{\cdots\varkappa}p_\varkappa,$$

be identically satisfied by any $v^\varkappa$, $\nabla_\lambda v^\varkappa$ and p_λ. Hence we have

$$(3.13)\qquad P_{\nu\mu\lambda}^{\cdots\varkappa} = 0,\quad P_{\nu\mu\lambda} = 0,$$

which shows that the space is a D_n. Thus we have

THEOREM 3.2. *In order that an A_n admit a group of projective motions of the maximum order $n^2 + 2n$, it is necessary and sufficient that the A_n be a D_n.*

§ 4. A group as group of projective motions.

We apply now Theorems 3.1, 3.2 and 3.3 of Ch. III to the case of groups of projective motions.

We consider an G_r in a X_n and we first suppose that the rank of $\underset{b}{v}^\varkappa$ in a neighbourhood is $r \leq n$. We choose a coordinate system with respect to which we have (3.2) of Ch. III. Then the equations, which determine a projective connexion $\overset{p}{\Gamma}{}^{\kappa}_{\mu\lambda}$, are

$$(4.1)\quad \underset{a}{\pounds}\overset{p}{\Gamma}{}^{\kappa}_{\mu\lambda} = \partial_\mu \partial_\lambda v^\kappa + \underset{a}{v}{}^\alpha \partial_\alpha \overset{p}{\Gamma}{}^{\kappa}_{\mu\lambda} - \overset{p}{\Gamma}{}^{\rho}_{\mu\lambda} \partial_\rho \underset{a}{v}{}^\kappa + \overset{p}{\Gamma}{}^{\kappa}_{\beta\lambda} \partial_\mu \underset{a}{v}{}^\beta - \overset{p}{\Gamma}{}^{\kappa}_{\mu\beta} \partial_\lambda \underset{a}{v}{}^\beta - \frac{2}{n+1} A^{\kappa}_{(\mu} \partial_{\lambda)} \partial_\alpha v^\alpha = 0$$

$$(4.2)\qquad \overset{p}{\Gamma}{}^{\kappa}_{[\mu\lambda]} = 0, \quad \overset{p}{\Gamma}{}^{\rho}_{\mu\rho} = 0,$$

and consequently, defining the functions $\Theta_{\alpha\mu\lambda}^{\ \ \ \ \kappa}(\overset{p}{\Gamma}, \xi)$ by

$$(4.3)\quad \underset{a}{v}{}^\alpha \Theta_{\alpha\mu\lambda}^{\ \ \ \ \kappa}(\overset{p}{\Gamma}, \xi) \overset{\text{def}}{=} - \partial_\mu \partial_\lambda v^\kappa + \overset{p}{\Gamma}{}^{\rho}_{\mu\lambda} \partial_\rho \underset{a}{v}{}^\kappa - \overset{p}{\Gamma}{}^{\kappa}_{\beta\lambda} \partial_\mu \underset{a}{v}{}^\beta - \overset{p}{\Gamma}{}^{\kappa}_{\mu\beta} \partial_\lambda \underset{a}{v}{}^\beta + \frac{2}{n+1} A^{\kappa}_{(\mu} \partial_{\lambda)} \partial_\alpha \underset{a}{v}{}^\alpha,$$

we obtain

$$(4.4)\qquad \underset{a}{\pounds}\overset{p}{\Gamma}{}^{\kappa}_{\mu\lambda} = \underset{a}{v}{}^\alpha [\partial_\alpha \overset{p}{\Gamma}{}^{\kappa}_{\mu\lambda} - \Theta_{\alpha\mu\lambda}^{\ \ \ \ \kappa}(\overset{p}{\Gamma}, \xi)] = 0$$

from which

$$(4.5)\qquad \partial_\alpha \overset{p}{\Gamma}{}^{\kappa}_{\mu\lambda} = \Theta_{\alpha\mu\lambda}^{\ \ \ \ \kappa}(\overset{p}{\Gamma}, \xi), \; \overset{p}{\Gamma}{}^{\kappa}_{[\mu\lambda]} = 0, \quad \overset{p}{\Gamma}{}^{\rho}_{\mu\rho} = 0.$$

By the same method as in § 3 of Ch. III, we can prove

$$(4.6)\qquad \Theta_{\beta\tau\sigma}^{\ \ \ \ \rho} \frac{\partial \Theta_{\alpha\mu\lambda}^{\ \ \ \ \kappa}}{\partial \overset{p}{\Gamma}{}^{\rho}_{\tau\sigma}} + \partial_\beta \Theta_{\alpha\mu\lambda}^{\ \ \ \ \kappa} = \Theta_{\alpha\tau\rho}^{\ \ \ \ \rho} \frac{\partial \Theta_{\beta\mu\lambda}^{\ \ \ \ \kappa}}{\partial \overset{p}{\Gamma}{}^{\rho}_{\tau\sigma}} + \partial_\alpha \Theta_{\beta\mu\lambda}^{\ \ \ \ \kappa}.$$

Moreover, we can easily see that the $\Theta_{\alpha\mu\lambda}^{\ \ \ \ \kappa}(\overset{p}{\Gamma}, \xi)$ satisfy the equations

$$(4.7)\qquad \begin{cases} \underset{a}{v}{}^\alpha \Theta_{\alpha[\mu\lambda]}^{\ \ \ \ \ \ \kappa} = \overset{p}{\Gamma}{}_{[\mu\lambda]}^{\ \ \ \ \rho} \partial_\rho \underset{a}{v}{}^\kappa \\ \underset{a}{v}{}^\alpha \Theta_{\alpha\mu\rho}^{\ \ \ \ \rho} = - \overset{p}{\Gamma}{}^{\rho}_{\alpha\rho} \partial_\mu \underset{a}{v}{}^\alpha. \end{cases}$$

The equations (4.6) and (4.7) show that the mixed system of partial differential equations (4.5) is completely integrable. Hence we have

THEOREM 5.1. *A G_r in an X_n such that the rank of $\underset{b}{v}{}^\kappa$ in a neighbourhood is $r \leq n$ can be regarded as a group of projective motions in an A_n whose components of projective connexion can contain $\frac{1}{2}n^2(n+1) - n$ arbitrary functions or constants.*

We next consider a G_r in an X_n such that the rank of $\underset{b}{v}{}^\kappa$ in a neighbourhood is $q < r, n$. We choose a coordinate system with respect to which (3.9) of Ch. III holds.

Then the equations, which determine a projective connexion $\overset{p}{\Gamma}{}^{\varkappa}_{\mu\lambda}$, are

$$
(4.8)\quad \begin{cases}
\underset{i}{\pounds}\overset{p}{\Gamma}{}^{\varkappa}_{\mu\lambda} = \partial_\mu \partial_\lambda \underset{i}{v}{}^{\varkappa} + \underset{i}{v}{}^{\alpha}\partial_\alpha \overset{p}{\Gamma}{}^{\varkappa}_{\mu\lambda} - \overset{p}{\Gamma}{}^{\rho}_{\mu\lambda}\partial_\rho \underset{i}{v}{}^{\varkappa} + \overset{p}{\Gamma}{}^{\varkappa}_{\alpha\lambda}\partial_\mu \underset{i}{v}{}^{\alpha} + \overset{p}{\Gamma}{}^{\varkappa}_{\mu\alpha}\partial_\lambda \underset{i}{v}{}^{\alpha} \\
\qquad\qquad - \dfrac{2}{n+1} A^{\ \varkappa}_{(\mu}\partial_{\lambda)}\partial_\alpha \underset{i}{v}{}^{\alpha} \\
\underset{u}{\pounds}\overset{p}{\Gamma}{}^{\varkappa}_{\mu\lambda} = \varphi^i_u \underset{i}{\pounds}\overset{p}{\Gamma}{}^{\varkappa}_{\mu\lambda} + (\partial_\mu\partial_\lambda \varphi^i_u)\underset{i}{v}{}^{\varkappa} + 2(\partial_{(\mu}\varphi^i_{|u|})(\partial_{\lambda)}\underset{i}{v}{}^{\varkappa}) \\
\quad - \overset{p}{\Gamma}{}^{\rho}_{\mu\lambda}(\partial_\rho \varphi^i_u)\underset{i}{v}{}^{\varkappa} + \overset{p}{\Gamma}{}^{\varkappa}_{\alpha\lambda}(\partial_\mu \varphi^i_u)\underset{i}{v}{}^{\alpha} + \overset{p}{\Gamma}{}^{\varkappa}_{\mu\alpha}(\partial_\lambda \varphi^i_u)\underset{i}{v}{}^{\alpha} \\
\quad - \dfrac{2}{n+1} A^{\varkappa}_{(\mu}\{(\partial_{\lambda)}\partial_\alpha \varphi^i_u)\underset{i}{v}{}^{\alpha} + (\partial_{\lambda)}\varphi^i_u)(\partial_\alpha \underset{i}{v}{}^{\alpha}) + (\partial_{|\alpha}\varphi^i_{u|})(\partial_{\lambda)}\underset{i}{v}{}^{\alpha})\} = 0 \\
\qquad\qquad \overset{p}{\Gamma}{}^{\varkappa}_{[\mu\lambda]} = 0, \quad \overset{p}{\Gamma}{}^{\rho}_{\mu\rho} = 0.
\end{cases}
$$

If we put

$$
\begin{aligned}
\underset{i}{v}{}^{\alpha}\Theta_{\alpha\mu\lambda}^{\ \ \varkappa}(\overset{p}{\Gamma}, \xi) \overset{\text{def}}{=} & - \partial_\mu\partial_\lambda \underset{i}{v}{}^{\varkappa} + \overset{p}{\Gamma}{}^{\rho}_{\mu\lambda}\partial_\rho \underset{i}{v}{}^{\varkappa} - \overset{p}{\Gamma}{}^{\varkappa}_{\alpha\lambda}\partial_\mu \underset{i}{v}{}^{\alpha} - \overset{p}{\Gamma}{}^{\varkappa}_{\mu\alpha}\partial_\lambda \underset{i}{v}{}^{\alpha} \\
& + \frac{2}{n+1} A^{\varkappa}_{(\mu}\partial_{\lambda)}\partial_\alpha \underset{i}{v}{}^{\alpha}.
\end{aligned}
$$

$$
\begin{aligned}
\Xi_{u\mu\lambda}^{\ \ \varkappa}(\overset{p}{\Gamma}, \xi) \overset{\text{def}}{=} & (\partial_\mu\partial_\lambda \varphi^i_u)\underset{i}{v}{}^{\varkappa} + 2(\partial_{(\mu}\varphi^i_{|u|})(\partial_{\lambda)}\underset{i}{v}{}^{\varkappa}) \\
& - \overset{p}{\Gamma}{}^{\rho}_{\mu\lambda}(\partial_\rho \varphi^i_u)\underset{i}{v}{}^{\varkappa} + \overset{p}{\Gamma}{}^{\varkappa}_{\alpha\lambda}(\partial_\mu \varphi^i_u)\underset{i}{v}{}^{\alpha} + \overset{p}{\Gamma}{}^{\varkappa}_{\mu\alpha}(\partial_\lambda \varphi^i_u)\underset{i}{v}{}^{\alpha} \\
& - \frac{2}{n+1} A^{\varkappa}_{(\mu}\{(\partial_{\lambda)}\partial_\alpha \varphi^i_u)\underset{i}{v}{}^{\alpha} + (\partial_{\lambda)}\varphi^i_u)(\partial_\alpha \underset{i}{v}{}^{\alpha}) + (\partial_{|\alpha}\varphi^i_{u|})(\partial_{\lambda)}\underset{i}{v}{}^{\alpha}),
\end{aligned}
$$

we can write (4.8) in the form

$$
\begin{cases}
\underset{i}{\pounds}\overset{p}{\Gamma}{}^{\varkappa}_{\mu\lambda} = \underset{i}{v}{}^{\alpha}[\partial_\alpha \overset{p}{\Gamma}{}^{\varkappa}_{\mu\lambda} - \Theta_{\alpha\mu\lambda}^{\ \ \varkappa}(\overset{p}{\Gamma}, \xi)] = 0, \\
\underset{u}{\pounds}\overset{p}{\Gamma}{}^{\varkappa}_{\mu\lambda} = \varphi^i_u \underset{i}{\pounds}\overset{p}{\Gamma}{}^{\varkappa}_{\mu\lambda} + \Xi_{u\mu\lambda}^{\ \ \varkappa}(\overset{p}{\Gamma}, \xi) = 0, \\
\qquad \overset{p}{\Gamma}{}^{\varkappa}_{[\mu\lambda]} = 0, \quad \overset{p}{\Gamma}{}^{\rho}_{\mu\rho} = 0,
\end{cases}
$$

or

$$(4.9)\qquad \begin{cases} \partial_\alpha \overset{p}{\Gamma}{}^{\varkappa}_{\mu\lambda} = \Theta_{\alpha\mu\lambda}^{\ \varkappa}(\overset{p}{\Gamma}, \xi), \\ \Xi_{u\mu\lambda}^{\ \varkappa}(\overset{p}{\Gamma}, \xi) = 0, \quad \overset{p}{\Gamma}{}^{\varkappa}_{[\mu\lambda]} = 0, \quad \overset{p}{\Gamma}{}^{\rho}_{\mu\rho} = 0. \end{cases}$$

By the same method as that used in § 3 of Ch. III, we can prove that, taking account of the last three equations of (4.9), we have

$$\Theta_{\beta\tau\sigma}^{\ \rho} \frac{\partial \Theta_{\alpha\mu\lambda}^{\ \varkappa}}{\partial \overset{p}{\Gamma}{}^{\rho}_{\tau\sigma}} + \partial_\beta \Theta_{\alpha\mu\lambda}^{\ \varkappa} = \Theta_{\alpha\tau\sigma}^{\ \rho} \frac{\partial \Theta_{\beta\mu\lambda}^{\ \varkappa}}{\partial \overset{p}{\Gamma}{}^{\rho}_{\tau\sigma}} + \partial_\alpha \Theta_{\beta\mu\lambda}^{\ \varkappa},$$

$$\Theta_{\alpha\tau\sigma}^{\ \rho} \frac{\partial \Xi_{u\mu\lambda}^{\ \varkappa}}{\partial \overset{p}{\Gamma}{}^{\rho}_{\tau\sigma}} + \partial_\alpha \Xi_{u\mu\lambda}^{\ \varkappa} = 0, \quad \Theta_{\alpha[\mu\lambda]}^{\ \varkappa} = 0, \quad \Theta_{\alpha\mu\rho}^{\ \rho} = 0,$$

which shows that the mixed system (4.9) is completely integrable. Thus we have

THEOREM 4.2. *Consider a G_r in an X_n such that the rank of $\underset{b}{v}^{\varkappa}$ in a neighbourhood is $q < r, n$. If, in the neighbourhood such that* (3.9) *of Ch.* III *holds, the equations $\Xi_{u\mu\lambda}^{\ \varkappa}(\overset{p}{\Gamma}, \xi) = 0$, $\overset{p}{\Gamma}{}^{\varkappa}_{[\mu\lambda]} = 0$, $\overset{p}{\Gamma}{}^{\rho}_{\mu\rho} = 0$ are compatible in $\overset{p}{\Gamma}{}^{\varkappa}_{\mu\lambda}$, then the group can be regarded as a group of projective motions in an A_n.*

A similar theorem holds for a multiply transitive group.

§ 5. The maximum order of a group of projevtice motions in an A_n with non vanishing projective curvature.

I. P. Egorov [1] and G. Vranceanu [2] have proved the following important

THEOREM 5.1. *If an A_n admits a group of projective motions of order greater than $n^2 - 2n + 5$, then the A_n is a P_n. An A_n admitting a group of projective motions of order $n^2 - 2n + 5$ exists for any n and the group is transitive in this case.*

We shall prove this theorem. We know that the integrability conditions

[1] EGOROV [6].
[2] VRANCEANU [3, 4].

of the equations $\underset{u}{\pounds}\Gamma^{\varkappa}_{\mu\lambda} = 2p_{(\mu}A^{\varkappa}_{\lambda)}$ are

$$(5.1)\qquad \begin{cases} \underset{u}{\pounds}P_{\nu\mu\lambda}^{\cdot\cdot\cdot\varkappa} = 0, \quad \underset{u}{\pounds}P_{\nu\mu\lambda} = -P_{\nu\mu\lambda}^{\cdot\cdot\cdot\varkappa}p_{\varkappa}, \\ \underset{u}{\pounds}\nabla_\omega P_{\nu\mu\lambda}^{\cdot\cdot\cdot\varkappa} = -2p_\omega P_{\nu\mu\lambda}^{\cdot\cdot\cdot\varkappa} + A^{\varkappa}_{\omega}P_{\nu\mu\lambda}^{\cdot\cdot\cdot\rho}p_\rho \\ \qquad\qquad - P_{\omega\mu\lambda}^{\cdot\cdot\cdot\varkappa}p_\nu - P_{\nu\omega\lambda}^{\cdot\cdot\cdot\varkappa}p_\mu - P_{\nu\mu\omega}^{\cdot\cdot\cdot\varkappa}p_\lambda, \\ \cdots\cdots\cdots \end{cases}$$

We consider the first equations of (5.1):

$$(5.2)\qquad \underset{v}{\pounds}P_{\nu\mu\lambda}^{\cdot\cdot\cdot\varkappa} = v^\sigma\nabla_\sigma P_{\nu\mu\lambda}^{\cdot\cdot\cdot\varkappa} - P_{\nu\mu\lambda}^{\cdot\cdot\cdot\rho}\nabla_\rho v^\varkappa + P_{\sigma\mu\lambda}^{\cdot\cdot\cdot\varkappa}\nabla_\nu v^\sigma + P_{\nu\sigma\lambda}^{\cdot\cdot\cdot\varkappa}\nabla_\mu v^\sigma + P_{\nu\mu\sigma}^{\cdot\cdot\cdot\varkappa}\nabla_\lambda v^\sigma = 0.$$

In these equations the coefficients of v^σ are given by $\nabla_\sigma P_{\nu\mu\lambda}^{\cdot\cdot\cdot\varkappa} = 0$ and those of $\nabla_\rho v^\sigma$ by

$$(5.3)\qquad S_{\nu\mu\lambda\cdot\sigma}^{\cdot\cdot\cdot\varkappa\cdot\rho} \overset{\text{def}}{=} A^\rho_\nu P_{\sigma\mu\lambda}^{\cdot\cdot\cdot\varkappa} + A^\rho_\mu P_{\nu\sigma\lambda}^{\cdot\cdot\cdot\varkappa} + A^\rho_\lambda P_{\nu\mu\sigma}^{\cdot\cdot\cdot\varkappa} - A^\varkappa_\sigma P_{\nu\mu\lambda}^{\cdot\cdot\cdot\rho}.$$

It should be noticed that the equations (5.2) do not contain p_λ.

We next consider the third equation of (5.1):

$$(5.4)\qquad \underset{v}{\pounds}\nabla_\omega P_{\nu\mu\lambda}^{\cdot\cdot\cdot\varkappa} + 2p_\omega P_{\nu\mu\lambda}^{\cdot\cdot\cdot\varkappa} - A^\varkappa_\omega P_{\nu\mu\lambda}^{\cdot\cdot\cdot\rho}p_\rho + P_{\omega\mu\lambda}^{\cdot\cdot\cdot\varkappa}p_\nu + P_{\nu\omega\lambda}^{\cdot\cdot\cdot\varkappa}p_\mu + P_{\nu\mu\omega}^{\cdot\cdot\cdot\varkappa}p_\lambda = 0.$$

In this equation the coefficients of p_ρ are given by

$$(5.5)\qquad U_{\omega\nu\mu\lambda}^{\cdot\cdot\cdot\cdot\varkappa\rho} \overset{\text{def}}{=} 2A^\rho_\omega P_{\nu\mu\lambda}^{\cdot\cdot\cdot\varkappa} + S_{\nu\mu\lambda\cdot\omega}^{\cdot\cdot\cdot\varkappa\cdot\rho}.$$

Thus denoting by T the matrix formed by the coefficients of $\nabla_\rho v^\sigma$ and p_ρ in the equations (5.2) and (5.4), we have

$$(5.6)\qquad T = \begin{pmatrix} S & 0 \\ * & U \end{pmatrix},$$

where S consists of n^2 columns and U of n columns.

In order to prove the first part of Theorem 5.1, we have only to prove that if the rank of the matrix T is less than $4n - 5$ [$= (n^2 + 2n) - (n^2 - 2n + 5)$], all the components of the projective curvature tensor $P_{\nu\mu\lambda}^{\cdot\cdot\cdot\varkappa}$ vanish.

We make a frequent use of the relations

$$P_{(\nu\mu)\lambda}^{\cdot\cdot\cdot\varkappa} = 0, \quad P_{[\nu\mu\lambda]}^{\cdot\cdot\cdot\varkappa} = 0, \quad P_{\varkappa\mu\lambda}^{\cdot\cdot\cdot\varkappa} = 0.$$

We shall prove a series of lemmas.

LEMMA 1. *If the rank of T is less than $4n-5$, then*

$$P^{\cdot\ \cdot\ \cdot\alpha_1}_{\alpha_3\alpha_2\alpha_2}=0. \tag{5.7}$$

We pick up the following $(3n-5)$-rowed square submatrix from S:

κ \ ρ $\nu\mu\lambda$ \ σ	α_1 α_j	α_p α_3	α_p α_2	α_3 α_2	α_1 α_1
α_i $\alpha_3\alpha_2\alpha_2$	$-\delta^i_j P^{\cdot\ \cdot\ \cdot\alpha_1}_{\alpha_3\alpha_2\alpha_2}$	*	*	*	*
α_1 $\alpha_q\alpha_2\alpha_2$	0	$\delta^p_q P^{\cdot\ \cdot\ \cdot\alpha_1}_{\alpha_3\alpha_2\alpha_2}$	*	*	*
α_1 $\alpha_q\alpha_3\alpha_2$	0	0	$-\delta^p_q P^{\cdot\ \cdot\ \cdot\alpha_1}_{\alpha_3\alpha_2\alpha_2}$	*	*
α_1 $\alpha_3\alpha_2\alpha_3$	0	0	0	$P^{\cdot\ \cdot\ \cdot\alpha_1}_{\alpha_3\alpha_2\alpha_2}$	*
α_1 $\alpha_3\alpha_2\alpha_2$	0	0	0	0	$-P^{\cdot\ \cdot\ \cdot\alpha_1}_{\alpha_3\alpha_2\alpha_2}$

and the following n-rowed square submatrix from U:

κ \ ρ $\omega\nu\mu\lambda$	α_2	α_3	α_p	α_1
α_1 $\alpha_2\alpha_3\alpha_2\alpha_2$	$4P^{\cdot\ \cdot\ \cdot\alpha_1}_{\alpha_3\alpha_2\alpha_2}$	0	0	0
α_1 $\alpha_3\alpha_3\alpha_2\alpha_2$	*	$3P^{\cdot\ \cdot\ \cdot\alpha_1}_{\alpha_3\alpha_2\alpha_2}$	0	0
α_1 $\alpha_q\alpha_3\alpha_2\alpha_2$	*	*	$2\delta^p_q P^{\cdot\ \cdot\ \cdot\alpha_1}_{\alpha_3\alpha_2\alpha_2}$	0
α_1 $\alpha_1\alpha_3\alpha_2\alpha_2$	*	*	*	$P^{\cdot\ \cdot\ \cdot\alpha_1}_{\alpha_3\alpha_2\alpha_2}$

$$2\leq i,j\leq n;\quad 3\leq k,l\leq n;\quad 4\leq p,q\leq n.$$

Since the rank of T is less than $4n - 5$, we conclude (5.7).

LEMMA 2. *If the rank of T is less than $4n - 5$, then*

$$P_{\alpha_2\alpha_1\alpha_1}^{\cdot\ \cdot\ \cdot\alpha_2} = 0. \tag{5.8}$$

Suppose that the $P_{\alpha_2\alpha_1\alpha_1}^{\cdot\ \cdot\ \cdot\alpha_2}$ were not zero. Since $P_{\varkappa\alpha_1\alpha_1}^{\cdot\ \cdot\ \cdot\varkappa} = 0$, there exists a subindex $k' \geq 3$ such that $P_{\alpha_2\alpha_1\alpha_1}^{\cdot\ \cdot\ \cdot\alpha_2} \neq P_{\alpha_{k'}\alpha_1\alpha_1}^{\cdot\ \cdot\ \cdot\alpha_{k'}}$. We denote the subindices satisfying this inequality by k' and l' and the other subindices satisfying the equality $P_{\alpha_2\alpha_1\alpha_1}^{\cdot\ \cdot\ \cdot\alpha_2} = P_{\alpha_a\alpha_1\alpha_1}^{\cdot\ \cdot\ \cdot\alpha_a}$ by a and $b \geq 2$. The number of the subindices such as a and b is denoted by λ. Then we have

$$1 \leq \lambda \leq n - 2. \tag{5.9}$$

Now, taking account of Lemma 1, we form the following square submatrix of S:

$\varkappa$ \ ρ ; $\nu\mu\lambda$ \ σ	α_a ; α_1	$\alpha_{k'}$; α_1	α_a ; $\alpha_{k'}$	$\alpha_{k'}$; α_a
α_1 ; $\alpha_b\alpha_1\alpha_1$	$-\delta_b^a P_{\alpha_a\alpha_1\alpha_1}^{\cdot\ \cdot\ \cdot\alpha_a}$	$*$	$*$	$*$
α_2 ; $\alpha_2\alpha_{l'}\alpha_1$	0	$\delta_{l'}^{k'} P_{\alpha_2\alpha_1\alpha_1}^{\cdot\ \cdot\ \cdot\alpha_2}$	$*$	$*$
$\alpha_{l'}$; $\alpha_b\alpha_1\alpha_1$	0	0	$\delta_b^a \delta_{k'}^{l'}(P_{\alpha_{k'}\alpha_1\alpha_1}^{\cdot\ \cdot\ \cdot\alpha_{k'}} - P_{\alpha_a\alpha_1\alpha_1}^{\cdot\ \cdot\ \cdot\alpha_a})$	$*$
α_b ; $\alpha_{l'}\alpha_1\alpha_1$	0	0	0	$\delta_a^b \delta_{l'}^{k'}(P_{\alpha_a\alpha_1\alpha_1}^{\cdot\ \cdot\ \cdot\alpha_a} - P_{\alpha_{k'}\alpha_1\alpha_1}^{\cdot\ \cdot\ \cdot\alpha_{k'}})$

The number of rows of this square matrix is

$$\lambda + (n - 1 - \lambda) + \lambda(n - 1 - \lambda) + \lambda(n - 1 - \lambda) = (2\lambda + 1)(n - 1) - 2\lambda^2.$$

Since

$$(2\lambda + 1)(n - 1) - 2\lambda^2 \geq 3n - 5,$$

the number of rows of this square matrix is greater than or equal to $3n - 5$.

Taking account of Lemma 1, we form the following n-rowed square submatrix of U:

$\overset{\rho}{\underset{\omega\nu\mu\lambda}{\varkappa}}$	α_k	α_2	α_1
$\overset{\alpha_2}{\alpha_l\alpha_2\alpha_1\alpha_1}$	$2\delta_l^k P_{\alpha_2\alpha_1\alpha_1}^{\cdot\ \cdot\ \cdot\alpha_2}$	$*$	$*$
$\overset{\alpha_2}{\alpha_2\alpha_2\alpha_1\alpha_1}$	0	$2P_{\alpha_2\alpha_1\alpha_1}^{\cdot\ \cdot\ \cdot\alpha_2}$	$*$
$\overset{\alpha_2}{\alpha_1\alpha_2\alpha_1\alpha_1}$	0	0	$4P_{\alpha_2\alpha_1\alpha_1}^{\cdot\ \cdot\ \cdot\alpha_2}$

$$3 \leq k,\ l \leq n.$$

Since the rank of the submatrix of T containing the last two matrices is greater than $3n - 5 + n = 4n - 5$, this is a contradiction. Hence we must have (5.8).

LEMMA 3. *If the rank of T is less than $4n - 5$, then*

$$P_{\alpha_4\alpha_3\alpha_2}^{\cdot\ \cdot\ \cdot\alpha_1} = 0. \tag{5.9}$$

Taking account of Lemmas 1 and 2, we form the following $(4n - 9)$-rowed square submatrix of S:

$\overset{\rho\ \sigma}{\underset{\nu\mu\lambda}{\varkappa}}$	$\alpha_1\ \alpha_1$	$\alpha_2\ \alpha_3$	$\alpha_2\ \alpha_4$	$\alpha_3\ \alpha_2$	$\alpha_r\ \alpha_2$	$\alpha_r\ \alpha_3$	$\alpha_r\ \alpha_4$	$\alpha_1\ \alpha_j$
$\overset{\alpha_1}{\alpha_4\alpha_3\alpha_2}$	$-P_{\alpha_4\alpha_3\alpha_2}^{\cdot\ \cdot\ \cdot\alpha_1}$	0	0	0	0	0	0	0
$\overset{\alpha_1}{\alpha_4\alpha_2\alpha_2}$	$*$	$P_{\alpha_4\alpha_3\alpha_2}^{\cdot\ \cdot\ \cdot\alpha_1}+P_{\alpha_4\alpha_2\alpha_3}^{\cdot\ \cdot\ \cdot\alpha_1}$	0	0	0	0	0	0
$\overset{\alpha_1}{\alpha_2\alpha_3\alpha_2}$	$*$	$*$	$P_{\alpha_4\alpha_3\alpha_2}^{\cdot\ \cdot\ \cdot\alpha_1}+P_{\alpha_2\alpha_3\alpha_4}^{\cdot\ \cdot\ \cdot\alpha_1}$	0	0	0	0	0
$\overset{\alpha_1}{\alpha_4\alpha_3\alpha_3}$	$*$	$*$	$*$	$P_{\alpha_4\alpha_2\alpha_3}^{\cdot\ \cdot\ \cdot\alpha_1}+P_{\alpha_4\alpha_3\alpha_2}^{\cdot\ \cdot\ \cdot\alpha_1}$	0	0	0	0
$\overset{\alpha_1}{\alpha_4\alpha_3\alpha_s}$	$*$	$*$	$*$	$*$	$\delta_s^r P_{\alpha_4\alpha_3\alpha_2}^{\cdot\ \cdot\ \cdot\alpha_1}$	0	0	0
$\overset{\alpha_1}{\alpha_4\alpha_s\alpha_2}$	$*$	$*$	$*$	$*$	$*$	$\delta_s^r P_{\alpha_4\alpha_3\alpha_2}^{\cdot\ \cdot\ \cdot\alpha_1}$	0	0
$\overset{\alpha_1}{\alpha_s\alpha_3\alpha_2}$	$*$	$*$	$*$	$*$	$*$	$*$	$\delta_s^r P_{\alpha_4\alpha_3\alpha_2}^{\cdot\ \cdot\ \cdot\alpha_1}$	0
$\overset{\alpha_i}{\alpha_4\alpha_3\alpha_2}$	$*$	$*$	$*$	$*$	$*$	$*$	$*$	$\delta_j^i P_{\alpha_4\alpha_3\alpha_2}^{\cdot\ \cdot\ \cdot\alpha_1}$

$$2 \leq i,\ j \leq n; \qquad 5 \leq r,\ s \leq n.$$

and also the following n-rowed square submatrix of U:

ρ / $\overset{\kappa}{\omega\nu\mu\lambda}$	α_2	α_3	α_4	α_r	α_1
$\overset{\alpha_1}{\alpha_3\alpha_4\alpha_2\alpha_2}$	$P^{\cdot\cdot\cdot\alpha_1}_{\alpha_4\alpha_3\alpha_2} + P^{\cdot\cdot\cdot\alpha_1}_{\alpha_4\alpha_2\alpha_3}$	0	0	0	0
$\overset{\alpha_1}{\alpha_2\alpha_4\alpha_3\alpha_3}$	*	$P^{\cdot\cdot\cdot\alpha_1}_{\alpha_4\alpha_2\alpha_3} + P^{\cdot\cdot\cdot\alpha_1}_{\alpha_4\alpha_3\alpha_2}$	0	0	0
$\overset{\alpha_1}{\alpha_2\alpha_4\alpha_3\alpha_4}$	*	*	$P^{\cdot\cdot\cdot\alpha_1}_{\alpha_2\alpha_3\alpha_4} + P^{\cdot\cdot\cdot\alpha_1}_{\alpha_4\alpha_3\alpha_2}$	0	0
$\overset{\alpha_1}{\alpha_2\alpha_4\alpha_3\alpha_s}$	*	*	*	$\delta^r_s P^{\cdot\cdot\cdot\alpha_1}_{\alpha_4\alpha_3\alpha_2}$	0
$\overset{\alpha_1}{\alpha_1\alpha_4\alpha_3\alpha_2}$	*	*	*	*	$P^{\cdot\cdot\cdot\alpha_1}_{\alpha_4\alpha_3\alpha_2}$

$$5 \leq r, \quad s \leq n.$$

Since the number of rows of the smallest square submatrix of T containing the last two submatrices is

$$(4n - 9) + n = (4n - 5) + (n - 4) \geq 4n - 5,$$

we must have

$$P^{\cdot\cdot\cdot\alpha_1}_{\alpha_4\alpha_3\alpha_2}(P^{\cdot\cdot\cdot\alpha_1}_{\alpha_4\alpha_3\alpha_2} + P^{\cdot\cdot\cdot\alpha_1}_{\alpha_4\alpha_2\alpha_3})(P^{\cdot\cdot\cdot\alpha_1}_{\alpha_4\alpha_3\alpha_2} + P^{\cdot\cdot\cdot\alpha_1}_{\alpha_2\alpha_3\alpha_4}) = 0,$$

from which

$$P^{\cdot\cdot\cdot\alpha_1}_{\alpha_4\alpha_3\alpha_2}[2(P^{\cdot\cdot\cdot\alpha_1}_{\alpha_4\alpha_3\alpha_2})^2 + P^{\cdot\cdot\cdot\alpha_1}_{\alpha_4\alpha_2\alpha_3}P^{\cdot\cdot\cdot\alpha_1}_{\alpha_2\alpha_3\alpha_4}] = 0,$$

$$2(P^{\cdot\cdot\cdot\alpha_1}_{\alpha_4\alpha_3\alpha_2})^3 = -P^{\cdot\cdot\cdot\alpha_1}_{\alpha_4\alpha_3\alpha_2}P^{\cdot\cdot\cdot\alpha_1}_{\alpha_3\alpha_2\alpha_4}P^{\cdot\cdot\cdot\alpha_1}_{\alpha_2\alpha_4\alpha_3}.$$

The last equation shows

$$P^{\cdot\cdot\cdot\alpha_1}_{\alpha_4\alpha_3\alpha_2} = P^{\cdot\cdot\cdot\alpha_1}_{\alpha_3\alpha_2\alpha_4} = P^{\cdot\cdot\cdot\alpha_1}_{\alpha_2\alpha_4\alpha_3}.$$

But the sum of these three is zero, from which we get (5.9).

LEMMA 4. *If the rank of T is less than $4n - 5$, then*

$$P^{\cdot\cdot\cdot\alpha_3}_{\alpha_3\alpha_2\alpha_1} = 0. \tag{5.10}$$

Suppose that $P_{\alpha_3\alpha_2\alpha_1}^{\cdot\,\cdot\,\cdot\,\alpha_3}$ were not zero. We denote by k and l the subindices satisfying

$$P_{\alpha_3\alpha_2\alpha_1}^{\cdot\,\cdot\,\cdot\,\alpha_3} = P_{\alpha_k\alpha_2\alpha_1}^{\cdot\,\cdot\,\cdot\,\alpha_k}$$

and $3 \leq k,\ l \leq n$, and by p and q the subindices satisfying

$$P_{\alpha_3\alpha_2\alpha_1}^{\cdot\,\cdot\,\cdot\,\alpha_3} = P_{\alpha_p\alpha_2\alpha_1}^{\cdot\,\cdot\,\cdot\,\alpha_p}$$

and $4 \leq p,\ q \leq n$.

We consider the following submatrices of S and U respectively:

S_1:

$\varkappa$ \ ρ $\nu\mu\lambda$ \ σ	α_p α_1	$\alpha_{k'}$ α_1	α_p α_2	α_k α_2
α_3 $\alpha_3\alpha_2\alpha_q$	$\delta_q^p P_{\alpha_3\alpha_2\alpha_1}^{\cdot\,\cdot\,\cdot\,\alpha_2}$	*	*	*
α_3 $\alpha_3\alpha_2\alpha_{l'}$	0	$\delta_{l'}^{k'} P_{\alpha_3\alpha_2\alpha_1}^{\cdot\,\cdot\,\cdot\,\alpha_3}$	*	*
α_3 $\alpha_3\alpha_q\alpha_1$	0	0	$\delta_q^p P_{\alpha_3\alpha_2\alpha_1}^{\cdot\,\cdot\,\cdot\,\alpha_3}$	*
α_2 $\alpha_l\alpha_2\alpha_1$	0	0	0	$\delta_l^k P_{\alpha_k\alpha_2\alpha_1}^{\cdot\,\cdot\,\cdot\,\alpha_k}$

$k',\ l' \neq 3.$

S_2:

$\varkappa$ \ ρ $\nu\mu\lambda$ \ σ	α_p α_l	α_k α_q	α_1 α_1
α_k $\alpha_q\alpha_2\alpha_1$	$\delta_q^p \delta_l^k (P_{\alpha_k\alpha_2\alpha_1}^{\cdot\,\cdot\,\cdot\,\alpha_k} - P_{\alpha_p\alpha_2\alpha_1}^{\cdot\,\cdot\,\cdot\,\alpha_p})$	*	*
α_p $\alpha_l\alpha_2\alpha_1$	0	$\delta_q^p \delta_l^k (P_{\alpha_p\alpha_2\alpha_1}^{\cdot\,\cdot\,\cdot\,\alpha_p} - P_{\alpha_k\alpha_2\alpha_1}^{\cdot\,\cdot\,\cdot\,\alpha_k})$	*
α_3 $\alpha_3\alpha_2\alpha_1$	0	0	$P_{\alpha_3\alpha_2\alpha_1}^{\cdot\,\cdot\,\cdot\,\alpha_3}$

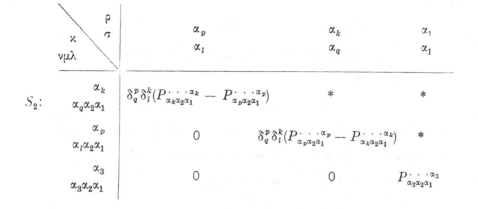

S_3:

$\varkappa$ / $\nu\mu\lambda$ \ ρ / σ	α_2 / α_1	α_1 / α_2
α_3 / $\alpha_3\alpha_2\alpha_2$	$P_{\alpha_3\alpha_1\alpha_2}^{\cdot\;\cdot\;\cdot\;\alpha_3} + P_{\alpha_3\alpha_2\alpha_1}^{\cdot\;\cdot\;\cdot\;\alpha_3}$	0
α_3 / $\alpha_3\alpha_1\alpha_1$	0	$P_{\alpha_3\alpha_2\alpha_1}^{\cdot\;\cdot\;\cdot\;\alpha_3} + P_{\alpha_3\alpha_1\alpha_2}^{\cdot\;\cdot\;\cdot\;\alpha_1}$

S_4:

$\varkappa$ / $\nu\mu\lambda$ \ ρ / σ	α_3 / α_1
α_1 / $\alpha_3\alpha_1\alpha_2$	$-P_{\alpha_3\alpha_1\alpha_2}^{\cdot\;\cdot\;\cdot\;\alpha_3}$

S_5:

$\varkappa$ / $\nu\mu\lambda$ \ ρ / σ	α_2 / α_2	α_1 / α_l
α_3 / $\alpha_3\alpha_2\alpha_1$	$P_{\alpha_3\alpha_2\alpha_1}^{\cdot\;\cdot\;\cdot\;\alpha_3}$	*
α_k / $\alpha_1\alpha_2\alpha_1$	0	$\delta_l^k(P_{\alpha_k\alpha_2\alpha_1}^{\cdot\;\cdot\;\cdot\;\sigma_k} + P_{\alpha_2\alpha_1\alpha_k}^{\cdot\;\cdot\;\cdot\;\alpha_k} - P_{\alpha_1\alpha_2\alpha_1}^{\cdot\;\cdot\;\cdot\;\alpha_1})$

U_1:

$\varkappa$ / $\omega\nu\mu\lambda$ \ ρ	α_1	α_2	α_3	α_r
α_3 / $\alpha_1\alpha_3\alpha_2\alpha_1$	$3P_{\alpha_3\alpha_2\alpha_1}^{\cdot\;\cdot\;\cdot\;\alpha_3}$	0	0	0
α_3 / $\alpha_2\alpha_3\alpha_2\alpha_1$	*	$3P_{\alpha_3\alpha_2\alpha_1}^{\cdot\;\cdot\;\cdot\;\alpha_3}$	0	0
α_2 / $\alpha_2\alpha_3\alpha_2\alpha_1$	*	*	$-P_{\alpha_3\alpha_2\alpha_1}^{\cdot\;\cdot\;\cdot\;\alpha_3}$	0
α_3 / $\alpha_1\alpha_3\alpha_2\alpha_s$	*	*	*	$\delta_s^r P_{\alpha_3\alpha_2\alpha_1}^{\cdot\;\cdot\;\cdot\;\alpha_3}$

$$4 \leq r,\ s \leq n.$$

Now if we denote the number of indices k such that $P_{\alpha_3\alpha_2\alpha_1}^{\cdot\,\cdot\,\cdot\alpha_3} = P_{\alpha_k\alpha_2\alpha_1}^{\cdot\,\cdot\,\cdot\alpha_k}$, by λ, then it is clear that $1 \leq \lambda \leq n-2$. We first assume $4 \leq n$ and consider

I. The case $1 \leq \lambda < n-2$.

Then the order of the square matrix

$$\begin{pmatrix} S_1 & * & * & * \\ 0 & S_2 & * & * \\ 0 & 0 & S_3 & * \\ 0 & 0 & 0 & U_1 \end{pmatrix}$$

is

$$[2(n-2)-1] + [2\lambda(n-2-\lambda)+1] + 2 + n > 4n-5.^{1}$$

Since the rank of T is less than $4n-5$, we should have

$$P_{\alpha_3\alpha_2\alpha_1}^{\cdot\,\cdot\,\cdot\alpha_3} + P_{\alpha_3\alpha_1\alpha_2}^{\cdot\,\cdot\,\cdot\alpha_3} = 0$$

from which

$$P_{\alpha_3\alpha_1\alpha_2}^{\cdot\,\cdot\,\cdot\alpha_3} = -P_{\alpha_3\alpha_2\alpha_1}^{\cdot\,\cdot\,\cdot\alpha_3} \neq 0.$$

Then the determinant

$$\begin{vmatrix} S_1 & * & * & * \\ 0 & S_2 & * & * \\ 0 & 0 & S_4 & * \\ 0 & 0 & 0 & U_1 \end{vmatrix}$$

is of order

$$2(n-2) - 1 + [2\lambda(n-2-\lambda)+1] + 1 + n \geq 4n-5$$

and does not vanish, which is a contradiction. We next consider

II. The case $\lambda = n-2$.

The order of the matrix

$$\begin{pmatrix} S_1 & * & * & * \\ 0 & S_3 & * & * \\ 0 & 0 & S_5 & * \\ 0 & 0 & 0 & U_1 \end{pmatrix}$$

is

$$[2(n-2)-1] + 2 + [1 + (n-2)] + n = 4n-4 > 4n-5.$$

---

[1] Note that $2\lambda(n-2-\lambda) - (n-3) = (n-3-\lambda)(2\lambda-1) + \lambda > 0$.

Consequently, we should have either

$$P^{\cdot\ \cdot\ \cdot\alpha_3}_{\alpha_3\alpha_2\alpha_1} + P^{\cdot\ \cdot\ \cdot\alpha_3}_{\alpha_3\alpha_1\alpha_2} = 0 \tag{5.11}$$

or

$$P^{\cdot\ \cdot\ \cdot\alpha_k}_{\alpha_k\alpha_2\alpha_1} + P^{\cdot\ \cdot\ \cdot\alpha_k}_{\alpha_1\alpha_2\alpha_k} - P^{\cdot\ \cdot\ \cdot\alpha_1}_{\alpha_1\alpha_2\alpha_1} = 0. \tag{5.12}$$

If (5.11) were valid, then we should have

$$P^{\cdot\ \cdot\ \cdot\alpha_3}_{\alpha_3\alpha_1\alpha_2} = -P^{\cdot\ \cdot\ \cdot\alpha_3}_{\alpha_3\alpha_2\alpha_1} \neq 0$$

but then the determinant

$$\begin{vmatrix} S_1 & * & * & * \\ 0 & S_4 & * & * \\ 0 & 0 & S_5 & * \\ 0 & 0 & 0 & U_1 \end{vmatrix}$$

is of order

$$[2(n-2)-1]+1+[1+(n-2)]+n = 4n-5$$

and does not vanish, which is a contradiction.

Thus we should have (5.12), from which

$$\Sigma_{k=3}^{n} P^{\cdot\ \cdot\ \cdot\alpha_k}_{\alpha_k\alpha_2\alpha_1} + \Sigma_{k=3}^{n} P^{\cdot\ \cdot\ \cdot\alpha_k}_{\alpha_1\alpha_2\alpha_k} - (n-2)P^{\cdot\ \cdot\ \cdot\alpha_1}_{\alpha_1\alpha_2\alpha_1} = 0,$$

$$-P^{\cdot\ \cdot\ \cdot\alpha_1}_{\alpha_1\alpha_2\alpha_1} - P^{\cdot\ \cdot\ \cdot\alpha_1}_{\alpha_1\alpha_2\alpha_1} - P^{\cdot\ \cdot\ \cdot\alpha_2}_{\alpha_1\alpha_2\alpha_2} - (n-2)P^{\cdot\ \cdot\ \cdot\alpha_1}_{\alpha_1\alpha_2\alpha_1} = 0,$$

$$nP^{\cdot\ \cdot\ \cdot\alpha_1}_{\alpha_1\alpha_2\alpha_1} - P^{\cdot\ \cdot\ \cdot\alpha_2}_{\alpha_2\alpha_1\alpha_2} = 0. \tag{5.13}$$

On the other hand, we have from (5.12)

$$P^{\cdot\ \cdot\ \cdot\alpha_3}_{\alpha_3\alpha_2\alpha_1} + P^{\cdot\ \cdot\ \cdot\alpha_2}_{\alpha_1\alpha_2\alpha_3} - P^{\cdot\ \cdot\ \cdot\alpha_1}_{\alpha_1\alpha_2\alpha_1} = 0,$$

$$2P^{\cdot\ \cdot\ \cdot\alpha_3}_{\alpha_3\alpha_2\alpha_1} - P^{\cdot\ \cdot\ \cdot\alpha_3}_{\alpha_3\alpha_1\alpha_2} - P^{\cdot\ \cdot\ \cdot\alpha_1}_{\alpha_1\alpha_2\alpha_1} = 0. \tag{5.14}$$

Substituting

$$P^{\cdot\ \cdot\ \cdot\alpha_1}_{\alpha_1\alpha_2\alpha_1} = -(n-2)P^{\cdot\ \cdot\ \cdot\alpha_3}_{\alpha_3\alpha_2\alpha_1} \tag{5.15}$$

in (5.14), we find

$$nP^{\cdot\ \cdot\ \cdot\alpha_1}_{\alpha_3\alpha_2\alpha_1} - P^{\cdot\ \cdot\ \cdot\alpha_3}_{\alpha_3\alpha_1\alpha_2} = 0$$

which shows that $P^{\cdot\ \cdot\ \cdot\alpha_3}_{\alpha_3\alpha_1\alpha_2} \neq 0$. Thus repeating the whole argument, we get

$$nP^{\cdot\ \cdot\ \cdot\alpha_2}_{\alpha_2\alpha_1\alpha_2} - P^{\cdot\ \cdot\ \cdot\alpha_1}_{\alpha_1\alpha_2\alpha_1} = 0. \tag{5.16}$$

From (5.13) and (5.16) we find

$$P^{\cdot\cdot\cdot\alpha_1}_{\alpha_1\alpha_2\alpha_1} = 0.$$

Thus, from (5.15) we obtain

$$P^{\cdot\cdot\cdot\alpha_3}_{\alpha_3\alpha_2\alpha_1} = 0,$$

which is a contradiction. Thus Lemma 4 is proved for $4 \leq n$.

When $n = 3$, we have $4n - 5 = 7$. We consider the following submatrix of S:

S_6:

$\varkappa$ \ ρ $\nu\mu\lambda$ \ σ	α_3 α_1	α_3 α_2	α_2 α_2	α_1 α_3
α_3 $\alpha_3\alpha_2\alpha_3$	$P^{\cdot\cdot\cdot\alpha_3}_{\alpha_1\alpha_2\alpha_3} + P^{\cdot\cdot\cdot\alpha_3}_{\alpha_3\alpha_2\alpha_1}$	*	*	*
α_2 $\alpha_3\alpha_2\alpha_1$	0	$P^{\cdot\cdot\cdot\alpha_3}_{\alpha_3\alpha_2\alpha_1}$	*	*
α_3 $\alpha_3\alpha_2\alpha_1$	0	0	$P^{\cdot\cdot\cdot\alpha_3}_{\alpha_3\alpha_2\alpha_1}$	*
α_3 $\alpha_1\alpha_2\alpha_1$	0	0	0	$P^{\cdot\cdot\cdot\alpha_3}_{\alpha_3\alpha_2\alpha_1} + P^{\cdot\cdot\cdot\alpha_3}_{\alpha_1\alpha_2\alpha_3}$

Since the rank of T is less than 7, the determinant

$$\begin{vmatrix} S_6 & 0 \\ * & U \end{vmatrix}$$

should vanish, and consequently

$$P^{\cdot\cdot\cdot\alpha_3}_{\alpha_3\alpha_2\alpha_1}(P^{\cdot\cdot\cdot\alpha_3}_{\alpha_3\alpha_2\alpha_1} + P^{\cdot\cdot\cdot\alpha_3}_{\alpha_1\alpha_2\alpha_3}) = 0.$$

Similarly we have

$$P^{\cdot\cdot\cdot\alpha_3}_{\alpha_3\alpha_1\alpha_2}(P^{\cdot\cdot\cdot\alpha_3}_{\alpha_3\alpha_1\alpha_2} + P^{\cdot\cdot\cdot\alpha_3}_{\alpha_2\alpha_1\alpha_3}) = 0.$$

Adding these two, we find

$$(P^{\cdot\cdot\cdot\alpha_3}_{\alpha_3\alpha_2\alpha_1})^2 + (P^{\cdot\cdot\cdot\alpha_3}_{\alpha_3\alpha_1\alpha_2})^2 + (P^{\cdot\cdot\cdot\alpha_3}_{\alpha_1\alpha_2\alpha_3})^2 = 0,$$

from which

$$P^{\cdot\cdot\cdot\alpha_3}_{\alpha_3\alpha_2\alpha_1} = P^{\cdot\cdot\cdot\alpha_3}_{\alpha_3\alpha_1\alpha_2} = P^{\cdot\cdot\cdot\alpha_3}_{\alpha_1\alpha_2\alpha_3} = 0.$$

Thus Lemma 4 is proved for $n = 3$.

LEMMA 5. *If the rank of T is less than $4n - 5$, then*

$$P_{\alpha_2\alpha_1\alpha_3}^{\cdot\cdot\cdot\alpha_3} = 0. \tag{5.17}$$

This follows from Lemma 4 and

$$P_{\alpha_3\alpha_2\alpha_1}^{\cdot\cdot\cdot\alpha_3} + P_{\alpha_2\alpha_1\alpha_3}^{\cdot\cdot\cdot\alpha_3} - P_{\alpha_3\alpha_1\alpha_2}^{\cdot\cdot\cdot\alpha_3} = 0.$$

From Lemmas 1, 2, 3, 4 and 5, we have

LEMMA 6. *If the rank of T is less than $4n - 5$, then*

$$P_{\nu\mu\lambda}^{\cdot\cdot\cdot\varkappa} = 0. \tag{5.18}$$

This last Lemma proves the first part of Theorem 5.1.

To prove the other part of Theorem 5.1, we give the following example. An A_n with

$$\Gamma_{32}^{1} = \Gamma_{23}^{1} = \xi^2,$$

the other $\Gamma_{\mu\lambda}^{\varkappa}$ being zero, or with

$$\overset{p}{\Gamma}{}_{32}^{1} = \overset{p}{\Gamma}{}_{23}^{1} = \xi^2$$

the other $\overset{p}{\Gamma}{}_{\mu\lambda}^{\varkappa}$ being zero, admits an $(n^2 - 2n + 5)$-parameter group of projective motions generated by

$$p_1,\ p_k,\ \xi^2 p_1,\ \xi^3 p_1,\ p_2 - \xi^2\xi^3 p_1,\ \xi^2 p_2 + 2\xi^1 p_1,$$
$$\xi^2 p_3 - \tfrac{1}{3}(\xi^2)^3 p_1,\ \xi^3 p_3 + \xi^1 p_1,\ \xi^p p_1,\ \xi^2 p_q,\ \xi^3 p_q,\ \xi^p p_q.$$

Calculating the projective curvature tensor of this A_n, we find

$$P_{232}^{\cdot\cdot\cdot 1} \neq 0,$$

which shows that the A_n is not a P_n.

If the order of a group of projective motions is $n^2 - 2n + 5$, then, as the above proof shows, the rank of the matrix T is equal to $4n - 5$, and consequently we can give the initial values of $v^\varkappa$ arbitrarily. Thus the group is transitive.

Thus the theorem is completely proved.

§ 6. An A_n admitting a complete group of affine motions of order greater than $n^2 - n + 1$.

We prove the following

THEOREM 6.1.[1] *Let an A_n, $n \geq 4$, admit a group of affine motions of order r.*

1°. *If $r > n^2 - n + 1$*

(a) *the A_n is a P_n,*

(b) *the Ricci tensor $R_{\mu\lambda}$ has the form $R_{\mu\lambda} = \varepsilon w_\mu w_\lambda$, where $\varepsilon = \pm 1$ and $w_\lambda = \partial_\lambda w$,*

(c) *the vector w_λ satisfies*

$$\nabla_\mu w_\lambda = \sigma w_\mu w_\lambda, \tag{6.1}$$

where σ is a function of w.

2°. *If* (a), (b) *and* (c) *in* 1° *hold, the curvature tensor of the space has the form*

$$R_{\nu\mu\lambda}^{\cdot\cdot\cdot\kappa} = c(w_\nu A_\mu^\kappa - w_\mu A_\nu^\kappa)w_\lambda; \qquad c = \text{constant}. \tag{6.2}$$

If $w_\lambda = 0$, the space is affinely Euclidean and $r = n^2 + n$.
If $w_\lambda \neq 0$ and $\sigma =$ constant, then $r = n^2$ and the group is transitive.
If $w_\lambda \neq 0$ and $\sigma \neq$ constant, then $r = n^2 - 1$ and the group is intransitive.

3°. *The conditions* (a), (b), (c) *in* 1° *are equivalent to the following which constitute a completely integrable system of partial differential equations.*

(α) *In a suitable coordinate system, we have*

$$\Gamma_{\mu\lambda}^\kappa = 2p_{(\mu} A_{\lambda)}^\kappa; \; p_\lambda = \partial_\lambda p. \tag{6.3}$$

$$(\beta) \qquad \partial_\mu p_\lambda = p_\mu p_\lambda - \frac{\varepsilon}{n-1} w_\mu w_\lambda,$$

$$(\gamma) \qquad \partial_\mu w_\lambda = \sigma(w) w_\mu w_\lambda + w_\mu p_\lambda + w_\lambda p_\mu.$$

PROOF.

1°. (a) Since a group of affine motions is a group of projective motions, by Theorem 5.1, we have (a).

(b) If we denote by $\underset{v}{\pounds} f = v^\kappa \partial_\kappa f$ an infinitesimal affine motion, then we have $\underset{v}{\pounds} R_{\mu\lambda} = 0$, from which $\underset{v}{\pounds} R_{[\mu\lambda]} = 0$, thus, denoting by $2k$ the rank of $R_{[\mu\lambda]}$, we have, by 2° of Theorem 12.1 of Ch. V,

$$n^2 - n + 1 < r \leq n^2 - (n-k)(2k-1),$$

from which

$$(k-1)[2(n-k) + 2k - 1] < 0.$$

[1] EGOROV [7].

Since $n \geqq 2k$, this inequality holds if and only if $k = 0$. This proves that the $R_{\mu\lambda}$ is symmetric.

We have $\underset{v}{\pounds} R_{\mu\lambda} = 0$ for a symmetric $R_{\mu\lambda}$. Consequently, denoting by m the rank of the matrix $R_{\mu\lambda}$ and applying Theorem 11.1 of Ch. v, we find

$$n^2 - n + 1 < r \leq n^2 + n - nm + \tfrac{1}{2}m(m-1),$$

from which

$$[(n-m) + (n-1)](m-2) < 0.$$

Since $n \geqq m$, this inequality holds if and only if $m = 0$ or 1. Thus

$$R_{\mu\lambda} = \varepsilon w_\mu w_\lambda; \quad \varepsilon = \pm 1.$$

From this equation and $\underset{v}{\pounds} R_{\mu\lambda} = 0$, we find $\underset{v}{\pounds} w_\lambda = 0$ and consequently $\underset{v}{\pounds} \nabla_\mu w_\lambda = 0$. Applying again 2° of Theorem 12.1 of Ch. v to $\nabla_{[\mu} w_{\lambda]}$, we get $\nabla_{[\mu} w_{\lambda]} = 0$, that is, $w_\lambda = \partial_\lambda w$, which proves (b).

(c) Substituting $R_{\mu\lambda} = \varepsilon w_\mu w_\lambda$ in the identity $\nabla_{[\nu} R_{\mu]\lambda} = 0$ which holds for a projectively Euclidean space, we find $w_{[\mu} \nabla_{\nu]} w_\lambda = 0$, from which

$$\nabla_\mu w_\lambda = \sigma w_\mu w_\lambda.$$

Since $\underset{v}{\pounds} \nabla_\mu w_\lambda = 0$, $\underset{v}{\pounds} w_\lambda = 0$, we find from the above equation $\underset{v}{\pounds} \sigma = 0$, from which $\underset{v}{\pounds} \nabla_\lambda \sigma = 0$ and consequently $\underset{v}{\pounds} w_{[\mu} \nabla_{\lambda]} \sigma = 0$. Thus applying 2° of Theorem 12.1 of Ch. v to $w_{[\mu} \nabla_{\lambda]} \sigma$, we find $w_{[\mu} \nabla_{\lambda]} \sigma = 0$, from which $\sigma = \sigma(w)$. This proves (c).

2°. The space is projectively Euclidean and the Ricci tensor is symmetric, and consequently the curvature tensor has the form

$$R_{\nu\mu\lambda}^{\cdot\cdot\cdot\varkappa} = \frac{2}{n-1} A^\varkappa_{[\nu} R_{\mu]\lambda}.$$

Substituting $R_{\mu\lambda} = \varepsilon w_\mu w_\lambda$ in this equation, we find

$$R_{\nu\mu\lambda}^{\cdot\cdot\cdot\varkappa} = c(w_\nu A^\varkappa_\mu - w_\mu A^\varkappa_\nu) w_\lambda, \qquad c = \text{constant}.$$

Thus, if $w_\lambda = 0$, then $R_{\nu\mu\lambda}^{\cdot\cdot\cdot\varkappa} = 0$ and, as is well-known, $r = n^2 + n$. If $w_\lambda \neq 0$, then by (a), the integrability conditions of $\underset{v}{\pounds} \Gamma^\varkappa_{\mu\lambda} = 0$ are given by $\underset{v}{\pounds} R_{\mu\lambda} = 0$, $\underset{v}{\pounds} \nabla_\nu R_{\mu\lambda} = 0$, ... But by (b) and (c), these are equivalent to $\underset{v}{\pounds} w_\lambda = 0$, $\underset{v}{\pounds} \sigma = 0$, $\underset{v}{\pounds} \nabla_\mu w_\lambda = 0$, $\underset{v}{\pounds} \nabla_\lambda \sigma = 0$, ...

Now suppose that $\sigma =$ constant, then these conditions are equivalent to $\underset{v}{\pounds} w_\lambda = 0$. Consequently we have $r \geq n^2$.

On the other hand, the rank of $R_{\mu\lambda}$ is 1. Consequently, by Theorem 11.1 of Ch. v, we have $r \leq n^2$, from which $r = n^2$. Moreover by 3° of Theorem 12.1 of Ch. v, the group is transitive.

Suppose next that $\sigma \neq$ constant, then we have $\underset{v}{\pounds}\sigma = 0$, and consequently the group is intransitive.

Since we have

$$\underset{v}{\pounds}\nabla_\lambda\sigma = \underset{v}{\pounds}\left(\frac{dw}{d\sigma}\,w_\lambda\right) = \frac{d^2\sigma}{dw^2}(\underset{v}{\pounds}w)w_\lambda + \frac{d\sigma}{dw}\,\underset{v}{\pounds}w_\lambda$$

$$= \frac{\dfrac{d^2\sigma}{dw^2}}{\dfrac{d\sigma}{dw}}(\underset{v}{\pounds}\sigma)w_\lambda + \frac{d\sigma}{dw}\,\underset{v}{\pounds}w_\lambda,$$

the integrability conditions of $\underset{v}{\pounds}\Gamma^\varkappa_{\mu\lambda} = 0$ are given by $\underset{v}{\pounds}w_\lambda = 0$ and $\underset{v}{\pounds}\sigma = 0$.

If $\underset{v}{\pounds}w_\lambda = 0$ and $\underset{v}{\pounds}\sigma = 0$ are not independent, then by 3° of Theorem 12.1 of Ch. v, the group becomes transitive, which is a contradiction. Thus $\underset{v}{\pounds}w_\lambda = 0$ and $\underset{v}{\pounds}\sigma = 0$ are independent and we have $r = n^2 - 1$.

3°. We assume (a), (b) and (c) of 1°. Since the space is projectively Euclidean and the Ricci tensor is symmetric, we have (α). From (α), we obtain

$$R_{\mu\lambda} = -(n-1)(\partial_\mu p_\lambda - p_\mu p_\lambda).$$

Substituting $R_{\mu\lambda} = \varepsilon w_\mu w_\lambda$ in this equation, we find

$$\partial_\mu p_\lambda = p_\mu p_\lambda - \frac{\varepsilon}{n-1}\,w_\mu w_\lambda$$

which proves (β).

Moreover, from

$$\nabla_\mu w_\lambda = \sigma(w) w_\mu w_\lambda,$$

we find

$$\partial_\mu w_\lambda = \sigma(w) w_\mu w_\lambda + w_\mu p_\lambda + w_\lambda p_\mu$$

which proves (γ).

It is easily to be seen that (α), (β), (γ) are equivalent to (a), (b), (c). The fact that (β) and (γ) form a completely integrable system of partial differential equations is verified by a straightforward calculation.

I. P. Egorov [1] proved also

THEOREM 6.2.

(1) *An A_n, not equi-affine, with maximal mobility, admits a transitive complete group of affine motions exactly of order $n^2 - n + 1$. Such a space is necessarily projectively Euclidean.*

(2) *A projectively Euclidean A_n, for which the rank of the skew-symmetric part of the Ricci tensor is $2k$, admits a transitive group of affine motions exactly of order $r = n^2 - (n-k)(2k-1)$.*

(3) *There are no A_n's, admitting a complete transitive group of affine motions of order r, with $n^2 - n + 1 < r < n^2$.*

(4) *The maximum order for an intransitive group of affine motions of an A_n is exactly $n^2 - 1$.*

(5) *There are no A_n's admitting an intransitive group of affine motions of order r, with $n^2 - n + 1 < r < n^2 - 1$.*

Y. Mutō [2] proved, by a method quite different from that of Egorov, the following theorems.

THEOREM 6.3. *An A_n with non-vanishing curvature tensor admits a complete group of affine motions of the maximum order if and only if the equations*

$$R_{\nu\mu\lambda}^{\cdot\cdot\cdot\varkappa} = \varepsilon(w_\nu A_\mu^\varkappa - w_\mu A_\nu^\varkappa)w_\lambda, \quad \nabla_\mu w_\lambda = a w_\mu w_\lambda, \quad \varepsilon = \pm 1, \quad a = \text{const.} \tag{6.4}$$

are satisfied. Then the order is n^2, and we can find a coordinate system with respect to which the components of the linear connexion are

$$\Gamma_{nn}^\alpha = -\varepsilon\xi^\alpha, \quad \Gamma_{nn}^n = -a, \quad \text{the other } \Gamma_{\mu\lambda}^\varkappa = 0; \quad \alpha = 1, 2, \ldots, n-1,$$

and the finite equations of the group are given by

$$\begin{cases} '\xi^\alpha = P_\beta^\alpha \xi^\beta + Q^\alpha e^{c_1\xi^n} + R^\alpha e^{c_2\xi^n} \\ '\xi^n = \xi^n + S \end{cases} \tag{6.5}$$

or

$$\begin{cases} '\xi^\alpha = P_\beta^\alpha \xi^\beta + (Q^\alpha + R^\alpha \xi^n) e^{c\xi^n} \\ '\xi^n = \xi^n + S \end{cases} \tag{6.6}$$

---

[1] EGOROV [7, 9].

[2] MUTŌ [3, 4].

according as the roots c_1 *and* c_2 *of the quadratic equation* $(\xi)^2 + a\xi - \varepsilon = 0$ *satisfy* $c_1 \neq c_2$ *or* $c_1 = c_2 = c$.

THEOREM 6.4.[1] *In order that a projectively Euclidean*[2] A_n *with non-vanishing curvature tensor admit a complete group of affine motions* G_r *of order* $r < n^2 - n$, *it is necessary and sufficient that the curvature tensor belong to one of the following three types* T_1, T_2 and T_3, *and the vectors appearing in the expressions of the curvature tensors satisfy the associated equations. Such linear connexions and groups actually exist.*

The curvature tensors,

$$T_1:\ R_{\nu\mu\lambda}^{\cdots\varkappa} = \varepsilon(w_\nu A_\mu^\varkappa - w_\mu A_\nu^\varkappa)w_\lambda,\quad \varepsilon = \pm 1,\quad w_\lambda \neq 0. \tag{6.7}$$

$$\begin{aligned} T_2:\ R_{\nu\mu\lambda}^{\cdots\varkappa} = {} & \varepsilon(w_\nu A_\mu^\varkappa - w_\mu A_\nu^\varkappa)w_\lambda \\ & + A_\nu^\varkappa(w_\mu x_\lambda - w_\lambda x_\mu) - A_\mu^\varkappa(w_\nu x_\lambda - w_\lambda x_\mu) \\ & - 2(w_\nu x_\mu - w_\mu x_\nu)A_\lambda^\varkappa, \end{aligned} \tag{6.8}$$

$\varepsilon = \pm 1$; w_λ and x_λ are linearly independent.

$$T_3:\ R_{\nu\mu\lambda}^{\cdots\varkappa} = \varepsilon_1(w_\nu A_\mu^\varkappa - w_\mu A_\nu^\varkappa)w_\lambda + \varepsilon_2(x_\nu A_\mu^\varkappa - x_\mu A_\nu^\varkappa)x_\lambda, \tag{6.9}$$

$\varepsilon_1, \varepsilon_2 = \pm 1$; w_λ and x_λ are linearly independent.

The associated equations

T_1:

$$\nabla_\mu w_\lambda = \alpha w_\mu w_\lambda;\quad \alpha = \alpha(w),\quad w_\lambda = \partial_\lambda w. \tag{6.10}$$

T_2:

$$\begin{cases} \nabla_\mu w_\lambda = \theta(2\varepsilon w_\mu w_\lambda - w_\mu x_\lambda + w_\lambda x_\mu) \\ \nabla_\mu x_\lambda = y_\mu x_\lambda + \theta(\varepsilon w_\mu x_\lambda - x_\mu x_\lambda) \end{cases} \tag{6.11}$$

$$\begin{aligned} \nabla_\mu y_\lambda - \nabla_\lambda y_\mu = {} & \varepsilon\theta(y_\mu w_\lambda - y_\lambda w_\mu) \\ & - 2\theta(x_\mu y_\lambda - x_\lambda y_\mu) - \varepsilon(w_\mu x_\lambda - w_\lambda x_\mu);\ \varepsilon\theta^2 = -1. \end{aligned} \tag{6.12}$$

T_3:

$$\nabla_\mu w_\lambda = -\varepsilon_1\varepsilon_2 y_\mu x_\lambda,\quad \nabla_\mu x_\lambda = y_\mu x_\lambda. \tag{6.13}$$

$$\nabla_\mu y_\lambda - \nabla_\lambda y_\mu = \varepsilon(x_\mu w_\lambda - x_\lambda w_\mu). \tag{6.14}$$

---

[1] MUTŌ [3, 4].

[2] For $n \geq 5$, we have

$$r > n^2 - n \geq n^2 - 2n + 5.$$

Consequently, according to Theorem 5.1, we do not need this assumption.

The groups

T_1:

(6.15) $$\pounds w_\lambda = 0, \ \pounds \alpha = 0.$$

If α is a constant, then $r = n^2$ and the group is transitive.
If α is not a constant, then $r = n^2 - 1$ and the group is intransitive.

T_2:

(6.16) $$\pounds w_\lambda = 0, \ \pounds x_\lambda = \beta w_\lambda; \quad \beta\text{: a scalar.}$$

$r = n^2 - n + 1$ and the group is transitive.

T_3:

(6.17) $$\pounds w_\lambda = -\varepsilon_1 \varepsilon_2 \beta x_\lambda, \ \pounds x_\lambda = \beta w_\lambda; \quad \beta\text{: a scalar.}$$

$r = n^2 - n + 1$ and the group is transitive.

Using his own method, Y. Mutō[1] studied also the A_n which admits a group G_r of affine motions of order $r < n^2 - 2n$.

§ 7. An L_n admitting an n^2-parameter of affine motions.

In § 10 of Ch. v, we have found that if an L_n admits an n^2-parameter group of affine motions, the connexion is semi-symmetric:

(7.1) $$\Gamma^{\varkappa}_{[\mu\lambda]} = A^{\varkappa}_{\mu} S_\lambda - A^{\varkappa}_{\lambda} S_\mu.$$

We denote by A_n the space with symmetric linear connexion $\Gamma^{\varkappa}_{(\mu\lambda)}$, and prove

THEOREM 7.1.[2] *If an L_n with a semi-symmetric linear connexion admits an n^2-parameter group of affine motions, then*

1°. (a) *the A_n is a P_n,*

(b) *the Ricci tensor $R_{\mu\lambda}$ of A_n has the form*

(7.2) $$R_{\mu\lambda} = c S_\mu S_\lambda \qquad c = \text{constant},$$

(c) *the vector S_λ satisfies*

$$\nabla_\mu S_\lambda = {}'c S_\mu S_\lambda \qquad {}'c = \text{constant}$$

2°. *The above three conditions* (a), (b), (c) *are equivalent to the following*

---

[1] MUTŌ [5].
[2] EGOROV [8].

equations:

$$(\alpha) \qquad \Gamma^{\varkappa}_{(\mu\lambda)} = A^{\varkappa}_{\mu} p_{\lambda} + A^{\varkappa}_{\lambda} p_{\mu},$$

$$(\beta) \qquad \partial_{\mu} p_{\lambda} = p_{\mu} p_{\lambda} + c S_{\mu} S_{\lambda},$$

$$(\gamma) \qquad \partial_{\mu} S_{\lambda} = {}'c S_{\mu} S_{\lambda} + S_{\mu} p_{\lambda} + S_{\lambda} p_{\mu}.$$

The last two constitute a completely integrable system of partial differential equations, and consequently, there exists actually a space satisfying all the conditions stated in 1°.

PROOF.

1°. Denoting by $\underset{v}{\pounds} f = v^{\varkappa}\partial_{\varkappa} f$ an infinitesimal operator of the group, we have $\underset{v}{\pounds}\Gamma^{\varkappa}_{\mu\lambda} = 0$, from which

$$\underset{v}{\pounds}\Gamma^{\varkappa}_{(\mu\lambda)} = 0, \quad \underset{v}{\pounds} S_{\lambda} = 0.$$

Thus the group is an n^2-parameter group of affine motions in the space A_n and consequently we have (a).

On the other hand, we know that

$$R_{\mu\lambda} = \varepsilon w_{\mu} w_{\lambda}, \quad \underset{v}{\pounds} w_{\lambda} = 0, \quad \nabla_{\mu} w_{\lambda} = \varepsilon w_{\mu} w_{\lambda}.$$

Thus, applying Theorem 12.1 of Ch. V to the tensor $S_{[\mu} w_{\lambda]}$, we find $S_{[\mu} w_{\lambda]} = 0$, from which $w_{\lambda} = \alpha S_{\lambda}$ and $\underset{v}{\pounds}\alpha = 0$. But the group is transitive and consequently $\alpha =$ constant. Thus we have (b) and (c).

2°. The (α) follows from the (a). The (β) follows from (b) and

$$R_{\mu\lambda} = -(n-1)(\partial_{\mu} p_{\lambda} - p_{\mu} p_{\lambda}).$$

The (γ) follows from (c) and (α).

The last statement can be proved by a straightforward calculation.

CHAPTER VII

GROUPS OF CONFORMAL MOTIONS

§ 1. Groups of conformal motions.

An infinitesimal conformal motion $\xi^\varkappa \to \xi^\varkappa + v^\varkappa dt$ is characterized by

$$\underset{v}{\pounds} g_{\mu\lambda} = 2\phi g_{\mu\lambda}, \tag{1.1}$$

or by

$$\underset{v}{\pounds} \mathfrak{G}_{\mu\lambda} = 0, \quad \mathfrak{G}_{\mu\lambda} \overset{\text{def}}{=} \mathfrak{g}^{-\frac{1}{n}} g_{\mu\lambda}; \quad \mathfrak{g} = |\text{Det}(g_{\mu\lambda})|. \tag{1.2}$$

If two vectors $v^\varkappa$ and $\rho v^\varkappa$ give conformal motions, we have

$$\nabla_{(\mu} v_{\lambda)} = \phi g_{\mu\lambda}; \quad \nabla_{(\mu}(\rho v_{\lambda)}) = \varphi g_{\mu\lambda},$$

from which

$$(\nabla_{(\mu} \rho) v_{\lambda)} = (\varphi - \rho\phi) g_{\mu\lambda},$$

and consequently we find $\nabla_\mu \rho = 0$, hence $\rho =$ constant. Thus we have

THEOREM 1.1.[1] *Two different infinitesimal conformal motions cannot have the same streamlines.*

The conformal fundamental tensor density $\mathfrak{G}_{\lambda\varkappa}$ is a linear differential geometric object. Thus, according to Theorems 2.1 and 2.2 of Ch. III, we have

THEOREM 1.2.[2] *If a V_n admits an infinitesimal conformal motion, it admits also a one-parameter group of conformal motions generated by this infinitesimal conformal motion.*

THEOREM 1.3.[3] *In order that a V_n admit a one-parameter group of conformal motions, it is necessary and sufficient that there exist a coordinate system with respect to which the components $\mathfrak{G}_{\lambda\varkappa}$ of the conformal fundamental tensor density are independent of one of the coordinates.*

[1], [2], [3] G. T., p. 51.

If the $\mathfrak{G}_{\lambda\varkappa}$ is independent of ξ^1, then $g_{\lambda\varkappa}$ has the form

$$g_{\lambda\varkappa} = \alpha(\xi^\nu) f_{\lambda\varkappa}(\xi^2, \ldots, \xi^n). \tag{1.3}$$

Conversely, if $g_{\lambda\varkappa}$ has the form (1.3), we can easily see that $\mathfrak{G}_{\lambda\varkappa}$ is independent of ξ^1. Thus

THEOREM 1.4.[1] *In order that a V_n admit a one-parameter group of conformal motions, it is necessary and sufficient that there exist a coordinate system with respect to which the fundamental tensor $g_{\lambda\varkappa}$ has the form* (1.3).

If we choose a coordinate system with respect to which $\xi^\varkappa = v^\varkappa$, then

$$\underset{v}{\pounds}\mathfrak{G}_{\lambda\varkappa} = \xi^\mu \partial_\mu \mathfrak{G}_{\lambda\varkappa} = 0,$$

from which we get

THEOREM 1.5.[2] *In order that a V_n admit a one-parameter group of conformal motions, it is necessary and sufficient that there exist a coordinate system with respect to which the components of the conformal fundamental tensor density are homogeneous functions of degree zero of the coordinates.*

Since the conformal tensor density $\mathfrak{G}_{\lambda\varkappa}$ is a linear differential geometric object, Theorems 2.3, 2.4, 2.5 and 2.6 of Ch. III also hold for conformal motions.

§ 2. Transformations carrying conformal circles into conformal circles.

A conformal circle[3] is defined as a curve which satisfies the differential equations

$$u^\varkappa \overset{\text{def}}{=} \frac{\delta^3\xi^\varkappa}{ds^3} + \frac{d\xi^\varkappa}{ds}\left(g_{\mu\lambda}\frac{\delta^2\xi^\mu}{ds^2}\frac{\delta^2\xi^\lambda}{ds^2} - \frac{1}{n-2}L_{\mu\lambda}\frac{d\xi^\mu}{ds}\frac{d\xi^\lambda}{ds}\right) + \frac{1}{n-2}L_{\lambda}^{\cdot\varkappa}\frac{d\xi^\lambda}{ds} = 0, \tag{2.1}$$

where

$$L_{\mu\lambda} \overset{\text{def}}{=} -K_{\mu\lambda} + \frac{1}{2(n-1)}Kg_{\mu\lambda}. \tag{2.2}$$

It is evident that a conformal motion carries every conformal circle into a conformal circle.

---

[1], [2] G. T., p. 51.

[3] YANO [1], SCHOUTEN [8], p. 331.

Conversely we assume that an infinitesimal transformation $\xi^\varkappa \to \xi^\varkappa + v^\varkappa dt$ carries every conformal circle into a conformal circle. Calculating $\underset{v}{\pounds} u^\varkappa$, we find [1]

$$\begin{aligned}
\pounds u^\varkappa = &- 3u^\varkappa \frac{\pounds ds}{ds} + 3(\pounds\{^{\ \varkappa}_{\mu\lambda}\}) \frac{\delta^2\xi^\mu}{ds^2}\frac{d\xi^\lambda}{ds} \\
&+ (\nabla_\nu \pounds\{^{\ \varkappa}_{\mu\lambda}\}) \frac{d\xi^\nu}{ds}\frac{d\xi^\mu}{ds}\frac{d\xi^\lambda}{ds} - 3\frac{\delta^2\xi^\varkappa}{ds^2}\frac{d}{ds}\frac{\pounds ds}{ds} \\
&+ \frac{d\xi^\varkappa}{ds}\left[(\pounds g_{\mu\lambda})\frac{\delta^2\xi^\mu}{ds^2}\frac{\delta^2\xi^\lambda}{ds} + 2g_{\tau\sigma}(\pounds\{^{\ \tau}_{\mu\lambda}\})\frac{\delta^2\xi^\sigma}{ds^2}\frac{d\xi^\mu}{ds}\frac{d\xi^\lambda}{ds}\right. \\
&\left. - \frac{1}{n-2}(\pounds L_{\mu\lambda})\frac{d\xi^\mu}{ds}\frac{d\xi^\lambda}{ds} - 2g_{\mu\lambda}\frac{\delta^2\xi^\mu}{ds^2}\frac{\delta^2\xi^\lambda}{ds^2} - \frac{d^2}{ds^2}\frac{\pounds ds}{ds}\right] \\
&+ \frac{1}{n-2}(\pounds L_\lambda^{\cdot\varkappa})\frac{d\xi^\lambda}{ds} - \frac{1}{n-2}L_\lambda^{\cdot\varkappa}\frac{d\xi^\lambda}{ds}\frac{\pounds ds}{ds},
\end{aligned}$$

from which, taking account of $u^\varkappa = 0$,

$$\begin{aligned}
g_{\mu\lambda}(\pounds u^\mu)\frac{\delta^2\xi^\lambda}{ds^2} = &\ 3g_{\tau\sigma}(\pounds\{^{\ \tau}_{\mu\lambda}\})\frac{\delta^2\xi^\sigma}{ds^2}\frac{d\xi^\mu}{ds}\frac{d\xi^\lambda}{ds} \\
&+ g_{\tau\sigma}(\nabla_\nu \pounds\{^{\ \tau}_{\mu\lambda}\})\frac{\delta^2\xi^\sigma}{ds^2}\frac{d\xi^\nu}{ds}\frac{d\xi^\mu}{ds}\frac{d\xi^\lambda}{ds} \\
&- 3g_{\tau\sigma}\frac{\delta^2\xi^\tau}{ds^2}\frac{\delta^2\xi^\sigma}{ds^2}\left[\tfrac{1}{2}(\nabla_\nu \pounds g_{\mu\lambda})\frac{d\xi^\nu}{ds}\frac{d\xi^\mu}{ds}\frac{d\xi^\lambda}{ds} + (\pounds g_{\mu\lambda})\frac{\delta^2\xi^\mu}{ds^2}\frac{d\xi^\lambda}{ds}\right] \\
+ \frac{1}{n-2}g_{\nu\mu}\pounds L_\lambda^{\cdot\nu}\frac{\delta^2\xi^\mu}{ds^2}\frac{d\xi^\lambda}{ds} &+ \frac{1}{n-2}(L_{\omega\nu}\pounds g_{\mu\lambda})\frac{\delta^2\xi^\omega}{ds^2}\frac{d\xi^\nu}{ds}\frac{d\xi^\mu}{ds}\frac{d\xi^\lambda}{ds}.
\end{aligned}$$

If an infinitesimal transformation $\xi^\varkappa \to \xi^\varkappa + v^\varkappa dt$ carries every conformal circle into a conformal circle, $g_{\mu\lambda}(\pounds u^\varkappa)\dfrac{\delta^2\xi^\mu}{ds^2}\dfrac{d\xi^\lambda}{ds}$ must vanish for an arbitrary unit vector $\dfrac{d\xi^\varkappa}{ds}$ and a vector $\dfrac{\delta^2\xi^\varkappa}{ds^2}$ satisfying $g_{\mu\lambda}\dfrac{\delta^2\xi^\mu}{ds^2}\dfrac{d\xi^\lambda}{ds} = 0$. Consequently, considering the coefficients of the term of the highest degree with respect to $\dfrac{\delta^2\xi^\varkappa}{ds^2}$, we can conclude

$$\pounds g_{\mu\lambda} = 2\phi g_{\mu\lambda}.$$

---

[1] YANO and TOMONAGA [1].

Thus we have

THEOREM 2.1.[1] *In order that an infinitesimal transformation in a* V_n *carry every conformal circle into a conformal circle, it is necessary and sufficient that the transformation be a conformal motion.*

§ 3. Integrability conditions of $\underset{v}{£} g_{\mu\lambda} = 2\phi g_{\mu\lambda}$.

We now consider the integrability conditions of

$$\underset{v}{£} g_{\mu\lambda} = 2\nabla_{(\mu} v_{\lambda)} = 2\phi g_{\mu\lambda}. \tag{3.1}$$

Substituting (3.1) in (cf. p. 52)

$$\underset{v}{£}\{{}^{\,\varkappa}_{\mu\lambda}\} = \tfrac{1}{2} g^{\varkappa\rho}[\nabla_\mu \underset{v}{£} g_{\lambda\rho} + \nabla_\lambda \underset{v}{£} g_{\mu\rho} - \nabla_\rho \underset{v}{£} g_{\mu\lambda}],$$

we find

$$\underset{v}{£}\{{}^{\,\varkappa}_{\mu\lambda}\} = A^\varkappa_\mu \phi_\lambda + A^\varkappa_\lambda \phi_\mu - \phi^\varkappa g_{\mu\lambda}, \tag{3.2}$$

where $\phi_\lambda = \partial_\lambda \phi$.

Substituting (3.2) into

$$\underset{v}{£} K_{\nu\mu\lambda}{}^{\varkappa} = 2\nabla_{[\nu} \underset{v}{£}\{{}^{\,\varkappa}_{\mu]\lambda}\},$$

we obtain

$$\underset{v}{£} K_{\nu\mu\lambda}{}^{\varkappa} = -\,2A^\varkappa_{[\nu} \nabla_{\mu]} \phi_\lambda - 2(\nabla_{[\nu} \phi^\varkappa) g_{\mu]\lambda}. \tag{3.3}$$

By contraction with respect to $\varkappa$ and ν, it follows from (3.3) that

$$\underset{v}{£} K_{\mu\lambda} = -\,(n-2)\nabla_\mu \phi_\lambda - g_{\mu\lambda} \nabla_\rho \phi^\rho. \tag{3.4}$$

Transvecting (3.4) with $g^{\mu\lambda}$, we find

$$g^{\mu\lambda} \underset{v}{£} K_{\mu\lambda} = -\,2(n-1)\nabla_\rho \phi^\rho,$$

from which

$$\underset{v}{£} K = -\,2\phi K - 2(n-1)\nabla_\rho \phi^\rho. \tag{3.5}$$

From (3.5), we get

$$\nabla_\rho \phi^\rho = -\frac{1}{2(n-1)}[\underset{v}{£} K + 2\phi K].$$

[1] YANO and TOMOGANA [1]; G. T., p. 50.

Substituting this in (3.4), we find

$$\frac{1}{n-2} \underset{v}{\pounds} L_{\mu\lambda} = \nabla_\mu \phi_\lambda, \tag{3.6}$$

from which

$$\frac{1}{n-2} \underset{v}{\pounds} L_{\lambda}^{\cdot\,\varkappa} + \frac{2}{n-2} \phi L_{\lambda}^{\cdot\,\varkappa} = \nabla_\lambda \phi^\varkappa. \tag{3.7}$$

Substituting (3.6) and (3.7) in (3.3), we find

$$\underset{v}{\pounds} C_{\nu\mu\lambda}^{\cdot\cdot\cdot\,\varkappa} = 0 \tag{3.8}$$

where the $C_{\nu\mu\lambda}^{\cdot\cdot\cdot\,\varkappa}$ is the conformal curvature tensor.

From (3.6) and the formula

$$2\nabla_{[\nu} \nabla_{\mu]} \phi_\lambda = - K_{\nu\mu\lambda}^{\cdot\cdot\cdot\,\varkappa} \phi_\varkappa,$$

it follows

$$\frac{2}{n-2} \nabla_{[\nu} (\underset{v}{\pounds} L_{\mu]\lambda}) = - K_{\nu\mu\lambda}^{\cdot\cdot\cdot\,\varkappa} \phi_\varkappa. \tag{3.9}$$

On the other hand, we have

$$\begin{aligned} \underset{v}{\pounds} \nabla_\nu L_{\mu\lambda} - \nabla_\nu \underset{v}{\pounds} L_{\mu\lambda} &= - (\underset{v}{\pounds} \{^{\,\rho}_{\nu\mu}\}) L_{\rho\lambda} - (\underset{v}{\pounds} \{^{\,\rho}_{\nu\lambda}\}) L_{\mu\rho} \\ &= - 2\phi_\nu L_{\mu\lambda} - 2L_{\nu(\mu} \phi_{\lambda)} + 2g_{\nu(\mu} L_{\lambda)}^{\cdot\,\varkappa} \phi_\varkappa, \end{aligned}$$

from which

$$\nabla_{[\nu} \underset{v}{\pounds} L_{\mu]\lambda} = \underset{v}{\pounds} (\nabla_{[\nu} L_{\mu]\lambda}) + (A_{[\nu}^{\varkappa} L_{\mu]\lambda} + L_{[\nu}^{\cdot\,\varkappa} g_{\mu]\lambda}) \phi_\varkappa. \tag{3.10}$$

The equations (3.9) and (3.10) give

$$\underset{v}{\pounds} C_{\nu\mu\lambda} = - C_{\nu\mu\lambda}^{\cdot\cdot\cdot\,\varkappa} \phi_\varkappa, \tag{3.11}$$

where

$$C_{\nu\mu\lambda} \overset{\text{def}}{=} \frac{2}{n-2} \nabla_{[\nu} L_{\mu]\lambda}. \tag{3.12}$$

To find further integrability conditions, we substitute (3.2) and (3.8) in the identity (cf. p. 16)

$$\begin{aligned} &\underset{v}{\pounds} \nabla_\omega C_{\nu\mu\lambda}^{\cdot\cdot\cdot\,\varkappa} - \nabla_\omega \underset{v}{\pounds} C_{\nu\mu\lambda}^{\cdot\cdot\cdot\,\varkappa} \\ &= (\underset{v}{\pounds} \{^{\,\varkappa}_{\omega\rho}\}) C_{\nu\mu\lambda}^{\cdot\cdot\cdot\,\rho} - (\underset{v}{\pounds} \{^{\,\rho}_{\omega\nu}\}) (C_{\rho\mu\lambda}^{\cdot\cdot\cdot\,\varkappa} - (\underset{v}{\pounds} \{^{\,\rho}_{\omega\mu}\}) C_{\nu\rho\lambda}^{\cdot\cdot\cdot\,\varkappa} - (\underset{v}{\pounds} \{^{\,\rho}_{\omega\lambda}\}) C_{\nu\mu\rho}^{\cdot\cdot\cdot\,\varkappa}, \end{aligned}$$

then we obtain

$$(3.13)\quad \underset{v}{\pounds}\nabla_\omega C_{\nu\mu\lambda}^{\cdot\cdot\cdot\varkappa} = -\,2\phi_\omega C_{\nu\mu\lambda}^{\cdot\cdot\cdot\varkappa} + A_\omega^\varkappa C_{\nu\mu\lambda}^{\cdot\cdot\cdot\rho}\phi_\rho - C_{\omega\mu\lambda}^{\cdot\cdot\cdot\varkappa}\phi_\nu - C_{\nu\omega\lambda}^{\cdot\cdot\cdot\varkappa}\phi_\mu - C_{\nu\mu\omega}^{\cdot\cdot\cdot\varkappa}\phi_\lambda - C_{\nu\mu\lambda\omega}\phi^\varkappa + \phi^\rho(g_{\omega\nu}C_{\rho\mu\lambda}^{\cdot\cdot\cdot\varkappa} + g_{\omega\mu}C_{\nu\rho\lambda}^{\cdot\cdot\cdot\varkappa} + g_{\omega\lambda}C_{\nu\mu\rho}^{\cdot\cdot\cdot\varkappa}).$$

We next substitute (3.2) and (3.11) in the identity

$$\underset{v}{\pounds}\nabla_\omega C_{\nu\mu\lambda} - \nabla_\omega \underset{v}{\pounds} C_{\nu\mu\lambda} = -\,(\underset{v}{\pounds}\{{}^{\,\rho}_{\omega\nu}\})C_{\rho\mu\lambda} - (\underset{v}{\pounds}\{{}^{\,\rho}_{\omega\mu}\})C_{\nu\rho\lambda} - (\underset{v}{\pounds}\{{}^{\,\rho}_{\omega\lambda}\})C_{\nu\mu\rho},$$

then we find

$$(3.14)\quad \underset{v}{\pounds}\nabla_\omega C_{\nu\mu\lambda} = -\,\frac{1}{n-2}(\underset{v}{\pounds}L_{\omega\rho})C_{\nu\mu\lambda}^{\cdot\cdot\cdot\rho} - \phi_\rho\nabla_\omega C_{\nu\mu\lambda}^{\cdot\cdot\cdot\rho} - 3\phi_\omega C_{\nu\mu\lambda} - C_{\omega\mu\lambda}\phi_\nu - C_{\nu\omega\lambda}\phi_\mu - C_{\nu\mu\omega}\phi_\lambda + \phi^\rho(g_{\omega\nu}C_{\rho\mu\lambda} + g_{\omega\mu}C_{\nu\rho\lambda} + g_{\omega\lambda}C_{\nu\mu\rho}).$$

We can continue this process as for as we wish. All equations contain only ϕ, ϕ_λ, $v^\varkappa$ and $\nabla_\lambda v^\varkappa$.

The above discussion shows the following. In order that a V_n admit an infinitesimal conformal motion $\xi^\varkappa \rightarrow \xi^\varkappa + v^\varkappa dt$, it is necessary and sufficient that the mixed system of partial differential equations

$$(3.15)\quad \begin{cases} v_{(\mu\lambda)} = \phi g_{\mu\lambda}, \\ \nabla_\lambda v^\varkappa = v_\lambda^{\cdot\varkappa}, \\ \nabla_\mu v_\lambda^{\cdot\varkappa} = -\,K_{\nu\mu\lambda}^{\cdot\cdot\cdot\varkappa}v^\nu + 2A_{(\mu}^\varkappa\phi_{\lambda)} - \phi^\varkappa g_{\mu\lambda}, \\ \nabla_\lambda\phi = \phi_\lambda, \\ \nabla_\mu\phi_\lambda = \dfrac{1}{n-2}[v^\rho\nabla_\rho L_{\mu\lambda} + L_{\rho\lambda}v_\mu^{\cdot\rho} + L_{\mu\rho}v_\lambda^{\cdot\rho}] \end{cases}$$

with $(n+1)^2$ unknowns ϕ, ϕ_λ, $v^\varkappa$, $v_\lambda^{\cdot\varkappa}$ admit solutions.

The integrability conditions of (3.15) are given by (3.8), (3.11), (3.13), (3.14) and the equations obtained in the same way. Thus we have

THEOREM 3.1.[1] *In order that a V_n admit a group of conformal motions, it is necessary and sufficient that the equations $v_{(\mu\lambda)} = \phi g_{\mu\lambda}$ and* (3.8), (3.11), (3.13), ... *be algebraically consistent with respect to ϕ, ϕ_λ, $v^\varkappa$ and $v_\lambda^{\cdot\varkappa}$. If there are, among the equations* (3.8), (3.11), (3.13), ..., *exactly s equa-*

[1] G. T., p. 55.

tions which are linearly independent among themselves and of $v_{(\mu\lambda)} = \phi g_{\mu\lambda}$, *then the space admits a* $\frac{1}{2}(n+1)(n+2) - s$ *parameter group of conformal motions.*

In order that a V_n admit a group of conformal motions of the maximum order $\frac{1}{2}(n+1)(n+2)$, it is necessary and sufficient that the equations

$$\underset{v}{\pounds} C_{\nu\mu\lambda}^{\cdots\kappa} = v^\sigma \nabla_\sigma C_{\nu\mu\lambda}^{\cdots\kappa} - C_{\nu\mu\lambda}^{\cdots\rho}\nabla_\rho v^\kappa + C_{\sigma\mu\lambda}^{\cdots\kappa}\nabla_\nu v^\sigma + C_{\nu\sigma\lambda}^{\cdots\kappa}\nabla_\mu v^\sigma + C_{\nu\mu\sigma}^{\cdots\kappa}\nabla_\lambda v^\sigma = 0,$$

and

$$\underset{v}{\pounds} C_{\nu\mu\lambda} = v^\sigma \nabla_\sigma C_{\nu\mu\lambda} + C_{\sigma\mu\lambda}\nabla_\nu v^\sigma + C_{\nu\sigma\lambda}\nabla_\mu v^\sigma + C_{\nu\mu\sigma}\nabla_\lambda v^\sigma = - C_{\nu\mu\lambda}^{\cdots\kappa}\phi_\kappa$$

be identically satisfied by any ϕ, ϕ_λ ,v^κ and $\nabla_\lambda v^\kappa$ such that

$$\nabla_{(\lambda} v_{\kappa)} = \phi g_{\lambda\kappa}.$$

From the arbitrariness of the ϕ_λ and v^κ, we find

$$\nabla_\sigma C_{\nu\mu\lambda} = 0 \quad \text{and} \quad C_{\nu\mu\lambda}^{\cdots\kappa} = 0.$$

In this case, the equation $\underset{v}{\pounds} C_{\nu\mu\lambda} = - C_{\nu\mu\lambda}^{\cdots\kappa}\phi_\kappa = 0$ can be written as

$$(3.14) \qquad (A_\nu^\rho C_{\sigma\mu\lambda} + A_\mu^\rho C_{\nu\sigma\lambda} + A_\lambda^\rho C_{\nu\mu\sigma}) g^{\sigma\tau}\nabla_\rho v_\tau = 0.$$

The equation (3.14) is of the form

$$(3.15) \qquad E_{\nu\mu\lambda}^{\cdots\rho\tau}\nabla_\rho v_\tau = 0,$$

where

$$(3.16) \qquad E_{\nu\mu\lambda}^{\cdots\rho\tau} \overset{\text{def}}{=} (A_\nu^\rho C_{\sigma\mu\lambda} + A_\mu^\rho C_{\nu\sigma\lambda} + A_\lambda^\rho C_{\nu\mu\sigma}) g^{\sigma\tau}.$$

Since the equation (3.15) can also be written as

$$(3.17) \qquad E_{\nu\mu\lambda}^{\cdots\rho\tau}\nabla_{(\rho} v_{\tau)} + E_{\nu\mu\lambda}^{\cdots\rho\tau}\nabla_{[\rho} v_{\nu]} = 0,$$

in order that (3.15) be satisfied for any $\nabla_\rho v_\tau$ satisfying $\nabla_{(\rho} v_{\tau)} = \phi g_\rho$ we must have

$$\phi E_{\nu\mu\lambda}^{\cdots\rho\tau} g_{\rho\tau} + E_{\nu\mu\lambda}^{\cdots\rho\tau}\nabla_{[\rho} v_{\tau]} = 0$$

for any $\nabla_{[\rho} v_{\tau]}$, from which it follows that

$$(3.18) \qquad E_{\nu\mu\lambda}^{\cdots\rho\tau} g_{\rho\tau} = 0, \quad E_{\nu\mu\lambda}^{\cdots[\rho\tau]} = 0.$$

Writing out these equations, we get

$$A_\nu^\rho C_{\rho\mu\lambda} + A_\mu^\rho C_{\nu\rho\lambda} + A_\lambda^\rho C_{\nu\mu\rho} = 0$$

and

$$(A_\nu^{[\rho} C_{\sigma\mu\lambda} + A_\mu^{[\rho} C_{\nu\sigma\lambda} + A_\lambda^{[\rho} C_{\nu\mu\sigma}) g^{\tau]\sigma} = 0,$$

and consequently $C_{\nu\mu\lambda} = 0$. Thus we have

THEOREM 3.2.[1] *In order that a V_n, $n \geq 3$, admit a group of conformal motions of the maximum order $\frac{1}{2}(n+1)(n+2)$, it is necessary and sufficient that the V_n be a C_n.*

§ 4. A group as group of conformal motions.

We apply now Theorems 3.1, 3.2 and 3.3 of Ch. III to the case of conformal motions. We consider a G_r in an X_n and denote the vectors generating the group by $\underset{b}{v}^\varkappa$. We first consider the case in which the rank of $\underset{b}{v}^\varkappa$ in a neighbourhood is $r \leq n$. We choose a coordinate system with respect to which we have (3.2) of Ch. III. Then the equations $\underset{a}{\pounds}\mathfrak{G}_{\lambda\varkappa} = 0$, $\mathfrak{G}_{[\lambda\varkappa]} = 0$, and $\mathrm{Det}(\mathfrak{G}_{\lambda\varkappa}) = 1$ become

$$\underset{a}{\pounds}\mathfrak{G}_{\lambda\varkappa} = \underset{a}{v}^\alpha \partial_\alpha \mathfrak{G}_{\lambda\varkappa} + \mathfrak{G}_{\alpha\varkappa} \partial_\lambda \underset{a}{v}^\alpha + \mathfrak{G}_{\lambda\alpha} \partial_\varkappa \underset{a}{v}^\alpha - \frac{2}{n} \mathfrak{G}_{\lambda\varkappa} \partial_\alpha \underset{a}{v}^\alpha = 0, \tag{4.1}$$

$$\mathfrak{G}_{[\lambda\varkappa]} = 0, \quad \mathrm{Det}(\mathfrak{G}_{\lambda\varkappa}) = 1. \tag{4.2}$$

and consequently, defining the functions $\Theta_{\alpha\lambda\varkappa}(\mathfrak{G}, \xi)$ by

$$\underset{a}{v}^\alpha \Theta_{\alpha\lambda\varkappa} = - \mathfrak{G}_{\alpha\varkappa} \partial_\lambda \underset{a}{v}^\alpha - \mathfrak{G}_{\lambda\alpha} \partial_\varkappa \underset{a}{v}^\alpha + \frac{2}{n} \mathfrak{G}_{\lambda\varkappa} \partial_\alpha \underset{a}{v}^\alpha, \tag{4.3}$$

we obtain

$$\underset{a}{\pounds}\mathfrak{G}_{\lambda\varkappa} = \underset{a}{v}^\alpha [\partial_\alpha \mathfrak{G}_{\lambda\varkappa} - \Theta_{\alpha\lambda\varkappa}(\mathfrak{G}, \xi)] = 0, \tag{4.4}$$

from which

$$\partial_\alpha \mathfrak{G}_{\lambda\varkappa} = \Theta_{\alpha\lambda\varkappa}(\mathfrak{G}, \xi); \quad \mathfrak{G}_{[\lambda\varkappa]} = 0, \quad \mathrm{Det}(\mathfrak{G}_{\lambda\varkappa}) = 1. \tag{4.5}$$

By the same method as was used in § 3 of Ch. III, we can prove

$$\Theta_{\gamma\sigma\rho} \frac{\partial \Theta_{\beta\lambda\varkappa}}{\partial \mathfrak{G}_{\sigma\rho}} + \partial_\gamma \Theta_{\beta\lambda\varkappa} = \Theta_{\beta\sigma\rho} \frac{\partial \Theta_{\gamma\lambda\varkappa}}{\partial \mathfrak{G}_{\sigma\rho}} + \partial_\beta \Theta_{\gamma\lambda\varkappa} \tag{4.6}$$

---

[1] SASAKI [1]; TAUB [1]; G. T., P. 56.

and also

$$(4.7)\qquad \begin{cases} \underset{a}{v}^{\alpha}\Theta_{\alpha[\lambda\varkappa]} = -\mathfrak{G}_{[\alpha\varkappa]}\partial_{\lambda}\underset{a}{v}^{\alpha} - \mathfrak{G}_{[\alpha\lambda]}\partial_{\varkappa}\underset{a}{v}^{\alpha} + \dfrac{2}{n}\mathfrak{G}_{[\lambda\varkappa]}\partial_{\alpha}\underset{a}{v}^{\alpha}, \\ \underset{a}{v}^{\alpha}\overset{-1}{\mathfrak{G}}{}^{\lambda\varkappa}\Theta_{\alpha\lambda\varkappa} = 0, \end{cases}$$

The equations (4.6) and (4.7) show that the mixed system (4.5) is completely integrable. Thus we have

THEOREM 4.1. *A G_r in an X_n for which the rank of $\underset{b}{v}^{\varkappa}$ in a neighbourhood is $r \leq n$ can be regarded as a group of conformal motions in a V_n whose fundamental tensor density can contain $\frac{1}{2}n(n+1) - 1$ arbitrary constants.*

We next consider a G_r in an X_n for which the rank of $\underset{b}{v}^{\varkappa}$ in a neighbourhood is $q < r, n$. We choose a coordinate system with respect to which (3.9) of Ch. III holds. Then we get

$$(4.8)\qquad \begin{cases} \underset{i}{\pounds}\mathfrak{G}_{\lambda\varkappa} = \underset{i}{v}^{\alpha}\partial_{\alpha}\mathfrak{G}_{\lambda\varkappa} + \mathfrak{G}_{\alpha\varkappa}\partial_{\lambda}\underset{i}{v}^{\alpha} + \mathfrak{G}_{\lambda\alpha}\partial_{\varkappa}\underset{i}{v}^{\alpha} - \dfrac{2}{n}\mathfrak{G}_{\lambda\varkappa}\partial_{\alpha}\underset{i}{v}^{\alpha} = 0, \\ \underset{u}{\pounds}\mathfrak{G}_{\lambda\varkappa} = \varphi^{i}_{u}\underset{i}{\pounds}\mathfrak{G}_{\lambda\varkappa} + \mathfrak{G}_{\alpha\varkappa}(\partial_{\lambda}\varphi^{i}_{u})\underset{i}{v}^{\alpha} + \mathfrak{G}_{\lambda\alpha}(\partial_{\varkappa}\varphi^{i}_{u})\underset{i}{v}^{\alpha} \\ \qquad\qquad - \dfrac{2}{n}\mathfrak{G}_{\lambda\varkappa}(\partial_{\alpha}\varphi^{i}_{u})\underset{i}{v}^{\alpha} = 0, \\ \mathfrak{G}_{[\lambda\varkappa]} = 0,\ \operatorname{Det}(\mathfrak{G}_{\lambda\varkappa}) = 1. \end{cases}$$

If we put

$$(4.9)\qquad \underset{i}{v}^{\alpha}\Theta_{\alpha\lambda\varkappa} \overset{\text{def}}{=} -\mathfrak{G}_{\alpha\varkappa}\partial_{\lambda}\underset{i}{v}^{\alpha} - \mathfrak{G}_{\lambda\alpha}\partial_{\varkappa}\underset{i}{v}^{\alpha} + \frac{2}{n}\mathfrak{G}_{\lambda\varkappa}\partial_{\alpha}\underset{i}{v}^{\alpha},$$

$$(4.10)\qquad \Xi_{u\lambda\varkappa} \overset{\text{def}}{=} \mathfrak{G}_{\alpha\varkappa}(\partial_{\lambda}\varphi^{i}_{u})\underset{i}{v}^{\alpha} + \mathfrak{G}_{\lambda\alpha}(\partial_{\varkappa}\varphi^{i}_{u})\underset{i}{v}^{\alpha} - \frac{2}{n}\mathfrak{G}_{\lambda\varkappa}(\partial_{\alpha}\phi^{i}_{u})\underset{i}{v}^{\alpha},$$

we can write (4.8) in the form

$$\begin{cases} \underset{i}{\pounds}\mathfrak{G}_{\lambda\varkappa} = \underset{i}{v}^{\alpha}[\partial_{\alpha}\mathfrak{G}_{\lambda\varkappa} - \Theta_{\alpha\lambda\varkappa}(\mathfrak{G}, \xi)] = 0, \\ \underset{u}{\pounds}\mathfrak{G}_{\lambda\varkappa} = \varphi^{i}_{u}\underset{i}{\pounds}\mathfrak{G}_{\lambda\varkappa} + \Xi_{u\lambda\varkappa}(\mathfrak{G}, \xi) = 0, \\ \mathfrak{G}_{[\lambda\varkappa]} = 0,\ \operatorname{Det}(\mathfrak{G}_{\lambda\varkappa}) = 1 \end{cases}$$

or

$$(4.11)\qquad \begin{cases} \partial_\alpha \mathfrak{G}_{\lambda\varkappa} = \Theta_{\alpha\lambda\varkappa}(\mathfrak{G}, \xi), \\ \Xi_{u\lambda\varkappa}(\mathfrak{G}, \xi) = 0,\ \mathfrak{G}_{[\lambda\varkappa]} = 0,\ \mathrm{Det}(\mathfrak{G}_{\lambda\varkappa}) = 1. \end{cases}$$

By the same method as was used in § 3 of Ch. III, we can prove

$$\Theta_{\gamma\sigma\rho} \frac{\partial \Theta_{\beta\lambda\varkappa}}{\partial \mathfrak{G}_{\sigma\rho}} + \partial_\gamma \Theta_{\beta\lambda\varkappa} = \Theta_{\beta\sigma\rho} \frac{\partial \Theta_{\gamma\lambda\varkappa}}{\partial \mathfrak{G}_{\sigma\rho}} + \partial_\beta \Theta_{\gamma\lambda\varkappa},$$

$$\Theta_{\alpha\sigma\rho} \frac{\partial \Xi_{u\lambda\varkappa}}{\partial \mathfrak{G}_{\sigma\rho}} + \partial_\alpha \Xi_{u\lambda\varkappa} = 0,$$

$$\underset{i}{v}^\alpha \Theta_{\alpha[\lambda\varkappa]} = - \mathfrak{G}_{[\alpha\varkappa]} \partial_\lambda \underset{i}{v}^\alpha - \mathfrak{G}_{[\alpha\lambda]} \partial_\varkappa \underset{i}{v}^\alpha + \frac{2}{n} \mathfrak{G}_{[\lambda\varkappa]} \partial_\alpha \underset{i}{v}^\alpha$$

$$\underset{a}{v}^\alpha \overset{-1}{\mathfrak{G}}{}^{\lambda\varkappa} \Theta_{\alpha\lambda\varkappa} = 0$$

which shows that the mixed system (4.11) is completely integrable. Thus we have

THEOREM 4.2. *Consider a G_r in an X_n for which the rank of $\underset{b}{v}^\varkappa$ in a neighbourhood is $q < r, n$. If, in a neighbourhood such that* (3.9) *of* Ch. III *holds, the equations $\Xi_{u\lambda\varkappa}(\mathfrak{G}, \xi) = 0$, $\mathfrak{G}_{[\lambda\varkappa]} = 0$ and $\mathrm{Det}(\mathfrak{G}_{\lambda\varkappa}) = 1$ are compatible at a point of the space, then the group can be regarded as a group of conformal motions in a V_n.*

A similar theorem holds for a multiply transitive group.

§ 5. Homothetic motions.[1]

Consider an infinitesimal transformation $'\xi^\varkappa = \xi^\varkappa + v^\varkappa dt$ in a V_n. If the square of the distance $ds^2 = g_{\mu\lambda}(\xi) d\xi^\mu d\xi^\lambda$ between $\xi^\varkappa$ and $\xi^\varkappa + d\xi^\varkappa$ and the square of the distance $d's^2 = g_{\mu\lambda}('\xi) d'\xi^\mu d'\xi^\lambda$ between $'\xi^\varkappa$ and $'\xi^\varkappa + d'\xi^\varkappa$ have always the same constant ratio, that is, if

$$(5.1)\qquad \underset{v}{\pounds} g_{\mu\lambda} = 2c g_{\mu\lambda}, \qquad c = \text{constant},$$

then the infinitesimal transformation is called a *homothetic motion*.

From (5.1) and the formula

$$(5.2)\qquad \underset{v}{\pounds} \{^{\,\varkappa}_{\mu\lambda}\} = \tfrac{1}{2} g^{\varkappa\rho} [\nabla_\mu \underset{v}{\pounds} g_{\lambda\rho} + \nabla_\lambda \underset{v}{\pounds} g_{\mu\rho} - \nabla_\rho \underset{v}{\pounds} g_{\mu\lambda}],$$

[1] SHANKS [1]; YANO [15].

we find $\underset{v}{\pounds}\{^{\varkappa}_{\mu\lambda}\} = 0$. Thus a homothetic motion is an affine motion. Conversely, if a conformal motion is an affine motion, then we have $\underset{v}{\pounds} g_{\mu\lambda} = 2\phi g_{\mu\lambda}$ and $\underset{v}{\pounds}\{^{\varkappa}_{\mu\lambda}\} = 0$, from which we conclude $\phi =$ constant. Thus we have

THEOREM 5.1. *In order that a transformation in a V_n be homothetic, it is necessary and sufficient that the transformation be conformal and affine at the same time.*

More generally, if a conformal motion is a projective motion, we have $\underset{v}{\pounds} g_{\mu\lambda} = 2\phi g_{\mu\lambda}$ and $\underset{v}{\pounds}\{^{\varkappa}_{\mu\lambda}\} = A^{\varkappa}_{\mu} p_{\lambda} + A^{\varkappa}_{\lambda} p_{\mu}$.

From these equations and (5.2), it follows that $\phi =$ constant and $p_\lambda = 0$. Thus we have

THEOREM 5.2. *In order that a transformation in a V_n be homothetic, it is necessary and sufficient that the transformation be conformal and projective at the same time.*

Applying the formula (4.9) of Ch. I to the fundamental tensor $g_{\mu\lambda}$, we obtain

$$\underset{v}{\pounds}\nabla_\nu g_{\mu\lambda} - \nabla_\nu \underset{v}{\pounds} g_{\mu\lambda} = -(\underset{v}{\pounds}\{^{\rho}_{\nu\mu}\}) g_{\rho\lambda} - (\underset{v}{\pounds}\{^{\rho}_{\nu\lambda}\}) g_{\mu\rho},$$

from which, for an affine motion,

$$\nabla_\nu \underset{v}{\pounds} g_{\mu\lambda} = 0. \tag{5.3}$$

If the metric of V_n is not decomposable, we obtain[1], from (5.3),

$$\underset{v}{\pounds} g_{\mu\lambda} = 2c g_{\mu\lambda}, \qquad c = \text{constant}.$$

Thus we have

THEOREM 5.3. *In a V_n whose metric is not decomposable, an affine motion is homothetic.*

When the constant c is zero, a homothetic motion reduces to a motion. We call a *proper* homothetic motion a homothetic motion for which $c \neq 0$ and *c the homothetic constant.*

Now if we consider an infinitesimal proper homothetic motion $\xi^\varkappa \to \xi^\varkappa + v^\varkappa dt$ whose streamlines are geodesics, we have $\underset{v}{\pounds} g_{\mu\lambda} = 2\nabla_{(\mu} v_{\lambda)}$

---

[1] T. Y. THOMAS [1, 2]; SCHOUTEN [8], p. 286.

$= 2cg_{\mu\lambda}$ and $v^\mu \nabla_\mu v_\lambda = \alpha v_\lambda$, where α is a scalar. Transvecting the latter equation with v^λ, we obtain $c = \alpha$ by virtue of the former. Transvecting next the former equation with v^λ, we find $cv_\mu = \nabla_\mu(\frac{1}{2}v_\lambda v^\lambda)$, which shows that v_λ is a gradient vector. Thus $\nabla_\mu v_\lambda$ is symmetric in μ and λ, and consequently

$$\nabla_\mu v_\lambda = cg_{\mu\lambda}, \tag{5.4}$$

that is, the $v^\varkappa$ is a concurrent vector field.[1] Since the converse is evident, we have

THEOREM 5.4. *In order that a V_n admit an infinitesimal proper homothetic motion whose streamlines are geodesics, it is necessary and sufficient that the V_n admit a concurrent vector field.*

In order that a V_n admit a concurrent vector field, it is necessary and sufficient that there exist a coordinate system with respect to which the linear element takes the form

$$ds^2 = (d\xi^1)^2 + (\xi^1)^2 f_{\zeta\eta}(\xi^2, \ldots, \xi^n) d\xi^\zeta d\xi^\eta \qquad (\eta, \zeta = 2, 3, \ldots, n). \tag{5.5}$$

Thus we have

THEOREM 5.5. *In order that a V_n admit an infinitesimal proper homothetic motion whose streamlines are geodesics, it is necessary and sufficient that there exist a coordinate system with respect to which the linear element of V_n takes the form* (5.5).

In order that an infinitesimal transformation be a projective (conformal) motion in a V_n, it is necessary and sufficient that the transformation carry every geodesic (conformal circle) into a geodesic (conformal circle). Thus from Theorem 5.2 we have

THEOREM 5.6. *In order that an infinitesimal transformation be homothetic, it is necessary and sufficient that the transformation carry every geodesic into a geodesic and every conformal circle into a conformal circle.*

If we take a coordinate system with respect to which $v^\varkappa = e^\varkappa_1$, then the equation $\underset{v}{\mathcal{L}} g_{\mu\lambda} = 2cg_{\mu\lambda}$ gives $\partial g_{\mu\lambda}/\partial\xi^1 = 2cg_{\mu\lambda}$, from which

$$g_{\mu\lambda} = e^{2c\xi^1} f_{\mu\lambda}(\xi^2, \ldots, \xi^n). \tag{5.6}$$

---

[1] YANO [6].

Conversely, if there exists a coordinate system with respect to which the fundamental tensor takes the form (5.6), then the space admits a one-parameter group of homothetic motions generated by $'\xi^\varkappa = \xi^\varkappa + \underset{1}{e}^\varkappa dt$. Thus we have

THEOREM 5.7. *If a V_n admits an infinitesimal homothetic motion, then the V_n admits also a one-parameter group of homothetic motions generated by the infinitesimal homothetic motion.*

THEOREM 5.8. *In order that a V_n admit a one-parameter group of homothetic motions with the homothetic constant c, it is necessary and sufficient that there exist a coordinate system with respect to which the fundamental tensor takes the form* (5.6).

If we take a coordinate system with respect to which $v^\varkappa = \xi^\varkappa$, then the equation $\underset{v}{£} g_{\mu\lambda} = 2cg_{\mu\lambda}$ becomes $\xi^\nu \partial_\nu g_{\mu\lambda} = 2(c-1)g_{\mu\lambda}$, from which we see that the $g_{\mu\lambda}$ are homogeneous functions of degree $2(c-1)$ with respect to $\xi^\varkappa$. Thus we have

THEOREM 5.8. *In order that a V_n admit a one-parameter group of homothetic motions with homothetic constant c, it is necessary and sufficient that there exist a coordinate system with respect to which the components of the fundamental tensor are homogeneous functions of degree* $2(c-1)$ *of the coordinates.*

Using (5.1), we can easily verify that Theorems 2.3 and 2.4 of Ch. III are also valid for a group of homothetic motions.

If $\underset{b}{£} f$ are generators of r one-parameter groups of transformations, then we have

$$(\underset{c}{£}\underset{b}{£}) g_{\mu\lambda} = \underset{cb}{£} g_{\mu\lambda}.$$

If $\underset{b}{£} f$ are generators of r one-parameter groups of homothetic motions, then we have $(\underset{c}{£}\underset{b}{£}) g_{\mu\lambda} = 0$ and consequently $\underset{cb}{£} g_{\mu\lambda} = 0$. Thus we have

THEOREM 5.9. *If $\underset{b}{£} f$ are generators of r one-parameter groups of homothetic motions, then $\underset{cb}{£} f$ are those of a one-parameter group of motions.*

If $\underset{b}{£} f$ are r generators of an r-parameter group of transformations, then we have

$$(\underset{c}{\pounds}\underset{b}{\pounds})g_{\mu\lambda} = c^a_{cb}\underset{a}{\pounds}g_{\mu\lambda},$$

where c^a_{cb} are the structural constants of the group. For an r-parameter group of homothetic motions, we have $(\underset{c}{\pounds}\underset{b}{\pounds})g_{\mu\lambda} = 0$ and $\underset{a}{\pounds}g_{\mu\lambda} = 2c_a g_{\mu\lambda}$ and consequently $c^a_{cb}c_a = 0$. Thus we have

THEOREM 5.10. *If $\underset{b}{\pounds}f$ are r generators of an r-parameter group of homothetic motions with homothetic constants c_a, then there exist the relations $c^a_{cb}c_a = 0$ between the structural constants c^a_{cb} and the homothetic constants c_a.*

Since $c^a_{cb}c_a = 0$ means that the first derived group is of order $\leq r - 1$, combining Theorems 5.9 and 5.10, we get

THEOREM 5.11. *The first derived group of a group of homothetic motions in a V_n is a group of motions of order $\leq r - 1$.*

Moreover we have

THEOREM 5.12. *If $\underset{b}{\pounds}f$ are generators of the complete set of r one-parameter groups of homothetic motions, they are generators of an r-parameter group G_r of homothetic motions. Moreover G_r must contain a complete set of one-parameter groups of motions, consequently, G_r contains a complete group of motions.*

§ 6. Homothetic motions in conformally related spaces.

Let a V_n admit an r-parameter group G_r of homothetic motions whose generators are $\underset{a}{\pounds}f = \underset{a}{v}^\varkappa\partial_\varkappa f : \underset{a}{\pounds}g_{\mu\lambda} = 2c_a g_{\mu\lambda}$. In order that a $'V_n$ conformal to V_n admit G_r as a group of homothetic motions with the same homothetic constants c_a, it is necessary and sufficient that there exist a function ρ such that $\underset{a}{\pounds}(\rho^2 g_{\mu\lambda}) = 2c_a\rho^2 g_{\mu\lambda}$, from which $\underset{a}{\pounds}\rho^2 = 0$. Now if we assume that the rank of $\underset{b}{v}^\varkappa$ is $r < n$, then the equations $\underset{a}{\pounds}\rho^2 = 0$ are completely integrable and admit $n - r$ functionally independent solutions. Thus we have

THEOREM 6.1. *If a V_n admits an r-parameter group G_r of homothetic motions such that the rank of the generators $\underset{b}{v}^\varkappa$ is $r < n$, then there exist spaces which are (not trivially) conformal to V_n and which admit G_r as a group of homothetic motions with the same homothetic constants.*

Let again a V_n admit an r-parameter group G_r of homothetic motions: $\underset{a}{\pounds} g_{\mu\lambda} = 2c_a g_{\mu\lambda}$. In order that a $'V_n$ conformal to V_n admit G_r as a group of motions, it is necessary and sufficient that there exist a function ρ such that $\underset{a}{\pounds}(\rho^2 g_{\mu\lambda}) = 0$, from which $\underset{a}{\pounds} \log \rho = - c_a$. If we assume that the rank of $\underset{b}{v}^{\varkappa}$ is $r < n$, then the equations $\underset{a}{\pounds} \log \rho = - c_a$ are completely integrable by virtue of $(\underset{c}{\pounds}\underset{b}{\pounds}) \log \rho = c_{cb}^{a} \underset{a}{\pounds} \log \rho$ and $c_{cb}^{a} c_a = 0$. Thus we have

THEOREM 6.2. *If a V_n admits an r-parameter group G_r of homothetic motions such that the rank of the generators $\underset{b}{v}^{\varkappa}$ is $r < n$, then there exists a $'V_n$ which is conformal to V_n and which admits G_r as a group of motions.*

§ 7. Subgroups of homothetic motions contained in a group of conformal motions or in a group of affine motions.

Let a V_n admit an r-parameter group G_r of conformal motions: $\underset{a}{\pounds} g_{\mu\lambda} = 2\phi_a g_{\mu\lambda}$, ϕ_a being r scalars. In order that the group G_r contain a subgroup of homothetic motions, it is necessary and sufficient that there exist constants c^a not all zero such that $c^a \phi_a =$ constant. By successive covariant differentiations of this equation, we get $c_a \nabla_\lambda \phi_a = 0$, $c^a \nabla_{\lambda_2\lambda_1} \phi_a = 0$, ... If we denote by $\alpha_1, \alpha_2, \ldots$ the a-ranks of the sets $\nabla_\lambda \phi_a$; $\nabla_\lambda \phi_a$, $\nabla_{\lambda_2\lambda_1} \phi_a$; .... respectively, then we have $\alpha_1 \leqq \alpha_2 \leqq \ldots.$ Since the equations $c^a \nabla_\lambda \phi_a = 0$, $c^a \nabla_{\lambda_2\lambda_1} \phi_a = 0$, ... admit a set of solutions which are not all zero, we must have $\alpha_1 \leqq \alpha_2 \leqq \ldots < r$. On the other hand, we can easily prove that, if $\alpha_p = \alpha_{p+1}$, then $\alpha_{p+1} = \alpha_{p+2}$. Thus the ranks of the matrices must satisfy

$$\alpha_1 \leqq \alpha_2 \leqq \ldots \leqq \alpha_p = \alpha_{p+1} = \ldots = s < r. \tag{7.1}$$

Conversely, if the ranks $\alpha_1, \alpha_2, \ldots$ satisfy the relation (7.1), then we can find sets of linearly independent solutions $f_A^a(\xi)$ $(A, B, C, \ldots = 1, 2, \ldots, s)$ which are not all zero and such that

$$f_A^a \nabla_\lambda \phi_a = 0,\ f_A^a \nabla_{\lambda_2\lambda_1} \phi_a = 0,\ \ldots,\ f_A^a \nabla_{\lambda_p \cdots \lambda_1} \phi_a = 0. \tag{7.2}$$

Differentiating these equations covariantly and taking account of the fact that the ranks satisfy (7.1), we find that $\nabla_\lambda f_A^a$ are also solutions of (7.2). Thus there must exist a set of functions $P_{\lambda B}^{A}(\xi)$ such that

$$\nabla_\lambda f_B^a = P_{\lambda B}^{A} f_A^a.$$

The integrability conditions of this equation are

$$\nabla_\mu P^A_{\lambda B} - \nabla_\lambda P^A_{\mu B} + P^A_{\mu E} P^E_{\lambda B} - P^A_{\lambda E} P^E_{\mu B} = 0. \tag{7.3}$$

These equations show that there exists a set of functions $h^A(\xi)$ such that $h^A f^a_A =$ constants. In fact, the equations $\nabla_\lambda(h^A f^a_A) = 0$ give $\nabla_\lambda h^A + P^A_{\lambda B} h^B = 0$, which are completely integrable because of (7.3). Thus putting $c^a = h^A f^a_A$, we obtain $c^a \nabla_\lambda \phi_a = 0$ and consequently $c^a \phi_a =$ constant. Thus we have

THEOREM 7.1. *In order that an r-parameter group of conformal motions in a V_n contain a subgroup of homothetic motions, it is necessary and sufficient that the a-ranks $\alpha_1, \alpha_2, \ldots$ of the sets $\nabla_\lambda \phi_a$; $\nabla_\lambda \phi_a$, $\nabla_{\lambda_2 \lambda_1} \phi_a$; ... satisfy the relation* (7.1).

Let a V_n admit an r-parameter group G_r of affine motions: $\underset{a}{\pounds}\{^{\,\varkappa}_{\mu\lambda}\} = 0$. In order that the group G_r contain a subgroup of homothetic motions, it is necessary and sufficient that there exist constants c^a not all zero and c such that $c^a \underset{a}{\pounds} g_{\mu\lambda} = 2c g_{\mu\lambda}$. Thus the $\mu\lambda$-rank of the set $\underset{a}{\pounds} g_{\mu\lambda}$, $g_{\mu\lambda}$ must be less than $r + 1$.

Conversely, if the $\mu\lambda$-rank s of the set $\underset{a}{\pounds} g_{\mu\lambda}$, $g_{\mu\lambda}$ is less than $r + 1$, then we can find $r + 1 - s$ linearly independent solutions $f^a_L(\xi)$, $f_L(\xi)$ of $c^a \underset{a}{\pounds} g_{\mu\lambda} = 2c g_{\mu\lambda}$ such that

$$f^a_L(\xi) \underset{a}{\pounds} g_{\mu\lambda} = 2 f_L(\xi) g_{\mu\lambda}, \tag{7.4}$$

$$(L, M, N = 1, 2, \ldots, r + 1 - s).$$

Differentiating (7.4) covariantly, we obtain

$$(\nabla_\nu f^a_L) \underset{a}{\pounds} g_{\mu\lambda} + f^a_L \nabla_\nu \underset{a}{\pounds} g_{\mu\lambda} = 2(\nabla_\nu f_L) g_{\mu\lambda},$$

from which

$$(\nabla_\nu f^a_L) \underset{a}{\pounds} g_{\mu\lambda} = 2(\nabla_\nu f_L) g_{\mu\lambda}$$

because of $\nabla_\nu \pounds g_{\mu\lambda} = 0$. Thus $\nabla_\nu f^a_L$, $\nabla_\nu f_L$ are also solutions of (7.4) and consequently there exist functions $P^L_{\nu M}$ such that

$$\nabla_\nu f^a_M = P^L_{\nu M} f^a_L, \quad \nabla_\nu f_M = P^L_{\nu M} f_L.$$

The integrability conditions of these equations are

$$\nabla_\omega P^L_{\nu M} - \nabla_\nu P^L_{\omega M} + P^L_{\omega N} P^N_{\nu M} - P^L_{\nu N} P^N_{\omega M} = 0. \tag{7.5}$$

This equation shows that there exists a set of functions $h^L(\xi)$ such that $h^L f^a_L =$ constants, $h^L f_L =$ constant. Thus we have

THEOREM 7.2. *In order that an r-parameter group of affine motions in a V_n contain a subgroup of homothetic motions, it is necessary and sufficient that the $\mu\lambda$-rank of the set $\underset{a}{\pounds} g_{\mu\lambda}$, $g_{\mu\lambda}$ be less than $r + 1$.*

§ 8. Integrability conditions of $\underset{v}{\pounds} g_{\mu\lambda} = 2cg_{\mu\lambda}$.

From the equation

$$\underset{v}{\pounds} g_{\mu\lambda} = 2\nabla_{(\mu} v_{\lambda)} = 2cg_{\mu\lambda},$$

we obtain

$$(8.1) \qquad v_{(\mu\lambda)} = \frac{1}{n} g^{\tau\sigma} v_{\tau\sigma} g_{\mu\lambda}, \quad v_{\mu\lambda} \overset{\text{def}}{=} \nabla_\mu v_\lambda.$$

The fact that the c is a constant can be expressed by $\nabla_\nu \underset{v}{\pounds} g_{\mu\lambda} = 0$, or by

$$\underset{v}{\pounds} \{^{\varkappa}_{\mu\lambda}\} = \nabla_\mu \nabla_\lambda v^\varkappa + K_{\nu\mu\lambda}{}^{\cdot\cdot\cdot\varkappa} v^\nu = 0,$$

from which

$$(8.2) \qquad \nabla_\lambda v^\varkappa = v_\lambda{}^{\cdot\varkappa}, \quad \nabla_\mu v_\lambda{}^{\cdot\varkappa} = - K_{\nu\mu\lambda}{}^{\cdot\cdot\cdot\varkappa} v^\nu.$$

Thus we have a mixed system of the partial differential equations (8.1) and (8.2). Since the integrability conditions of this mixed system are

$$(8.3) \qquad \underset{v}{\pounds} K_{\nu\mu\lambda}{}^{\cdot\cdot\cdot\varkappa} = 0, \quad \underset{v}{\pounds} \nabla^1_\omega K_{\nu\mu\lambda}{}^{\cdot\cdot\cdot\varkappa} = 0, \quad \ldots .$$

we have

THEOREM 8.1. *In order that a V_n admit a group of homothetic motions, it is necessary and sufficient that there exist a positive integer N such that the first N sets of equations in* (8.1) *and* (8.3) *are algebraically consistent in $v^\varkappa$ and $v_\lambda{}^{\cdot\varkappa}$ and all $v^\varkappa$ and $v_\lambda{}^{\cdot\varkappa}$ satisfying these equations satisfy the $(N + 1)$st set of equations.*

The complete integrability condition of the mixed system is that $\underset{a}{\pounds} K_{\nu\mu\lambda}{}^{\cdot\cdot\cdot\varkappa} = 0$ be identically satisfied by any $v^\varkappa$ and $v_\lambda{}^{\cdot\varkappa}$ satisfying (8.1), the number of $v^\varkappa$ and $v_\lambda{}^{\cdot\varkappa}$ which can be given arbitrarily being $n^2 + n - [\frac{1}{2}n(n + 1) - 1] = \frac{1}{2}n(n + 1) + 1$.

From this we have

$$K_{\nu\mu\lambda}^{\cdots\varkappa} = \frac{K}{n(n-1)}(A_\nu^\varkappa g_{\mu\lambda} - A_\mu^\varkappa g_{\nu\lambda})$$

and consequently, from $\underset{v}{\pounds} K_{\nu\mu\lambda}^{\cdots\varkappa} = 0$ and $\underset{v}{\pounds} g_{\mu\lambda} = 2cg_{\mu\lambda}$, we find

$$\underset{v}{\pounds} K_{\nu\mu\lambda}^{\cdots\varkappa} = 2cK_{\nu\mu\lambda}^{\cdots\varkappa} = 0.$$

Thus we have

THEOREM 8.2. *In order that a V_n admit a group of homothetic motions of the maximum order $\frac{1}{2}n(n+1)+1$, it is necessary and sufficient that the V_n be Euclidean.*

If $\underset{v}{\pounds} f = v^\varkappa \partial_\varkappa f$ is a generator of a one-parameter group of homothetic motions, then we have $\underset{v}{\pounds} g_{\mu\lambda} = 2cg_{\mu\lambda}$ and $\underset{v}{\pounds} K_{\nu\mu\lambda}^{\cdots\varkappa} = 0$ and consequently $\underset{v}{\pounds} K_{\mu\lambda} = 0$.

Thus, if the V_n is an Einstein space (or an S_n), i.e., if $K_{\mu\lambda} = \frac{1}{n} Kg_{\mu\lambda}$, K being a constant, then we have

$$\underset{v}{\pounds} K_{\mu\lambda} = \frac{1}{n} K \underset{v}{\pounds} g_{\mu\lambda} = \frac{1}{n} Kcg_{\mu\lambda} = 0.$$

Thus if $K \neq 0$, then $c = 0$ and consequently we have

THEOREM 8.3. *If an Einstein space (or an S_n) with non-vanishing curvature scalar admits a homothetic motion, it is a motion. Consequently an Einstein space (or an S_n) with non-vanishing curvature scalar cannot admit a proper homothetic motion.*

§ 9. A group as group of homothetic motions.

We apply now Theorems 3.1, 3.2 and 3.3 of Ch. III to the case of homothetic motions. We consider a G_r in an X_n and we suppose that there exist r constants c_a not all zero such that $c_{cb}^a c_a = 0$, c_{cb}^a being structural constants of the G_r. Denoting by $\underset{b}{v}^\varkappa$ r vectors generating the group, we first consider the case in which the rank of $\underset{b}{v}^\varkappa$ in a neighbourhood is $r \leq n$. We choose a coordinate system with respect to which we have (3.2) of Ch. III. Then the equations, which determine the $g_{\mu\lambda}$, are

$$\underset{a}{\pounds} g_{\mu\lambda} = \underset{a}{v}^\alpha \partial_\alpha g_{\mu\lambda} + g_{\alpha\lambda} \partial_\mu \underset{a}{v}^\alpha + g_{\mu\alpha} \partial_\lambda \underset{a}{v}^\alpha = 2c_a g_{\mu\lambda} \tag{9.1}$$

and

$$g_{[\mu\lambda]} = 0. \tag{9.2}$$

We define the functions $\Theta_{\alpha\mu\lambda}(g, \xi)$ by

$$\underset{a}{v}^{\alpha}\Theta_{\alpha\mu\lambda}(g, \xi) \overset{\text{def}}{=} -g_{\alpha\lambda}\partial_{\mu}\underset{a}{v}^{\alpha} - g_{\mu\alpha}\partial_{\lambda}\underset{a}{v}^{\alpha} + 2c_a g_{\mu\lambda} \tag{9.3}$$

then we obtain

$$\underset{a}{\pounds} g_{\mu\lambda} - 2c_a g_{\mu\lambda} = \underset{a}{v}^{\alpha}[\partial_{\alpha} g_{\mu\lambda} - \Theta_{\alpha\mu\lambda}(g, \xi)] = 0, \tag{9.4}$$

from which

$$\partial_{\alpha} g_{\mu\lambda} = \Theta_{\alpha\mu\lambda}(g, \xi), \; g_{[\mu\lambda]} = 0. \tag{9.5}$$

By the same method as was used in § 3 of Ch. III, we can prove

$$\Theta_{\gamma\sigma\rho}\frac{\partial\Theta_{\beta\mu\lambda}}{\partial g_{\sigma\rho}} + \partial_{\gamma}\Theta_{\beta\mu\lambda} = \Theta_{\beta\sigma\rho}\frac{\partial\Theta_{\gamma\mu\lambda}}{\partial g_{\sigma\rho}} + \partial_{\beta}\Theta_{\gamma\mu\lambda} \tag{9.6}$$

and

$$\underset{a}{v}^{\alpha}\Theta_{\alpha[\mu\lambda]} = -g_{[\alpha\lambda]}\partial_{\mu}\underset{a}{v}^{\alpha} - g_{[\mu\alpha]}\partial_{\lambda}\underset{a}{v}^{\alpha} + 2c_a g_{[\mu\lambda]}. \tag{9.7}$$

The equations (9.6) and (9.7) show that the mixed system (9.5) is completely integrable. Thus we have

THEOREM 9.1. *A G_r in an X_n, such that the rank of $\underset{b}{v}^{\varkappa}$ in a neighbourhood is $r \leq n$ and that there exist r constants c_a not all zero satisfying $c_{cb}^{a} c_a = 0$, can be regarded as a group of homothetic motions with homothetic constants c_a in a V_n whose fundamental tensor can contain $\frac{1}{2}n(n+1)$ arbitrary functions of $n - r$ variables.*

We next consider a G_r in an X_n for which the rank of $\underset{b}{v}^{\varkappa}$ in a neighbourhood is $q < r, n$. We choose a coordinate system with respect to which (3.9) of Ch. III holds. Then the equations, which determine the $g_{\mu\lambda}$, are

$$\underset{i}{\pounds} g_{\mu\lambda} = \underset{i}{v}^{\alpha}\partial_{\alpha} g_{\mu\lambda} + g_{\alpha\lambda}\partial_{\mu}\underset{i}{v}^{\alpha} + g_{\mu\alpha}\partial_{\lambda}\underset{i}{v}^{\alpha} = 2c_i g_{\mu\lambda}, \tag{9.8}$$

$$\underset{u}{\pounds} g_{\mu\lambda} = \varphi_u^i \underset{i}{\pounds} g_{\mu\lambda} + g_{\alpha\lambda}(\partial_{\mu}\varphi_u^i)\underset{i}{v}^{\alpha} + g_{\mu\alpha}(\partial_{\lambda}\varphi_u^i)\underset{i}{v}^{\alpha} = 2c_u g_{\mu\lambda}, \tag{9.9}$$

$$g_{[\mu\lambda]} = 0. \tag{9.10}$$

Thus, if we put

$$\underset{i}{v}^{\alpha}\Theta_{\alpha\mu\lambda} \overset{\text{def}}{=} - g_{\alpha\lambda}\partial_{\mu}\underset{i}{v}^{\alpha} - g_{\mu\alpha}\partial_{\lambda}\underset{i}{v}^{\alpha} + 2c_{i}g_{\mu\lambda}, \tag{9.11}$$

$$\Xi_{u\mu\lambda} \overset{\text{def}}{=} g_{\alpha\lambda}(\partial_{\mu}\varphi^{i}_{u})\underset{i}{v}^{\alpha} + g_{\mu\alpha}(\partial_{\lambda}\varphi^{i}_{u})\underset{i}{v}^{\alpha} + 2(\varphi^{i}_{u}c_{i} - c_{u})g_{\mu\lambda}, \tag{9.12}$$

we can write (9.8), (9.9) and (9.10) in the form

$$\begin{cases} \underset{i}{\pounds} g_{\mu\lambda} - 2c_{i}g_{\mu\lambda} = \underset{i}{v}^{\alpha}[\partial_{\alpha}g_{\mu\lambda} - \Theta_{\alpha\mu\lambda}(g, \xi)] = 0, \\ \underset{u}{\pounds} g_{\mu\lambda} - 2c_{u}g_{\mu\lambda} = \varphi^{i}_{u}(\underset{i}{\pounds} g_{\mu\lambda} - 2c_{i}g_{\mu\lambda}) + \Xi_{u\mu\lambda}(g, \xi) = 0, \\ g_{[\mu\lambda]} = 0 \end{cases}$$

from which

$$\begin{cases} \partial_{\alpha}g_{\mu\lambda} = \Theta_{\alpha\mu\lambda}(g, \xi), \\ \Xi_{u\mu\lambda}(g, \xi) = 0, \quad g_{[\mu\lambda]} = 0. \end{cases} \tag{9.13}$$

By the same method as was used in § 3 of Ch. III, we can prove

$$\Theta_{\gamma\sigma\rho}\frac{\partial\Theta_{\beta\mu\lambda}}{\partial g_{\sigma\rho}} + \partial_{\gamma}\Theta_{\beta\mu\lambda} = \Theta_{\beta\sigma\rho}\frac{\partial\Theta_{\gamma\mu\lambda}}{\partial g_{\sigma\rho}} + \partial_{\beta}\Theta_{\gamma\mu\lambda},$$

$$\Theta_{\alpha\sigma\rho}\frac{\partial\Xi_{u\mu\lambda}}{\partial g_{\sigma\rho}} + \partial_{\alpha}\Xi_{u\mu\lambda} = 0,$$

$$\underset{i}{v}^{\alpha}\Theta_{\alpha[\mu\lambda]} = - g_{[\alpha\lambda]}\partial_{\mu}\underset{i}{v}^{\alpha} - g_{[\mu\alpha]}\partial_{\lambda}\underset{i}{v}^{\alpha} + 2c_{i}g_{[\mu\lambda]}$$

which shows that the mixed system (9.13) is completely integrable. Thus we have

THEOREM 9.2. *Consider a G_r in an X_n such that the rank of $\underset{b}{v}^{\varkappa}$ in a neighbourhood is $q < r, n$ and that there exist r constants c_a satisfying $c^{a}_{cb}c_{a} = 0$. If, in a neighbourhood such that* (3.9) *of Ch.* III *holds, the equations $\Xi_{u\mu\lambda}(g, \xi) = 0$, $g_{[\mu\lambda]} = 0$, $\mathrm{Det}(g_{\mu\lambda}) \neq 0$ are compatible at a point of the space, then the group can be regarded as a group of homothetic motions in a V_n.*

A similar theorem holds for a multiply transitive group.

CHAPTER VIII

GROUPS OF TRANSFORMATIONS IN GENERALIZED SPACES

§ 1. Finsler spaces.

Let us consider an n-dimensional space of class C^r ($r \geqq 3$) in which is given a function $L(\xi^\varkappa, \dot{\xi}^\varkappa)$ of $2n$ independent variables $\xi^\varkappa$ and $\dot{\xi}^\varkappa$, positively homogeneous of degree one with respect to the variables $\dot{\xi}^\varkappa$:

$$L(\xi^\varkappa, \dot{\xi}^\varkappa) \geqq 0; \; L(\xi^\varkappa, \rho\dot{\xi}^\varkappa) = |\rho| \, L(\xi^\varkappa, \dot{\xi}^\varkappa), \tag{1.1}$$

and in which the length of an arc $\xi^\varkappa = \xi^\varkappa(t)$, $t_1 \leqq t \leqq t_2$, is defined as

$$s = \int_{t_1}^{t_2} L(\xi^\varkappa(t), \dot{\xi}^\varkappa(t))dt; \quad \dot{\xi}^\varkappa = d\xi^\varkappa/dt. \tag{1.2}$$

Such a space is called a *Finsler space*[1] and the function $L(\xi^\varkappa, \dot{\xi}^\varkappa)$ its *fundamental function*.

A coordinate transformation in a Finsler space is of the form

$$\xi^{\varkappa'} = \xi^{\varkappa'}(\xi^\nu), \quad \dot{\xi}^{\varkappa'} = A_\varkappa^{\varkappa'} \dot{\xi}^\varkappa. \tag{1.3}$$

The fundamental function $L(\xi, \dot{\xi})$ is assumed to be invariant under coordinate transformations.

Putting

$$F(\xi, \dot{\xi}) \stackrel{\text{def}}{=} \tfrac{1}{2} L^2(\xi, \dot{\xi}), \tag{1.4}$$

$$g_{\lambda\varkappa} \stackrel{\text{def}}{=} \dot{\partial}_\lambda \dot{\partial}_\varkappa F(\xi, \dot{\xi}); \quad \dot{\partial}_\varkappa \stackrel{\text{def}}{=} \partial/\partial\dot{\xi}^\varkappa, \tag{1.5}$$

we see that $g_{\lambda\varkappa}$ is a symmetric covariant tensor and that

$$L^2(\xi, \dot{\xi}) = g_{\lambda\varkappa}(\xi, \dot{\xi})\dot{\xi}^\lambda \dot{\xi}^\varkappa. \tag{1.6}$$

We assume that $g_{\lambda\varkappa}$ has the rank n and we use $g_{\lambda\varkappa}$ and its inverse $g^{\lambda\varkappa}$ for the lowering and the raising of indices. The $g_{\lambda\varkappa}$ and $g^{\lambda\varkappa}$ are called the *fundamental tensors* of the Finsler space.

---

[1] FINSLER [1]; E. CARTAN [10].

Now we put [1]

(1.7) $$\{^{\varkappa}_{\mu\lambda}\} \overset{\text{def}}{=} \tfrac{1}{2}g^{\varkappa\rho}(\partial_\mu g_{\lambda\rho} + \partial_\lambda g_{\mu\rho} - \partial_\rho g_{\mu\lambda}),$$

(1.8) $$\Gamma^{\varkappa} \overset{\text{def}}{=} \tfrac{1}{2}\{^{\varkappa}_{\mu\lambda}\}\dot{\xi}^\mu \dot{\xi}^\lambda,$$

(1.9) $$\Gamma^{\varkappa}_{\lambda} \overset{\text{def}}{=} \dot{\partial}_\lambda \Gamma^{\varkappa},$$

(1.10) $$C_{\mu\lambda\varkappa} \overset{\text{def}}{=} \tfrac{1}{2}\dot{\partial}_\mu g_{\lambda\varkappa} = \tfrac{1}{2}\dot{\partial}_\mu \dot{\partial}_\lambda \dot{\partial}_\varkappa F(\xi, \dot{\xi})$$

and

(1.11) $$\Gamma^{\varkappa}_{\mu\lambda} \overset{\text{def}}{=} \{^{\varkappa}_{\mu\lambda}\} - C_{\mu\rho}^{\cdot\cdot\varkappa}\Gamma^{\rho}_{\lambda} - C_{\lambda\rho}^{\cdot\cdot\varkappa}\Gamma^{\rho}_{\mu} + C_{\mu\lambda\rho}\Gamma^{\rho}_{\sigma} g^{\sigma\varkappa}.$$

The following relations can easily be verified:

(1.12) $$\Gamma^{\varkappa}_{[\mu\lambda]} = 0, \; C_{\mu\lambda\varkappa} = C_{(\mu\lambda\varkappa)},$$

(1.13) $$C_{\mu\lambda\varkappa}\dot{\xi}^\mu = 0, \; C_{\mu\lambda\varkappa}\dot{\xi}^\lambda = 0, \; C_{\mu\lambda\varkappa}\dot{\xi}^\varkappa = 0,$$

(1.14) $$\Gamma^{\varkappa}_{\mu\lambda}\dot{\xi}^\mu = \Gamma^{\varkappa}_{\lambda\mu}\dot{\xi}^\mu = \Gamma^{\varkappa}_{\lambda},$$

(1.15) $$\Gamma^{\varkappa}_{\mu\lambda}\dot{\xi}^\mu \dot{\xi}^\lambda = \Gamma^{\varkappa}_{\lambda}\dot{\xi}^\lambda = 2\Gamma^{\varkappa}.$$

Under a coordinate transformation (1.3), the $\Gamma^{\varkappa}$, $\Gamma^{\varkappa}_{\lambda}$, $C_{\mu\lambda\varkappa}$ and $\Gamma^{\varkappa}_{\mu\lambda}$ have respectively the following transformation laws:

(1.16) $$\Gamma^{\varkappa'} = A^{\varkappa'}_{\varkappa}\Gamma^{\varkappa} + \tfrac{1}{2}A^{\varkappa'}_{\lambda}(\partial_{\mu'} A^{\lambda}_{\lambda'})\dot{\xi}^{\mu'}\dot{\xi}^{\lambda'},$$

(1.17) $$\Gamma^{\varkappa'}_{\lambda'} = A^{\varkappa'\lambda}_{\varkappa\lambda'}\Gamma^{\varkappa}_{\lambda} + A^{\varkappa'}_{\lambda}(\partial_{\mu'} A^{\lambda}_{\lambda'})\dot{\xi}^{\mu'},$$

(1.18) $$C_{\mu'\lambda'\varkappa'} = A^{\mu\lambda\varkappa}_{\mu'\lambda'\varkappa'} C_{\mu\lambda\varkappa},$$

and

(1.19) $$\Gamma^{\varkappa'}_{\mu'\lambda'} = A^{\varkappa'}_{\varkappa}(A^{\mu\lambda}_{\mu'\lambda'}\Gamma^{\varkappa}_{\mu\lambda} + \partial_{\mu'} A^{\varkappa}_{\lambda'}).$$

Hence the $C_{\mu\lambda\varkappa}$ is a covariant tensor and the $\Gamma^{\varkappa}_{\mu\lambda}$ is a linear connexion. The covariant differential of a contravariant vector field $v^\varkappa(\xi, \dot{\xi})$ is defined by

(1.20) $$\delta v^\varkappa \overset{\text{def}}{=} dv^\varkappa + (\Gamma^{\varkappa}_{\mu\lambda} d\xi^\mu + C_{\mu\lambda}^{\cdot\cdot\varkappa}\delta\dot{\xi}^\mu)v^\lambda$$

where

(1.21) $$\delta\dot{\xi}^\varkappa \overset{\text{def}}{=} d\dot{\xi}^\varkappa + \Gamma^{\varkappa}_{\lambda} d\xi^\lambda$$

---

[1] In E. CARTAN [10], the $\Gamma^{\varkappa}$, $\Gamma^{\varkappa}_{\lambda}$ and $\Gamma^{\varkappa}_{\mu\lambda}$ introduced here are denoted by $G^{\varkappa}$, $G^{\varkappa}_{\lambda}$ and $\Gamma^{*\varkappa}_{\mu\lambda}$ respectively. Cf. BERWALD [1], SYNGE [1], TAYLOR [1].

and the covariant derivatives are given by

$$\nabla_\mu v^\varkappa \overset{\text{def}}{=} \partial_\mu v^\varkappa - \Gamma^\rho_\mu \dot\partial_\rho v^\varkappa + \Gamma^\varkappa_{\mu\lambda} v^\lambda, \tag{1.22}$$

$$\dot\nabla_\mu v^\varkappa \overset{\text{def}}{=} \dot\partial_\mu v^\varkappa \qquad\qquad + C_{\mu\lambda}^{\cdot\cdot\,\varkappa} v^\lambda. \tag{1.23}$$

For the covariant derivatives of the direction element $\dot\xi^\varkappa$, we find

$$\nabla_\mu \dot\xi^\varkappa = 0, \quad \dot\nabla_\mu \dot\xi^\varkappa = A^\varkappa_{\mu\cdot}. \tag{1.24}$$

We can easily verify that the linear connexion introduced here is metric:

$$\nabla_\mu g_{\lambda\varkappa} = 0, \quad \dot\nabla_\mu g_{\lambda\varkappa} = 0. \tag{1.25}$$

§ 2. The Lie derivative of the fundamental tensor.

Consider a point transformation

$$'\xi^\varkappa = f^\varkappa(\xi^\nu) \tag{2.1}$$

in a Finsler space. By this point transformation, the direction element $\dot\xi^\varkappa$ undergoes the transformation

$$'\dot\xi^\varkappa = (\partial_\lambda f^\varkappa)\dot\xi^\lambda. \tag{2.2}$$

Combining (2.1) and (2.2), we call it an *extended point transformation.*

We introduce now a coordinate transformation

$$\xi^{\varkappa'} = {}'\xi^\varkappa = f^\varkappa(\xi^\nu), \quad \dot\xi^{\varkappa'} = (\partial_\lambda f^\varkappa)\dot\xi^\lambda \tag{2.3}$$

and define a new tensor field which has the components

$$'g_{\lambda'\varkappa'}(\xi, \dot\xi) \overset{\text{def}}{=} g_{\lambda\varkappa}('\xi, '\dot\xi) \tag{2.4}$$

with respect to the coordinate system $(\varkappa')$ and the components

$$'g_{\lambda\varkappa}(\xi, \dot\xi) = (\partial_\lambda f^\sigma)(\partial_\varkappa f^\rho) g_{\sigma\rho}('\xi, '\dot\xi) \tag{2.5}$$

with respect to the coordinate system $(\varkappa)$. We call this tensor the *deformed tensor* of the original tensor $g_{\lambda\varkappa}$ under the extended point transformation $('\xi, '\dot\xi) \to (\xi, \dot\xi)$ and $'g_{\lambda\varkappa}(\xi, \dot\xi) - g_{\lambda\varkappa}(\xi, \dot\xi)$ the *Lie difference* of the tensor under the extended point transformation (2.1).

In the case in which (2.1) is an infinitesimal extended point transformation:

$$'\xi^\varkappa = \xi^\varkappa + v^\varkappa(\xi)dt, \quad '\dot\xi^\varkappa = \dot\xi^\varkappa + (\partial_\lambda v^\varkappa)\dot\xi^\lambda dt, \tag{2.6}$$

we find

$$'g_{\lambda\varkappa} = g_{\lambda\varkappa} + (\underset{v}{\pounds} g_{\lambda\varkappa})dt, \tag{2.7}$$

where

$$\underset{v}{\pounds} g_{\lambda\varkappa} = v^{\rho}\partial_{\rho} g_{\lambda\varkappa} + (\dot{\xi}^{\sigma}\partial_{\sigma} v^{\rho})\dot{\partial}_{\rho} g_{\lambda\varkappa} + g_{\rho\varkappa}\partial_{\lambda} v^{\rho} + g_{\lambda\rho}\partial_{\varkappa} v^{\rho} \tag{2.8}$$

is called the *Lie derivative* of the fundamental tensor with respect to the infinitesimal extended point transformation (2.6). Since $(\underset{v}{\pounds} g_{\lambda\varkappa})dt$ is the difference of two tensors, $\underset{v}{\pounds} g_{\lambda\varkappa}$ is also a tensor. In fact, we can put $\underset{v}{\pounds} g_{\lambda\varkappa}$ in the following tensorial form

$$\underset{v}{\pounds} g_{\lambda\varkappa} = 2\nabla_{(\lambda} v_{\varkappa)} + (\dot{\xi}^{\nu}\nabla_{\nu} v^{\mu})C_{\mu\lambda\varkappa}. \tag{2.9}$$

The Lie derivative of a general tensor, say $T_{\mu\lambda}{}^{\cdot\cdot\varkappa}$ is constructed in the same way:

$$\underset{v}{\pounds} T_{\mu\lambda}{}^{\cdot\cdot\varkappa} = v^{\rho}\nabla_{\rho} T_{\mu\lambda}{}^{\cdot\cdot\varkappa} + (v^{\sigma}\nabla_{\sigma} v^{\rho})\dot{\nabla}_{\rho} T_{\mu\lambda}{}^{\cdot\cdot\varkappa} - T_{\mu\lambda}{}^{\cdot\cdot\rho}\nabla_{\rho} v^{\varkappa} + T_{\rho\lambda}{}^{\cdot\cdot\varkappa}\nabla_{\mu} v^{\rho} + T_{\mu\rho}{}^{\cdot\cdot\varkappa}\nabla_{\lambda} v^{\rho}. \tag{2.10}$$

From (2.9) we can easily derive the formula

$$(\underset{c}{\pounds}\underset{b}{\pounds}) g_{\lambda\varkappa} = \underset{cb}{\pounds} g_{\lambda\varkappa}, \quad a, b, c = 1, 2, \ldots, r \tag{2.11}$$

where $\underset{b}{\pounds}$ denotes Lie derivative with respect to the vector $\underset{b}{v}{}^{\varkappa}$ and $\underset{cb}{\pounds}$ the Lie derivative with respect to the vector

$$\underset{c}{\pounds}\underset{b}{v}{}^{\varkappa} = -\underset{b}{\pounds}\underset{c}{v}{}^{\varkappa}. \tag{2.12}$$

If the $\underset{b}{v}{}^{\varkappa}$ generate an r-parameter group of transformations, we have

$$\underset{c}{\pounds}\underset{b}{v}{}^{\varkappa} = c_{cb}^{a}\underset{a}{v}{}^{\varkappa} \tag{2.13}$$

and consequently the equation (2.11) becomes

$$(\underset{c}{\pounds}\underset{b}{\pounds}) g_{\lambda\varkappa} = c_{cb}^{a}\underset{a}{\pounds} g_{\lambda\varkappa}. \tag{2.14}$$

§ 3. Motions in a Finsler space.

When the extended point transformation (2.3) does not change the fundamental function $L(\xi, \dot{\xi})$ of a Finsler space, that is, when we have

$$L('\xi, '\dot{\xi}) = L(\xi, \dot{\xi}), \tag{3.1}$$

we call this extended point transformation a *motion* in a Finsler space. Differentiating

$$F('\xi, '\dot{\xi}) = F(\xi, \dot{\xi}) \tag{3.2}$$

twice with respect to $\dot{\xi}$, we find

$$(\partial_\lambda f^\sigma)(\partial_\varkappa f^\rho) g_{\sigma\rho}('\xi, '\dot{\xi}) = g_{\lambda\varkappa}(\xi, \dot{\xi}). \tag{3.3}$$

Because of the homogeneity property of the function $F(\xi, \dot{\xi})$, the equations (3.2) and (3.3) are equivalent. Comparing (2.5) with (3.3), we obtain

THEOREM 3.1. *In order that* (2.1) *be a motion in a Finsler space, it is necessary and sufficient that the point transformation do not deform the fundamental tensor.*

Consequently, from (2.7) we get

THEOREM 3.2. *In order that an infinitesimal extended point transformation* (2.6) *be a motion in a Finsler space, it is necessary and sufficient that the Lie derivative of the fundamental tensor with respect to the transformation vanish.*

The equation

$$\underset{v}{\pounds} g_{\lambda\varkappa} = 0 \tag{3.4}$$

is called the *equation of Killing* in a Finsler space.

Making use of (2.8), (2.11) and (2.14), we can prove theorems corresponding to Theorems 2.1, 2.2, 2.3, 2.4, 2.5, 2.6 of Ch. III and Theorem 1.5 of Ch. IV.

The motions in a Finsler space have been studied by Davies,[1] Knebelman [2], B. Laptev, [3] Nakae, [4] Soós, [5] Su [6] and Wang [7] and the groups of homothetic transformations in a Finsler space by Hiramatu. [8]

Davies [9] and Su [10] have studied the motions in a so-called Cartan space. [11]

---

[1] DAVIES [5].
[2] KNEBELMANN [2].
[3] B. LAPTEV [1, 2].
[4] NAKAE [1].
[5] SOÓS [1].
[6] SU [4].
[7] WANG [1].
[8] HIRAMATU [3, 4].
[9] DAVIES [8, 11, 12].
[10] SU [8].
[11] E. CARTAN [9].

§ 4. Finsler spaces with completely integrable equations of Killing.

We may try to discuss the motions in a Finsler space following the arguments used in Ch. IV, but because of the fact that all the components of geometric objects appearing in a Finsler space are not only functions of the coordinates $\xi^\varkappa$ but also of the direction element $\dot{\xi}^\varkappa$, the equations which express the integrability conditions of the equations of Killing are so complicated that it is almost impossible to discuss them in a way analogous to that followed in Ch. IV.

Now by quite another method H. C. Wang[1] succeeded in determining the Finsler space with completely integrable equations of Killing. Wang proved

THEOREM 4.1. *If an n-dimensional Finsler space, $n \neq 4$, admits a group G_r of motions depending on $r > \frac{1}{2}n(n-1)+1$ essential parameters, the space is a Riemannian space of constant curvature.*

Here follows the proof. Let a Finsler space admit a group G_r of motions

$$'\xi^\varkappa = f^\varkappa(\xi, \eta) = f^\varkappa(\xi^1, \ldots, \xi^n; \eta^1, \ldots, \eta^r) \tag{4.1}$$

depending on r essential parameters $\eta^1, \ldots, \eta^r$. With (4.1), we associate

$$'\dot{\xi}^\varkappa = (\partial_\lambda f^\varkappa)\dot{\xi}^\lambda. \tag{4.2}$$

The equations (4.1) and (4.2) define again an r-parameter group. We take an arbitrary point $P(\overset{0}{\xi}{}^\varkappa)$ in the space and we consider all the motions leaving invariant the point P. These form the isotropy subgroup $G(P)$ at P:

$$T_\zeta:\quad '\xi^\varkappa = h^\varkappa(\xi, \zeta) = h^\varkappa(\xi^1, \ldots, \xi^n; \zeta^1, \ldots, \zeta^{r_0}) \tag{4.3}$$

with the property $\overset{0}{\xi}{}^\varkappa = h^\varkappa(\overset{0}{\xi}; \zeta)$. The isotropy subgroup $G(P)$ depends on $r_0 \geq r - n$ essential parameters if $r \geq n$. To each motion T_ζ of $G(P)$, corresponds a linear transformation

$$\tilde{T}_\zeta:\quad '\dot{\xi}^\varkappa = [\partial_\lambda h^\varkappa(\overset{0}{\xi}; \zeta)]\dot{\xi}^\lambda. \tag{4.4}$$

We know that all the $\tilde{T}_\zeta$ form a linear group $\tilde{G}(P)$ and that the two groups $G(P)$ and $\tilde{G}(P)$ are isomorphic in the sense of topological groups. Thus the group $\tilde{G}(P)$ has the same order $r_0 \geq r - n$ as $G(P)$.

---

[1] WANG [1].

Now since the group G_r is a group of motions in a Finsler space, we have

$$L^2('\xi^\varkappa, '\dot{\xi}^\varkappa) = L^2(\xi^\varkappa, \dot{\xi}^\varkappa).$$

Putting $'\xi^\varkappa = \xi^\varkappa = \underset{0}{\xi}^\varkappa$ and substituting (4.4) in this equation, we find

$$L^2(\underset{0}{\xi}^\varkappa, h^\varkappa_\lambda \dot{\xi}^\lambda) = L^2(\underset{0}{\xi}^\varkappa, \dot{\xi}^\varkappa), \tag{4.5}$$

where

$$h^\varkappa_\lambda \overset{\text{def}}{=} \partial_\lambda h^\varkappa(\underset{0}{\xi}; \zeta).$$

If we put

$$L^2(\dot{\xi}^\varkappa) \overset{\text{def}}{=} L^2(\underset{0}{\xi}^\varkappa, \dot{\xi}^\varkappa), \tag{4.6}$$

then the equation (4.5) becomes

$$L^2(h^\varkappa_\lambda \dot{\xi}^\lambda) = L^2(\dot{\xi}^\varkappa)$$

and this shows that the function $L^2(\dot{\xi}^\varkappa)$ is an absolute invariant of the linear group $\tilde{G}(P)$. Thus we obtain

LEMMA 1. *If a Finsler space admits an r-parameter group G_r of motions, the function $L^2(\dot{\xi}^\varkappa) = L^2(\underset{0}{\xi}^\varkappa, \dot{\xi}^\varkappa)$ is left invariant by a linear isotropy group $\tilde{G}(P)$ at $P(\underset{0}{\xi}^\varkappa)$ of order $r_0 \geq r - n$.*

Now we shall show that, but for a change of basis, the group $\tilde{G}(P)$ consists of orthogonal transformations only. To prove this we need the following lemmas:

LEMMA 2. *The set K of vectors $\dot{\xi}^\varkappa$ at $\underset{0}{\xi}^\varkappa$ satisfying the equations*

$$L^2(\dot{\xi}^\varkappa) = c^2;\ c^2 = \text{constant} > 0, \tag{4.6}$$

is bounded.

Let us denote by $N(\dot{\xi})$ the "norm" $\sqrt{\Sigma_{\varkappa=1}^n \dot{\xi}^\varkappa \dot{\xi}^\varkappa}$ of the vector $\dot{\xi}^\varkappa$ and let us consider the values of $L^2(\dot{\xi})$ as $\dot{\xi}^\varkappa$ varies on the hypersphere S defined by $N(\dot{\xi}) = 1$. Since $L^2(\dot{\xi}) > 0$ on S, we have

$$0 < \varepsilon^2 < L^2(\dot{\xi}), \qquad \dot{\xi}^\varkappa \in S, \tag{4.7}$$

where ε is a positive constant.

Let now $\dot{\xi}^\varkappa$ be any vector satisfying (4.5), then we have, from the homogeneity property of $L^2(\dot{\xi}^\varkappa)$,

$$c^2 = L^2(\dot{\xi}) = N^2(\dot{\xi})L^2\left(\frac{\dot{\xi}}{N(\dot{\xi})}\right). \tag{4.8}$$

Since the vector $\dot{\xi}^\varkappa/N(\dot{\xi}) \varepsilon S$, (4.7) and (4.8) imply

$$N(\dot{\xi}) \leq \frac{c}{\varepsilon}.$$

Hence K is bounded and Lemma 2 is proved.

LEMMA 3. *If a suitable basis is chosen, all linear transformations leaving $L^2(\dot{\xi})$ invariant are orthogonal.*

Let H be the set of motions $h^\varkappa_\lambda$ leaving $L^2(\dot{\xi})$ invariant, i.e.

$$L^2(h^\varkappa_\lambda \dot{\xi}^\lambda) = L^2(\dot{\xi}^\varkappa). \tag{4.9}$$

Since $L^2(\dot{\xi}) = 0$ if and only if all $\dot{\xi}^\varkappa$ vanish, no matrix of H can be singular. From this we can easily verify that H forms a group.

Now we shall prove that H is compact. For this purpose, we put in the equation (4.9), for each λ, $\dot{\xi}^\varkappa = 1$ whenever $\varkappa = \lambda$ and $\dot{\xi}^\varkappa = 0$ otherwise. Then we obtain

$$L^2(h^1_\lambda, h^2_\lambda, \ldots, h^n_\lambda) = L^2(0, \ldots, 0, 1, 0, \ldots, 0).$$

It follows from Lemma 2 that $h^1_\lambda, h^2_\lambda, \ldots, h^n_\lambda$ are bounded. As λ is arbitrary, H is bounded as well. Moreover the set H is defined by (4.9) in which all functions involved are continuous, so that H is a closed set in the space of all n-rowed square matrices. Hence the boundedness implies the compactness.

By a well-known theorem of Weyl[1], H leaves invariant a positive definite quadratic form $u(\dot{\xi})$, and we can choose a suitable basis such that

$$u(\dot{\xi}) = \Sigma^n_{\varkappa=1} \dot{\xi}^\varkappa \dot{\xi}^\varkappa.$$

Hence the matrices $h^\varkappa_\lambda$ are orthogonal and Lemma 3 is proved.

Now with the aid of the above lemmas, we can prove Theorem 4.1 without difficulty. In fact, if a Finsler space admits a group G_r of motions depending on $r > \frac{1}{2}n(n-1)+1$ parameters, then by Lemma 1 the function $L^2(\dot{\xi})$ is left invariant by a linear isotropy group $\tilde{G}(P)$ of the order

$$r_0 \geq r - n > \tfrac{1}{2}(n-1)(n-2).$$

[1] WEYL [1].

Lemma 3 tells us that $\tilde{G}(P)$ is a subgroup of the orthogonal group $O(n)$. As there is no proper subgroup of $O(n)$ of an order greater than $\frac{1}{2}(n-1)(n-2)$ for $n \neq 4$,[1] we conclude that $\tilde{G}(P)$ coincides with $O(n)$. Thus the function $L^2(\xi)$ is a scalar invariant of the orthogonal group and therefore the $L^2(\dot{\xi})$ takes the form

$$L^2(\dot{\xi}) = h(u).$$

From the homogeneity property of $L^2(\dot{\xi})$, we have $h(\lambda u) = \lambda h(u)$, which implies $h(u) = cu$, c being independent of $\dot{\xi}^\varkappa$. Thus

$$L^2(\dot{\xi}) = L^2(\underset{0}{\xi}, \dot{\xi}) = c(\underset{0}{\xi}) \textstyle\sum_{\varkappa=1}^{n} \dot{\xi}^\varkappa \dot{\xi}^\varkappa.$$

Since the $\underset{0}{\xi}^\varkappa$ are arbitrary and

$$\dot{\partial}_\mu g_{\lambda\varkappa}(\underset{0}{\xi}, \dot{\xi}) = \tfrac{1}{2}\dot{\partial}_\mu \dot{\partial}_\lambda \dot{\partial}_\varkappa L^2(\underset{0}{\xi}, \dot{\xi}) = 0,$$

the metric tensor $g_{\lambda\varkappa}$ depends only on the position. Thus the space is Riemannian. Hence by Theorem 8.2 of Ch. IV, we can conclude that the space is of constant curvature.

§ 5. General affine spaces of geodesics.[2]

Consider an n-dimensional space in which a system of curves called geodesics (or paths) is given by a system of ordinary differential equations

$$\text{(5.1)} \qquad \frac{d\dot{\xi}^\varkappa}{dt} + \Gamma^\varkappa(\xi, \dot{\xi}) = 0; \; \dot{\xi}^\varkappa = \frac{d\xi^\varkappa}{dt},$$

where $\Gamma^\varkappa(\xi, \dot{\xi})$ are functions of the $2n$ independent variables $\xi^\varkappa$ and $\dot{\xi}^\varkappa$, homogeneous of degree 2 with respect to $\dot{\xi}^\varkappa$ and t is a scalar parameter determined up to an affine transformation. Such a space is called a *general affine space of geodesics* (or *paths*) and its geometry the *general affine geometry of geodesics* (or *paths*).

We assume that the left-hand side of (5.1) is a contravariant vector. Then under a coordinate transformation

$$\text{(5.2)} \qquad \xi^{\varkappa'} = \xi^{\varkappa'}(\xi^\nu); \; \dot{\xi}^{\varkappa'} = A_\varkappa^{\varkappa'} \dot{\xi}^\varkappa,$$

the functions $\Gamma^\varkappa(\xi, \dot{\xi})$ are transformed into

$$\text{(5.3)} \qquad \Gamma^{\varkappa'} = A_\varkappa^{\varkappa'} \Gamma^\varkappa - (\partial_\mu A_\lambda^{\varkappa'}) \dot{\xi}^\mu \dot{\xi}^\lambda.$$

---

[1] MONTGOMERY and SAMELSON [1].

[2] DOUGLAS [1].

From (5.2) we find by partial differentiation with respect to $\dot{\xi}^{\lambda'}$,

$$\Gamma^{\kappa'}_{\lambda'} = A^{\kappa'\lambda}_{\kappa\lambda'} \Gamma^{\kappa}_{\lambda} - A^{\lambda}_{\lambda'}(\partial_{\mu} A^{\kappa'}_{\lambda})\dot{\xi}^{\mu}, \tag{5.4}$$

where

$$\Gamma^{\kappa'}_{\lambda'} \overset{\text{def}}{=} \tfrac{1}{2}\partial_{\lambda'} \Gamma^{\kappa'}, \quad \Gamma^{\kappa}_{\lambda} \overset{\text{def}}{=} \tfrac{1}{2}\partial_{\lambda} \Gamma^{\kappa}. \tag{5.5}$$

From (5.4), by partial differentiation with respect to $\dot{\xi}^{\mu'}$, we obtain

$$\Gamma^{\kappa'}_{\mu'\lambda'} = A^{\kappa'\mu\lambda}_{\kappa\mu'\lambda'} \Gamma^{\kappa}_{\mu\lambda} - A^{\mu\lambda}_{\mu'\lambda'} \partial_{\mu} A^{\kappa'}_{\lambda} \tag{5.6}$$

or

$$\Gamma^{\kappa'}_{\mu'\lambda'} = A^{\kappa'}_{\kappa}(A^{\mu\lambda}_{\mu'\lambda'} \Gamma^{\kappa}_{\mu\lambda} + \partial_{\mu'} A^{\kappa}_{\lambda'}), \tag{5.7}$$

where

$$\Gamma^{\kappa'}_{\mu'\lambda'} \overset{\text{def}}{=} \partial_{\mu'} \Gamma^{\kappa'}_{\lambda'}, \quad \Gamma^{\kappa}_{\mu\lambda} \overset{\text{def}}{=} \partial_{\mu} \Gamma^{\kappa}_{\lambda}. \tag{5.8}$$

Hence the $\Gamma^{\kappa}_{\mu\lambda}(\xi, \dot{\xi})$ are components of a symmetric linear connexion. By the homogeneity property of $\Gamma^{\kappa}(\xi, \dot{\xi})$, we get

$$\Gamma^{\kappa}_{\mu\lambda} \dot{\xi}^{\mu} = \Gamma^{\kappa}_{\lambda\mu} \dot{\xi}^{\mu} = \Gamma^{\kappa}_{\lambda} \tag{5.9}$$

and

$$\Gamma^{\kappa}_{\mu\lambda} \dot{\xi}^{\mu} \dot{\xi}^{\lambda} = \Gamma^{\kappa}_{\lambda} \dot{\xi}^{\lambda} = \Gamma^{\kappa}. \tag{5.10}$$

Thus the equation of the geodesics can be written as

$$\frac{d^2\xi^{\kappa}}{dt^2} + \Gamma^{\kappa}_{\mu\lambda} \frac{d\xi^{\mu}}{dt} \frac{d\xi^{\lambda}}{dt} = 0. \tag{5.11}$$

We now define the covariant differential of $\dot{\xi}^{\kappa}$ by

$$\delta\dot{\xi}^{\kappa} \overset{\text{def}}{=} d\dot{\xi}^{\kappa} + \Gamma^{\kappa}_{\mu\lambda} d\xi^{\mu} \dot{\xi}^{\lambda}$$

or by

$$\delta\dot{\xi}^{\kappa} \overset{\text{def}}{=} d\dot{\xi}^{\kappa} + \Gamma^{\kappa}_{\mu} d\xi^{\mu}, \tag{5.12}$$

and the covariant differential of a contravariant vector v^{κ} by

$$\delta v^{\kappa} \overset{\text{def}}{=} dv^{\kappa} + \Gamma^{\kappa}_{\mu\lambda} d\xi^{\mu} v^{\lambda}. \tag{5.13}$$

The covariant derivatives of the contravariant vector field v^{κ} are then given by

$$\nabla_{\mu} v^{\kappa} \overset{\text{def}}{=} \partial_{\mu} v^{\kappa} - \Gamma^{\rho}_{\mu} \dot{\partial}_{\rho} v^{\kappa} + \Gamma^{\kappa}_{\mu\lambda} v^{\lambda}, \tag{5.14}$$

$$\dot{\nabla}_{\mu} v^{\kappa} \overset{\text{def}}{=} \dot{\partial}_{\mu} v^{\kappa}. \tag{5.15}$$

For the covariant derivatives of the vector $\dot{\xi}^{\varkappa}$, we obtain

$$(5.16) \qquad \nabla_{\mu}\dot{\xi}^{\varkappa} = 0, \; \dot{\nabla}_{\mu}\dot{\xi}^{\varkappa} = A_{\mu}^{\varkappa}.$$

Now as generalizations of Ricci identities, we find

$$(5.17) \qquad (\nabla_{\nu}\nabla_{\mu} - \nabla_{\nu}\nabla_{\mu})v^{\varkappa} = R_{\nu\mu\lambda}^{\cdots\varkappa}v^{\lambda} - R_{\nu\mu\lambda}^{\cdots\rho}\dot{\xi}^{\lambda}\dot{\nabla}_{\rho}v^{\varkappa},$$

$$(5.18) \qquad (\dot{\nabla}_{\nu}\nabla_{\mu} - \nabla_{\mu}\dot{\nabla}_{\nu})v^{\varkappa} = T_{\nu\mu\lambda}^{\cdots\varkappa}v^{\lambda},$$

$$(5.19) \qquad (\dot{\nabla}_{\nu}\dot{\nabla}_{\mu} - \dot{\nabla}_{\mu}\dot{\nabla}_{\nu})v^{\varkappa} = 0,$$

where

$$(5.20) \qquad R_{\nu\mu\lambda}^{\cdots\varkappa} \overset{\text{def}}{=} (\partial_{\nu}\Gamma_{\mu\lambda}^{\varkappa} - \Gamma_{\nu}^{\rho}\dot{\partial}_{\rho}\Gamma_{\mu\lambda}^{\varkappa}) - (\partial_{\mu}\Gamma_{\nu\lambda}^{\varkappa} - \Gamma_{\mu}^{\rho}\dot{\partial}_{\rho}\Gamma_{\nu\lambda}^{\varkappa})$$
$$+ \Gamma_{\nu\rho}^{\varkappa}\Gamma_{\mu\lambda}^{\rho} - \Gamma_{\mu\rho}^{\varkappa}\Gamma_{\nu\lambda}^{\rho}$$

and

$$(5.21) \qquad T_{\nu\mu\lambda}^{\cdots\varkappa} \overset{\text{def}}{=} \dot{\partial}_{\nu}\Gamma_{\mu\lambda}^{\varkappa} = \tfrac{1}{2}\dot{\partial}_{\nu}\dot{\partial}_{\mu}\dot{\partial}_{\lambda}\Gamma^{\varkappa}$$

are curvature tensors of the space. The Bianchi identities for the curvature tensors take the form

$$(5.22) \qquad R_{[\nu\mu\lambda]}^{\cdots\varkappa} = 0,$$

$$(5.23) \qquad \nabla_{[\omega}R_{\nu\mu]\lambda}^{\cdots\varkappa} + R_{[\omega\nu|\sigma}^{\cdots\rho}\dot{\xi}^{\sigma}T_{\rho|\mu]\lambda}^{\cdots\varkappa} = 0.$$

Furthermore, we have the following identities:

$$(5.24) \qquad \dot{\nabla}_{\omega}R_{\nu\mu\lambda}^{\cdots\varkappa} = 2\nabla_{[\nu}T_{\mu]\omega\lambda}^{\cdots\varkappa},$$

$$(5.25) \qquad \dot{\nabla}_{\omega}T_{\nu\mu\lambda}^{\cdots\varkappa} = \dot{\nabla}_{(\omega}T_{\nu\mu)\lambda}^{\cdots\varkappa}.$$

From (5.24), we get

$$(5.26) \qquad \dot{\nabla}_{[\omega}R_{\nu\mu]\lambda}^{\cdots\varkappa} = 0,$$

from which

$$(5.27) \qquad \dot{\nabla}_{[\omega}V_{\nu\mu]} = 0,$$

where

$$(5.28) \qquad V_{\mu\nu} \overset{\text{def}}{=} R_{\nu\mu\lambda}^{\cdots\lambda} = -(R_{\nu\mu} - R_{\mu\nu})$$

$$(5.29) \qquad R_{\nu\mu} \overset{\text{def}}{=} R_{\rho\nu\mu}^{\cdots\rho}.$$

§ 6. Lie derivatives in a general affine space of geodesics.

Consider an extended point transformation

$$(6.1) \qquad '\xi^\varkappa = f^\varkappa(\xi^\nu), \quad '\dot{\xi}^\varkappa = (\partial_\lambda f^\varkappa)\dot{\xi}^\lambda$$

in a general affine space of geodesics.

We introduce a coordinate transformation

$$(6.2) \qquad \xi^{\varkappa'} = '\xi^\varkappa = f^\varkappa(\xi), \quad \dot{\xi}^{\varkappa'} = '\dot{\xi}^\varkappa = (\partial_\lambda f^\varkappa)\dot{\xi}^\lambda$$

and define a new linear connexion which has the components

$$(6.3) \qquad '\Gamma^{\varkappa'}_{\mu'\lambda'}(\xi, \dot{\xi}) \stackrel{\text{def}}{=} \Gamma^\varkappa_{\mu\lambda}('\xi, '\dot{\xi})$$

with respect to the coordinate system $(\varkappa')$. The components $'\Gamma^\varkappa_{\mu\lambda}(\xi, \dot{\xi})$ of the new linear connexion with respect to the coordinate system $(\varkappa)$ are given by

$$(6.4) \qquad (\partial_\mu {'\xi^\tau})(\partial_\lambda {'\xi^\sigma})\Gamma^\rho_{\tau\sigma}('\xi, '\dot{\xi}) = (\partial_\varkappa {'\xi^\rho}) '\Gamma^\varkappa_{\mu\lambda}(\xi, \dot{\xi}) - \partial_\mu \partial_\lambda {'\xi^\rho}.$$

Now we call $'\Gamma^\varkappa_{\mu\lambda}(\xi, \dot{\xi})$ the *deformed linear connexion* of the original linear connexion $\Gamma^\varkappa_{\mu\lambda}(\xi, \dot{\xi})$ under the transformation $('\xi^\varkappa, '\dot{\xi}^\varkappa) \rightarrow (\xi^\varkappa, \dot{\xi}^\varkappa)$ and

$$'\Gamma^\varkappa_{\mu\lambda}(\xi, \dot{\xi}) - \Gamma^\varkappa_{\mu\lambda}(\xi, \dot{\xi})$$

the *Lie difference*. The Lie difference of a linear connexion is a mixed tensor of the valence three.

When (6.1) is an infinitesimal transformation

$$(6.5) \qquad '\xi^\varkappa = \xi^\varkappa + v^\varkappa(\xi)dt, \quad '\dot{\xi}^\varkappa = \dot{\xi}^\varkappa + (\partial_\lambda v^\varkappa)\dot{\xi}^\lambda dt,$$

we find

$$(6.6) \qquad '\Gamma^\varkappa_{\mu\lambda} = \Gamma^\varkappa_{\mu\lambda} + \underset{v}{\pounds}\Gamma^\varkappa_{\mu\lambda}\, dt,$$

where

$$(6.7) \qquad \underset{v}{\pounds}\Gamma^\varkappa_{\mu\lambda} = \partial_\mu \partial_\lambda v^\varkappa + v^\nu \partial_\nu \Gamma^\varkappa_{\mu\lambda} + \dot{\xi}^\nu (\partial_\nu v^\rho)\dot{\partial}_\rho \Gamma^\varkappa_{\mu\lambda} - \Gamma^\rho_{\mu\lambda}\partial_\rho v^\varkappa + \Gamma^\varkappa_{\rho\lambda}\partial_\mu v^\rho + \Gamma^\varkappa_{\mu\rho}\partial_\lambda v^\rho$$

is the *Lie derivative* of the linear connexion $\Gamma^\varkappa_{\mu\lambda}$ with respect to the infinitesimal transformation (6.5). We can put $\underset{v}{\pounds}\Gamma^\varkappa_{\mu\lambda}$ in the following tensorial form

$$(6.8) \qquad \underset{v}{\pounds}\Gamma^\varkappa_{\mu\lambda} = \nabla_\mu \nabla_\lambda v^\varkappa + R_{\nu\mu\lambda}{}^{\varkappa} v^\nu + T_{\nu\mu\lambda}{}^{\varkappa} \dot{\xi}^\rho \nabla_\rho v^\nu.$$

We mention here some important formulae which contain the Lie derivatives and which will be useful later on.

$$\underset{v}{\pounds} u^{\varkappa} = v^{\mu}\partial_{\mu}u^{\varkappa} + (\dot{\xi}^{\rho}\partial_{\rho}v^{\mu})\dot{\partial}_{\mu}u^{\varkappa} - u^{\rho}\partial_{\rho}v^{\varkappa}$$
$$= v^{\mu}\nabla_{\mu}u^{\varkappa} + (\dot{\xi}^{\rho}\nabla_{\rho}v^{\mu})\dot{\nabla}_{\mu}u^{\varkappa} - u^{\rho}\nabla_{\rho}v^{\varkappa}, \tag{6.9}$$

$$\underset{v}{\pounds} w_{\lambda} = v^{\mu}\partial_{\mu}w_{\lambda} + (\dot{\xi}^{\rho}\partial_{\rho}v^{\mu})\dot{\partial}_{\mu}w_{\lambda} + w_{\rho}\partial_{\lambda}v^{\rho}$$
$$= v^{\mu}\nabla_{\mu}w_{\lambda} + (\dot{\xi}^{\rho}\nabla_{\rho}v^{\mu})\dot{\nabla}_{\mu}w_{\lambda} + w_{\rho}\nabla_{\lambda}v^{\rho}, \tag{6.10}$$

$$\underset{v}{\pounds} T_{\lambda}^{\cdot\varkappa} = v^{\mu}\partial_{\mu}T_{\lambda}^{\cdot\varkappa} + (\dot{\xi}^{\rho}\partial_{\rho}v^{\mu})\dot{\partial}_{\mu}T_{\lambda}^{\cdot\varkappa} - T_{\lambda}^{\cdot\rho}\partial_{\rho}v^{\varkappa} + T_{\rho}^{\cdot\varkappa}\partial_{\lambda}v^{\rho}$$
$$= v^{\mu}\nabla_{\mu}T_{\lambda}^{\cdot\varkappa} + (\dot{\xi}^{\rho}\nabla_{\rho}v^{\mu})\dot{\nabla}_{\mu}T_{\lambda}^{\cdot\varkappa} - T_{\lambda}^{\cdot\rho}\nabla_{\rho}v^{\varkappa} + T_{\rho}^{\cdot\varkappa}\nabla_{\lambda}v^{\rho}, \tag{6.11}$$

where $u^{\varkappa}$, w_{λ} and $T_{\lambda}^{\cdot\varkappa}$ are respectively a contravariant vector, a covariant vector and a mixed tensor.

If we apply the operator $\underset{v}{\pounds}$ to $\dot{\xi}^{\varkappa}$, we find

$$\underset{v}{\pounds}\dot{\xi}^{\varkappa} = 0. \tag{6.12}$$

Moreover, applying the operators $\underset{v}{\pounds}\nabla_{\mu} - \nabla_{\mu}\underset{v}{\pounds}$ and $\underset{v}{\pounds}\dot{\nabla}_{\mu} - \dot{\nabla}_{\mu}\underset{v}{\pounds}$ to an arbitrary tensor $T_{\lambda}^{\cdot\varkappa}$, we find

$$(\underset{v}{\pounds}\nabla_{\mu} - \nabla_{\mu}\underset{v}{\pounds})T_{\lambda}^{\cdot\varkappa} = (\underset{v}{\pounds}\Gamma_{\mu\rho}^{\varkappa})T_{\lambda}^{\cdot\rho} - (\underset{v}{\pounds}\Gamma_{\mu\lambda}^{\rho})T_{\rho}^{\cdot\varkappa} - (\underset{v}{\pounds}\Gamma_{\mu\sigma}^{\rho})\dot{\xi}^{\sigma}\dot{\nabla}_{\rho}T_{\lambda}^{\cdot\varkappa}, \tag{6.13}$$

and

$$(\underset{v}{\pounds}\dot{\nabla}_{\mu} - \dot{\nabla}_{\mu}\underset{v}{\pounds})T_{\lambda}^{\cdot\varkappa} = 0 \tag{6.14}$$

respectively. On the other hand, we have

$$\nabla_{\nu}\underset{v}{\pounds}\Gamma_{\mu\lambda}^{\varkappa} - \nabla_{\mu}\underset{v}{\pounds}\Gamma_{\nu\lambda}^{\varkappa} = \underset{v}{\pounds}R_{\nu\mu\lambda}^{\cdots\varkappa} + (\underset{v}{\pounds}\Gamma_{\nu\sigma}^{\rho})\dot{\xi}^{\sigma}T_{\rho\mu\lambda}^{\cdots\varkappa} - (\underset{v}{\pounds}\Gamma_{\mu\sigma}^{\rho})\dot{\xi}^{\sigma}T_{\rho\nu\lambda}^{\cdots\varkappa}. \tag{6.15}$$

$$\dot{\nabla}_{\nu}\underset{v}{\pounds}\Gamma_{\mu\lambda}^{\varkappa} = \underset{v}{\pounds}T_{\nu\mu\lambda}^{\cdots\varkappa}. \tag{6.16}$$

If for r vectors $\underset{b}{v}^{\varkappa}$, $a, b, c, \ldots = 1, 2, \ldots, r$, the Lie derivative with respect to $\underset{b}{v}^{\varkappa}$ is denoted by $\underset{b}{\pounds}$, we can easily verify

$$(\underset{c}{\pounds}\underset{b}{\pounds})\Gamma_{\mu\lambda}^{\varkappa} = \underset{cb}{\pounds}\Gamma_{\mu\lambda}^{\varkappa}, \tag{6.17}$$

where $\underset{cb}{\pounds}$ denotes the Lie derivative with respect to the vector

$$\underset{c}{\pounds}\underset{b}{v}^{\varkappa} = -\underset{b}{\pounds}\underset{c}{v}^{\varkappa}.$$

If r vectors generate an r-parameter group, we have $\underset{c}{\pounds}\underset{b}{v}^{\varkappa} = c_{cb}^{a}\underset{a}{v}^{\varkappa}$ and (6.17) becomes

$$(6.18) \qquad (\underset{c}{\pounds}\underset{b}{\pounds})\Gamma_{\mu\lambda}^{\varkappa} = c_{cb}^{a}\underset{a}{\pounds}\Gamma_{\mu\lambda}^{\varkappa}.$$

§ 7. Affine motions in a general affine space of geodesics.[1]

When the point transformation (6.1) transforms every geodesic into a geodesic and the affine parameter on it into an affine parameter on the deformed geodesic, we call the transformation an *affine motion* in a general affine space of geodesics.

The condition for the extended point transformation (6.1) to be an affine motion is given by

$$(7.1) \qquad \Gamma^{\rho}('\xi, '\dot{\xi}) = (\partial_{\varkappa}{'\xi^{\rho}})\Gamma^{\varkappa}(\xi, \dot{\xi}) - (\partial_{\mu}\partial_{\lambda}{'\xi^{\rho}})\dot{\xi}^{\mu}\dot{\xi}^{\lambda}$$

which is equivalent to

$$(7.2) \qquad (\partial_{\mu}{'\xi^{\tau}})(\partial_{\lambda}{'\xi^{\sigma}})\Gamma_{\tau\sigma}^{\rho}('\xi, '\dot{\xi}) = (\partial_{\varkappa}{'\xi^{\rho}})\Gamma_{\mu\lambda}^{\varkappa}(\xi, \dot{\xi}) - \partial_{\mu}\partial_{\lambda}{'\xi^{\rho}}.$$

Thus comparing (6.4) with (7.2), we can state

THEOREM 7.1. *In order that* (6.1) *be an affine motion in a general affine space of geodesics, it is necessary and sufficient that the point transformation do not deform the linear connexion.*

Consequently, from (6.6) we obtain

THEOREM 7.2. *In order that the infinitesimal extended point transformation* (6.5) *be an affine motion in a general affine space of geodesics, it is necessary and sufficient that the Lie derivative of the linear connexion with respect to the extended point transformation vanish.*

Using the formula (6.13), we can state a theorem corresponding to Theorem 4.2 of Ch. I.

Making use of the expression (6.7) for the Lie derivative $\underset{v}{\pounds}\Gamma_{\mu\lambda}^{\varkappa}$ of the linear connexion and of the formulae (6.17) and (6.18), we can prove theorems corresponding to Theorems 2.1, 2.2, 2.3, 2.4, 2.5, 2.6 of Ch. III and to Theorem 1.3 of Ch. V.

§ 8. Integrability conditions of the equations $\underset{v}{\pounds}\Gamma_{\mu\lambda}^{\varkappa} = 0$.

We now consider the integrability conditions of the equations

$$(8.1) \qquad \underset{v}{\pounds}\Gamma_{\mu\lambda}^{\varkappa} = \nabla_{\mu}\nabla_{\lambda}v^{\varkappa} + R_{\nu\mu\lambda}^{\cdot\cdot\cdot\varkappa}v^{\nu} + T_{\nu\mu\lambda}^{\cdot\cdot\cdot\varkappa}\dot{\xi}^{\rho}\nabla_{\rho}v^{\nu} = 0.$$

[1] KNEBELMANN [2, 4].

Since the $v^\varkappa$ are functions of the $\xi^\varkappa$ only, we get

$$\dot{\nabla}_\lambda v^\varkappa = 0$$

hence, taking account of (5.18), we find

$$\nabla_\mu \nabla_\lambda v^\varkappa = T_{\nu\mu\lambda}^{\cdot\cdot\cdot\varkappa} v^\nu.$$

From these equations we get

$$(8.2)\qquad \begin{cases} \text{(i)} \quad \nabla_\lambda v^\varkappa = v_\lambda^{\cdot\varkappa}, \qquad \text{(ii)} \quad \dot{\nabla}_\lambda v^\varkappa = 0, \\ \text{(iii)} \quad \nabla_\mu v_\lambda^{\cdot\varkappa} = -R_{\nu\mu\lambda}^{\cdot\cdot\cdot\varkappa} v^\nu - T_{\nu\mu\lambda}^{\cdot\cdot\cdot\varkappa} \dot{\xi}^\rho v_\rho^{\cdot\nu}, \\ \text{(iv)} \quad \dot{\nabla}_\mu v_\lambda^{\cdot\varkappa} = T_{\nu\mu\lambda}^{\cdot\cdot\cdot\varkappa} v^\nu, \end{cases}$$

which constitute a system of partial differential equations with unknown functions $v^\varkappa$ and $v_\lambda^{\cdot\varkappa}$. We shall study the integrability conditions of this system.

First we can easily verify that the integrability conditions which are obtained by substituting (8.2, i) and (8.2, ii) in (5.17), (5.18) and (5.19) are automatically satisfied because of (8.2, iii), (8.2, iv), the Bianchi identity (5.22) and the symmetry of $T_{\nu\mu\lambda}^{\cdot\cdot\cdot\varkappa}$ in the three lower indices.

Consequently we have only to consider the integrability conditions which are obtained by substituting (8.2, iii) and (8.2, iv) into

$$(8.3)\quad (\nabla_\omega \nabla_\mu - \nabla_\mu \nabla_\omega) v_\lambda^{\cdot\varkappa} = R_{\omega\mu\rho}^{\cdot\cdot\cdot\varkappa} v_\lambda^{\cdot\rho} - R_{\omega\mu\lambda}^{\cdot\cdot\cdot\rho} v_\rho^{\cdot\varkappa} - R_{\omega\mu\sigma}^{\cdot\cdot\cdot\rho} \dot{\xi}^\sigma \dot{\nabla}_\rho v_\lambda^{\cdot\varkappa},$$

$$(8.4)\qquad (\dot{\nabla}_\omega \nabla_\mu - \nabla_\mu \dot{\nabla}_\omega) v_\lambda^{\cdot\varkappa} = T_{\omega\mu\rho}^{\cdot\cdot\cdot\varkappa} v_\lambda^{\cdot\rho} - T_{\omega\mu\lambda}^{\cdot\cdot\cdot\rho} v_\rho^{\cdot\varkappa},$$

$$(8.5)\qquad (\dot{\nabla}_\omega \dot{\nabla}_\mu - \dot{\nabla}_\mu \dot{\nabla}_\omega) v_\lambda^{\cdot\varkappa} = 0.$$

But the equation which is obtained from (8.3) is equivalent to the equation obtained from (6.15) by putting $\underset{v}{\pounds}\Gamma^{\varkappa}_{\mu\lambda} = 0$:

$$(8.6)\qquad \underset{v}{\pounds} R_{\nu\mu\lambda}^{\cdot\cdot\cdot\varkappa} = 0.$$

Similarly the equation which is obtained from (8.4) is equivalent to the equation obtained from (6.16) by putting $\underset{v}{\pounds}\Gamma^{\varkappa}_{\mu\lambda} = 0$:

$$(8.7)\qquad \underset{v}{\pounds} T_{\nu\mu\lambda}^{\cdot\cdot\cdot\varkappa} = 0.$$

Thus the integrability conditions of the system (8.21) are given by (8.6), (8.7) and the equations which are obtained from (8.6) and (8.7) by successive covariant differentiations with respect to $\xi^\varkappa$ and $\dot{\xi}^\varkappa$, the terms $\nabla_\mu v_\lambda^{\cdot\varkappa}$ and $\dot{\nabla}_\mu v_\lambda^{\cdot\varkappa}$ being eliminated by the use of (8.2, iii) and (8.2, iv).

We first consider the equations obtained from (8.6) by successive covariant differentiations. From (8.6) we obtain

$$\nabla_\omega(\underset{v}{\pounds} R_{\nu\mu\lambda}^{\cdot\cdot\cdot\varkappa}) = 0, \ \dot{\nabla}_\omega(\underset{v}{\pounds} R_{\nu\mu\lambda}^{\cdot\cdot\cdot\varkappa}) = 0.$$

But from (5.24), we get

$$\dot{\nabla}_\omega(\underset{v}{\pounds} R_{\nu\mu\lambda}^{\cdot\cdot\cdot\varkappa}) = \underset{v}{\pounds}(\dot{\nabla}_\omega R_{\nu\mu\lambda}^{\cdot\cdot\cdot\varkappa}) = 2\underset{v}{\pounds}(\nabla_{[\nu} T_{\mu]\omega\lambda}^{\cdot\cdot\cdot\varkappa}) = 2\nabla_{[\nu}(\underset{v}{\pounds} T_{\mu]\omega\lambda}^{\cdot\cdot\cdot\varkappa}),$$

which shows that $\dot{\nabla}_\omega(\underset{v}{\pounds} R_{\nu\mu\lambda}^{\cdot\cdot\cdot\varkappa}) = 0$ is automatically satisfied, if the equations obtained from (8.7) by successive covariant differentiations are satisfied.

We next consider

$$\nabla_{\omega_2}\nabla_{\omega_1}(\underset{v}{\pounds} R_{\nu\mu\lambda}^{\cdot\cdot\cdot\varkappa}) = 0, \ \dot{\nabla}_\pi\nabla_\omega(\underset{v}{\pounds} R_{\nu\mu\lambda}^{\cdot\cdot\cdot\varkappa}) = 0.$$

But from (5.18) applied to $\underset{v}{\pounds} R_{\nu\mu\lambda}^{\cdot\cdot\cdot\varkappa}$, we find

$$\dot{\nabla}_\pi\nabla_\omega(\underset{v}{\pounds} R_{\nu\mu\lambda}^{\cdot\cdot\cdot\varkappa}) = \nabla_\omega\dot{\nabla}_\pi(\underset{v}{\pounds} R_{\nu\mu\lambda}^{\cdot\cdot\cdot\varkappa}) + T_{\pi\omega\rho}^{\cdot\cdot\cdot\varkappa}(\underset{v}{\pounds} R_{\nu\mu\lambda}^{\cdot\cdot\cdot\rho})$$
$$- T_{\pi\omega\nu}^{\cdot\cdot\cdot\rho}(\underset{v}{\pounds} R_{\rho\mu\lambda}^{\cdot\cdot\cdot\varkappa}) - T_{\pi\omega\mu}^{\cdot\cdot\cdot\rho}(\underset{v}{\pounds} R_{\nu\rho\lambda}^{\cdot\cdot\cdot\varkappa}) - T_{\pi\omega\lambda}^{\cdot\cdot\cdot\rho}(\underset{v}{\pounds} R_{\nu\mu\rho}^{\cdot\cdot\cdot\varkappa}),$$

which shows that $\dot{\nabla}_\pi\nabla_\omega(\underset{v}{\pounds} R_{\nu\mu\lambda}^{\cdot\cdot\cdot\varkappa}) = 0$ is automatically satisfied if the equations $\underset{v}{\pounds} R_{\nu\mu\lambda}^{\cdot\cdot\cdot\varkappa} = 0$ and the equations obtained from (8.7) by successive covariant differentiations are satisfied.

Repeating this process, we see that, as integrability conditions obtained from (8.6) by successive covariant differentiations, we have only to consider the equations

$$\nabla_{\omega_r \cdots \omega_2\omega_1}(\underset{v}{\pounds} R_{\nu\mu\lambda}^{\cdot\cdot\cdot\varkappa}) = 0, \qquad r = 1, 2, \ldots \tag{8.8}$$

which can also be written as

$$\underset{v}{\pounds}(\nabla_{\omega_r \cdots \omega_2\omega_1} R_{\nu\mu\lambda}^{\cdot\cdot\cdot\varkappa}) = 0, \qquad r = 1, 2, \ldots \tag{8.9}$$

We now consider the equations which are obtained from (8.7) by successive covariant differentiations. But the equation (5.18) applied to $\underset{v}{\pounds} T_{\nu\mu\lambda}^{\cdot\cdot\cdot\varkappa}$ shows that the conditions obtained from (8.7) by applying first the covariant differentiations with respect to $\xi^\varkappa$ and next the covariant differentiations with respect to $\xi^\varkappa$ are equivalent to the conditions obtained from (8.7) by applying the covariant differentiations in the reverse way.

Thus we shall consider first the conditions obtained from (8.7) by applying only the covariant differentiations with respect to $\dot{\xi}^\varkappa$:

$$\dot{\nabla}_{\omega_s \cdots \omega_2\omega_1}(\underset{v}{\pounds} T_{\nu\mu\lambda}^{\cdot\cdot\cdot\varkappa}) = 0, \qquad s = 1, 2, \ldots \tag{8.10}$$

But by virtue of the homogeneity property of $T_{\nu\mu\lambda}^{\cdot\cdot\cdot\varkappa}$ with respect to the $\dot{\xi}^\varkappa$, the s-th equation of (8.10) contains the preceding equations and consequently the equation (8.10) can be written as

$$\dot{\nabla}_{\omega_s \cdots \omega_2\omega_1}(\underset{v}{\pounds} T_{\nu\mu\lambda}^{\cdot\cdot\cdot\varkappa}) = 0. \tag{8.11}$$

or

$$\underset{v}{\pounds}(\dot{\nabla}_{\omega_s \cdots \omega_2\omega_1} T_{\nu\mu\lambda}^{\cdot\cdot\cdot\varkappa}) = 0. \tag{8.12}$$

From (8.12), by successive covariant differentiation with respect to ξ^π, we obtain

$$\underset{v}{\pounds}(\nabla_{\pi_t \cdots \pi_2\pi_1} \dot{\nabla}_{\omega_s \cdots \omega_2\omega_1} T_{\nu\mu\lambda}^{\cdot\cdot\cdot\varkappa}) = 0, \qquad t = 1, 2, \ldots \tag{8.13}$$

Gathering these results, we obtain

THEOREM 8.1. *In order that a general affine space of geodesics admit a group of affine motions, it is necessary and sufficient that, for a certain value of s, there exist a positive integer N such that the first N sets of the equations*

$$(8.14)\qquad \begin{aligned} &\underset{v}{\pounds}(\nabla_{\omega_r \cdots \omega_2\omega_1} R_{\nu\mu\lambda}^{\cdot\cdot\cdot\varkappa}) = 0 \\ &\underset{v}{\pounds}(\nabla_{\pi_t \cdots \pi_2\pi_1} \dot{\nabla}_{\omega_s \cdots \omega_2\omega_1} T_{\nu\mu\lambda}^{\cdot\cdot\cdot\varkappa}) = 0; \qquad r, t = 0, 1, 2, \ldots \end{aligned}$$

be algebraically consistent in $v^\varkappa$ and $v_\lambda^{\cdot\varkappa}$ and that all their solutions satisfy the $(N+1)$st set of the equations. If there exist $n^2 + n - r$ linearly independent equations in the first N sets, then the space admits an r-parameter complete group of affine motions.

If the system (8.2) of partial differential equations is completely integrable, then

$$(8.15)\qquad \begin{aligned} \underset{v}{\pounds} R_{\nu\mu\lambda}^{\cdot\cdot\cdot\varkappa} = {} & v^\omega \nabla_\omega R_{\nu\mu\lambda}^{\cdot\cdot\cdot\varkappa} - (\dot{\xi}^\rho \nabla_\rho v^\omega)\dot{\nabla}_\omega R_{\nu\mu\lambda}^{\cdot\cdot\cdot\varkappa} - R_{\nu\mu\lambda}^{\cdot\cdot\cdot\rho} \nabla_\rho v^\varkappa \\ & + R_{\rho\mu\lambda}^{\cdot\cdot\cdot\varkappa} \nabla_\nu v^\rho + R_{\nu\rho\lambda}^{\cdot\cdot\cdot\varkappa} \nabla_\mu v^\rho + R_{\nu\mu\rho}^{\cdot\cdot\cdot\varkappa} \nabla_\lambda v^\rho = 0 \end{aligned}$$

and

(8.16) $$\underset{v}{\pounds} T_{\nu\mu\lambda}^{\cdots\varkappa} = v^{\omega}\nabla_{\omega} T_{\nu\mu\lambda}^{\cdots\varkappa} - (\dot{\xi}^{\rho}\nabla_{\rho} v^{\omega})\dot{\nabla}_{\omega} T_{\nu\mu\lambda}^{\cdots\varkappa} - T_{\nu\mu\lambda}^{\cdots\rho}\nabla_{\rho} v^{\varkappa} + T_{\rho\mu\lambda}^{\cdots\varkappa}\nabla_{\nu} v^{\rho} + T_{\nu\rho\lambda}^{\cdots\varkappa}\nabla_{\mu} v^{\rho} + T_{\nu\mu\rho}^{\cdots\varkappa}\nabla_{\lambda} v^{\rho} = 0$$

must be satisfied identically for any $v^{\varkappa}$ and $\nabla_{\mu} v^{\varkappa}$ and hence we must have

(8.17) $$R_{\nu\mu\lambda}^{\cdots\varkappa} = 0, \quad T_{\nu\mu\lambda}^{\cdots\varkappa} = 0.$$

The equation $T_{\nu\mu\lambda}^{\cdots\varkappa} = 0$ shows that $\Gamma_{\mu\lambda}^{\varkappa}$ does not depend on $\dot{\xi}^{\varkappa}$ and the equation $R_{\nu\mu\lambda}^{\cdots\varkappa} = 0$ shows that the space is locally an E_n. Thus we have

THEOREM 8.2. *In order that a general affine space of geodesics admit a group of affine motions of the maximum order $n^2 + n$, it is necessary and sufficient that the geodesics be given by the equations of the form*

(8.18) $$\frac{d^2\xi^{\varkappa}}{dt^2} + \Gamma_{\mu\lambda}^{\varkappa}(\xi)\frac{d\xi^{\mu}}{dt}\frac{d\xi^{\lambda}}{dt} = 0$$

and the space be locally an E_n.

§ 9. General projective spaces of geodesics. [1]

The equations of geodesics

(9.1) $$\frac{d\dot{\xi}^{\varkappa}}{dt} + \Gamma^{\varkappa}(\xi, \dot{\xi}) = 0$$

can be written also in the form

(9.2) $$\dot{\xi}^{\omega}\left(\frac{d\dot{\xi}^{\varkappa}}{dt} + \Gamma^{\varkappa}\right) - \dot{\xi}^{\varkappa}\left(\frac{d\dot{\xi}^{\omega}}{dt} + \Gamma^{\omega}\right) = 0.$$

The equation (9.2) is a tensorial equation and does not change its form under an arbitrary transformation of the parameter t. Next to (9.2) we consider another equation of geodesics

(9.3) $$\dot{\xi}^{\omega}\left(\frac{d\dot{\xi}^{\varkappa}}{dt} + {}'\Gamma^{\varkappa}\right) - \dot{\xi}^{\varkappa}\left(\frac{d\dot{\xi}^{\omega}}{dt} + {}'\Gamma^{\omega}\right) = 0,$$

and we ask for the necessary and sufficient condition that the equations (9.2) and (9.3) define the same system of geodesics.

From (9.2) and (9.3), we find

$$\dot{\xi}^{\omega}({}'\Gamma^{\varkappa} - \Gamma^{\varkappa}) - \dot{\xi}^{\varkappa}({}'\Gamma^{\omega} - \Gamma^{\omega}) = 0$$

[1] BERWALD [3]; DOUGLAS [1]; KNEBELMAN [2, 4]; YANO [5].

or

$$(9.4) \qquad (A_\lambda^\omega T^\varkappa - A_\lambda^\varkappa T^\omega)\dot{\xi}^\lambda = 0,$$

where

$$(9.5) \qquad T^\varkappa \overset{\text{def}}{=} {}'\Gamma^\varkappa - \Gamma^\varkappa$$

are components of a contravariant vector, and are homogeneous functions of degree two with respect to $\dot{\xi}^\varkappa$.

Differentiating (9.4) with respect to $\dot{\xi}^\omega$, contracting with respect to ω and taking account of the homogeneity property of $T^\varkappa$, we find

$$(9.6) \qquad T^\varkappa = p\dot{\xi}^\varkappa,$$

where

$$(9.7) \qquad p \overset{\text{def}}{=} \frac{1}{n+1} \dot{\partial}_\omega T^\omega$$

is a homogeneous scalar function of degree one with respect to $\dot{\xi}^\varkappa$.

From (9.5) and (9.6) we obtain

$$(9.8) \qquad {}'\Gamma^\varkappa = \Gamma^\varkappa + p\dot{\xi}^\varkappa,$$

from which, by partial differentiation with respect to $\dot{\xi}^\lambda$,

$$(9.9) \qquad {}'\Gamma^\varkappa_{\mu\lambda} = \Gamma^\varkappa_{\mu\lambda} + p_\mu A_\lambda^\varkappa + p_\lambda A_\mu^\varkappa + p_{\mu\lambda}\dot{\xi}^\varkappa,$$

where

$$(9.10) \qquad p_\lambda \overset{\text{def}}{=} \dot{\nabla}_\lambda p, \quad p_{\mu\lambda} \overset{\text{def}}{=} \dot{\nabla}_\mu \dot{\nabla}_\lambda p.$$

Conversely, if two linear connexions are related by an equation of the form (9.8) or (9.9), the equations (9.2) and (9.3) define the same system of geodesics. Thus we obtain

THEOREM 9.1. *Two linear connexions* ${}'\Gamma^\varkappa_{\mu\lambda}$ *and* $\Gamma^\varkappa_{\mu\lambda}$ *give the same system of geodesics if and only if they are related by an equation of the form* (9.9).

The equation (9.9) gives the so-called *projective change* of the linear connexion $\Gamma^\varkappa_{\mu\lambda}$. The study of the properties of geodesics which are invariant under a projective change of $\Gamma^\varkappa_{\mu\lambda}$ is called the *general projective geometry of geodesics.*

It is known that the projective geometry of geodesics can be studied as a theory of the space of elements $(\xi, \dot{\xi})$ with normal projective connexion, whose family of geodesics is given by (9.1). The components $\Pi_{\mu\lambda}$, $C_{\mu\lambda}$ and $\Pi^{\varkappa}_{\mu\lambda}$ of this normal projective connexion with respect to a semi-natural frame of reference are given by [1]

(9.11)

$$
\begin{cases}
\Pi_{\mu\lambda} \overset{\text{def}}{=} \Pi^{\rho}_{\mu\sigma}\dot{\xi}^{\sigma}C_{\rho\lambda} - \dfrac{1}{n^2-1}\,(nN_{\mu\lambda} + N_{\lambda\mu}), \\[2mm]
C_{\mu\lambda} \overset{\text{def}}{=} \dfrac{1}{n+1}\,\dot{\partial}_{\mu}\Pi^{\rho}_{\lambda\rho} = \dfrac{1}{n+1}\,\dot{\partial}_{\mu}\Gamma^{\rho}_{\lambda\rho}, \\[2mm]
\Pi^{\varkappa}_{\mu\lambda} \overset{\text{def}}{=} \Gamma^{\varkappa}_{\mu\lambda} - C_{\mu\lambda}\dot{\xi}^{\varkappa}
\end{cases}
$$

where

(9.12) $N^{\cdots\varkappa}_{\nu\mu\lambda} \overset{\text{def}}{=} (\partial_{\nu}\Pi^{\varkappa}_{\mu\lambda} - \Pi^{\rho}_{\nu\sigma}\dot{\xi}^{\sigma}\dot{\partial}_{\rho}\Pi^{\varkappa}_{\mu\lambda}) - (\partial_{\mu}\Pi^{\varkappa}_{\nu\lambda} - \Pi^{\rho}_{\mu\sigma}\dot{\xi}^{\sigma}\dot{\partial}_{\rho}\Pi^{\varkappa}_{\nu\lambda})$

$$+ \Pi^{\varkappa}_{\nu\rho}\Pi^{\rho}_{\mu\lambda} - \Pi^{\varkappa}_{\mu\rho}\Pi^{\rho}_{\nu\lambda},$$

(9.13) $$N_{\mu\lambda} \overset{\text{def}}{=} N^{\cdots\rho}_{\rho\mu\lambda}.$$

The $\Pi_{\mu\lambda}$ and $\Pi^{\varkappa}_{\mu\lambda}$ are homogeneous functions of degree zero and the $C_{\mu\lambda}$ are homogeneous functions of degree -1 with respect to $\dot{\xi}^{\varkappa}$. It is easily verified that the $C_{\mu\lambda}$ are components of a tensor and the $\Pi^{\varkappa}_{\mu\lambda}$ are components of a linear connexion. Hence the $N^{\cdots\varkappa}_{\nu\mu\lambda}$ and $N_{\mu\lambda}$ are homogeneous functions of degree zero with respect to $\dot{\xi}^{\varkappa}$ and are components of tensors.

If we define the covariant derivatives of a tensor, say $T^{\cdot\varkappa}_{\lambda}$ by

(9.14)

$$
\begin{cases}
\nabla_{\mu}T^{\cdot\varkappa}_{\lambda} = \partial_{\mu}T^{\cdot\varkappa}_{\lambda} - \Pi^{\rho}_{\mu\sigma}\dot{\xi}^{\sigma}\dot{\partial}_{\rho}T^{\cdot\varkappa}_{\lambda} + \Pi^{\varkappa}_{\mu\rho}T^{\cdot\rho}_{\lambda} - \Pi^{\rho}_{\mu\lambda}T^{\cdot\varkappa}_{\rho}, \\[2mm]
\dot{\nabla}_{\mu}T^{\cdot\varkappa}_{\lambda} = \dot{\partial}_{\mu}T^{\cdot\varkappa}_{\lambda},
\end{cases}
$$

then by straightforward calculation, we can prove the following formulae:

(9.15) $\nabla_{\mu}\dot{\xi}^{\varkappa} = 0, \quad \dot{\nabla}_{\mu}\dot{\xi}^{\varkappa} = A^{\varkappa}_{\mu},$

(9.16) $(\nabla_{\nu}\nabla_{\mu} - \nabla_{\mu}\nabla_{\nu})T^{\cdot\varkappa}_{\lambda} = N^{\cdots\varkappa}_{\nu\mu\rho}T^{\cdot\rho}_{\lambda} - N^{\cdots\rho}_{\nu\mu\lambda}T^{\cdot\varkappa}_{\rho} - N^{\cdots\rho}_{\nu\mu\sigma}\dot{\xi}^{\sigma}\dot{\nabla}_{\rho}T^{\cdot\varkappa}_{\lambda},$

(9.17) $(\dot{\nabla}_{\nu}\nabla_{\mu} - \nabla_{\mu}\dot{\nabla}_{\nu})T^{\cdot\varkappa}_{\lambda} = U^{\cdots\varkappa}_{\nu\mu\rho}T^{\cdot\rho}_{\lambda} - U^{\cdots\rho}_{\nu\mu\lambda}T^{\cdot\varkappa}_{\rho} - U^{\cdots\rho}_{\nu\mu\sigma}\dot{\xi}^{\sigma}\dot{\nabla}_{\rho}T^{\cdot\varkappa}_{\lambda},$

(9.18) $\nabla_{[\omega}N^{\cdots\varkappa}_{\nu\mu]\lambda} + N^{\cdots\rho}_{[\omega\nu|\sigma}\dot{\xi}^{\sigma}U^{\cdots\varkappa}_{\rho|\mu]\lambda} = 0,$

(9.19) $\dot{\nabla}_{\omega}N^{\cdots\varkappa}_{\nu\mu\lambda} = 2\nabla_{[\nu}U^{\cdots\varkappa}_{|\omega|\mu]\lambda} - 2U^{\cdots\rho}_{\omega[\nu|\sigma}\dot{\xi}^{\sigma}U^{\cdots\varkappa}_{\rho|\mu]\lambda},$

---

[1] YANO [5]. Since

$$N_{\mu\lambda} = R_{\mu\lambda} - \dot{\xi}^{\rho}\nabla_{\rho}C_{\mu\lambda},$$

$$\dot{\xi}^{\rho}\dot{\nabla}_{\mu}(R_{\rho\lambda} - R_{\lambda\rho}) = (n+1)\dot{\xi}^{\rho}\nabla_{\rho}C_{\mu\lambda},$$

the equation (7.13) in YANO [5] coincides with the first equation of (9.11).

where

$$U_{\nu\mu\lambda}^{\cdot\cdot\cdot\varkappa} \overset{\text{def}}{=} \dot{\partial}_\nu \Pi_{\mu\lambda}^{\varkappa} \tag{9.20}$$

are homogeneous functions of degree -1 with respect to $\dot{\xi}^\varkappa$ and are components of a tensor.

The curvature tensors of the projective connexion are given by

$$P_{\nu\mu\lambda} \overset{\text{def}}{=} \nabla_\nu M_{\mu\lambda} - \nabla_\mu M_{\nu\lambda} + N_{\nu\mu\sigma}^{\cdot\cdot\cdot\rho} \dot{\xi}^\sigma C_{\rho\lambda}, \tag{9.21}$$

$$Q_{\nu\mu\lambda} \overset{\text{def}}{=} \dot{\nabla}_\nu M_{\mu\lambda} - \nabla_\mu C_{\nu\lambda} + U_{\nu\mu\sigma}^{\cdot\cdot\cdot\rho} \dot{\xi}^\sigma C_{\rho\lambda}, \tag{9.22}$$

$$P_{\nu\mu\lambda}^{\cdot\cdot\cdot\varkappa} \overset{\text{def}}{=} N_{\nu\mu\lambda}^{\cdot\cdot\cdot\varkappa} + A_\nu^\varkappa M_{\mu\lambda} - A_\mu^\varkappa M_{\nu\lambda} - (M_{\nu\mu} - M_{\mu\nu}) A_\lambda^\varkappa, \tag{9.23}$$

$$Q_{\nu\mu\lambda}^{\cdot\cdot\cdot\varkappa} \overset{\text{def}}{=} U_{\nu\mu\lambda}^{\cdot\cdot\cdot\varkappa} - C_{\nu\mu} A_\lambda^\varkappa - C_{\nu\lambda} A_\mu^\varkappa, \tag{9.24}$$

where

$$M_{\mu\lambda} \overset{\text{def}}{=} \Pi_{\mu\lambda} - \Pi_{\mu\sigma}^{\rho} \dot{\xi}^\sigma C_{\rho\lambda} = -\frac{1}{n^2 - 1}(nN_{\mu\lambda} + N_{\lambda\mu}) \tag{9.25}$$

is a tensor. The $P_{\nu\mu\lambda}$, $P_{\nu\mu\lambda}^{\cdot\cdot\cdot\varkappa}$ and $M_{\mu\lambda}$ are homogeneous functions of degree zero and the $Q_{\nu\mu\lambda}$ and $Q_{\nu\mu\lambda}^{\cdot\cdot\cdot\varkappa}$ are homogeneous functions of degree -1 with respect to $\dot{\xi}^\varkappa$.

Using the relations

$$C_{\mu\lambda} = C_{\lambda\mu}, \; C_{\mu\lambda} \dot{\xi}^\lambda = 0, \; \dot{\partial}_\rho \Pi_{\mu\lambda}^{\rho} = 2C_{\mu\lambda}, \; U_{\nu\mu\lambda}^{\cdot\cdot\cdot\varkappa} \dot{\xi}^\nu = 0, \tag{9.26}$$

we can easily verify that the projective curvature tensors satisfy the relations

$$\begin{cases} Q_{\nu\mu\lambda} \dot{\xi}^\nu = 0, \; Q_{\nu\mu\lambda}^{\cdot\cdot\cdot\varkappa} \dot{\xi}^\nu = 0, \; Q_{\nu\mu\lambda}^{\cdot\cdot\cdot\varkappa} = Q_{\nu\lambda\mu}^{\cdot\cdot\cdot\varkappa}, \\ P_{\rho\mu\lambda}^{\cdot\cdot\cdot\rho} = 0, \; P_{\nu\rho\lambda}^{\cdot\cdot\cdot\rho} = 0, \; P_{\nu\mu\rho}^{\cdot\cdot\cdot\rho} = 0, \; Q_{\rho\mu\lambda}^{\cdot\cdot\cdot\rho} = 0, \; Q_{\nu\rho\lambda}^{\cdot\cdot\cdot\rho} = 0, \\ Q_{\nu\mu\rho}^{\cdot\cdot\cdot\rho} = 0. \end{cases} \tag{9.27}$$

We remark here that the tensor $Q_{\nu\mu\lambda}^{\cdot\cdot\cdot\varkappa}$ can be written also in the form

$$Q_{\nu\mu\lambda}^{\cdot\cdot\cdot\varkappa} = \tfrac{1}{2} \dot{\partial}_\nu \dot{\partial}_\mu \dot{\partial}_\lambda \left[\Gamma^\varkappa - \frac{1}{n+1} (\dot{\partial}_\rho \Gamma^\rho) \dot{\xi}^\varkappa \right], \tag{9.28}$$

which shows that the $Q_{\nu\mu\lambda}^{\cdot\cdot\cdot\varkappa}$ is symmetric in the three lower indices.

Under a projective change (9.9) of $\Gamma_{\mu\lambda}^{\varkappa}$, the functions $\Pi_{\mu\lambda}$ $C_{\mu\lambda}$ and $\Pi_{\mu\lambda}^{\varkappa}$ are transformed into $'\Pi_{\mu\lambda}$, $'C_{\mu\lambda}$ and $'\Pi_{\mu\lambda}^{\varkappa}$ respectively following the formulae:

$$(9.29)\qquad \begin{cases} '\Pi_{\mu\lambda} = \Pi_{\mu\lambda} + \partial_\mu p_\lambda - \Pi^{\varkappa}_{\mu\lambda} p_\varkappa - p_\mu p_\lambda, \\ 'C_{\mu\lambda} = C_{\mu\lambda} + \dot{\partial}_\mu p_\lambda, \\ '\Pi^{\varkappa}_{\mu\lambda} = \Pi^{\varkappa}_{\mu\lambda} + p_\mu A^{\varkappa}_\lambda + p_\lambda A^{\varkappa}_\mu \end{cases}$$

and consequently the projective curvature tensors $P_{\nu\mu\lambda}$, $Q_{\nu\mu\lambda}$, $P_{\nu\mu\lambda}^{\cdot\cdot\cdot\varkappa}$ and $Q_{\nu\mu\lambda}^{\cdot\cdot\cdot\varkappa}$ are transformed as follows:

$$(9.30)\qquad \begin{cases} 'P_{\nu\mu\lambda} = P_{\nu\mu\lambda} - P_{\nu\mu\lambda}^{\cdot\cdot\cdot\varkappa} p_\varkappa, \\ 'Q_{\nu\mu\lambda} = Q_{\nu\mu\lambda} - Q_{\nu\mu\lambda}^{\cdot\cdot\cdot\varkappa} p_\varkappa, \\ 'P_{\nu\mu\lambda}^{\cdot\cdot\cdot\varkappa} = P_{\nu\mu\lambda}^{\cdot\cdot\cdot\varkappa},\ 'Q_{\nu\mu\lambda}^{\cdot\cdot\cdot\varkappa} = Q_{\nu\mu\lambda}^{\cdot\cdot\cdot\varkappa}. \end{cases}$$

We derive here some formulae which are useful in the discussions which follow. From (9.23), we have

$$(9.31)\qquad \dot{\nabla}_\omega P_{\nu\mu\lambda}^{\cdot\cdot\cdot\varkappa} = \dot{\nabla}_\omega N_{\nu\mu\lambda}^{\cdot\cdot\cdot\varkappa} + A^{\varkappa}_\nu \dot{\nabla}_\omega M_{\mu\lambda} - A^{\varkappa}_\mu \dot{\nabla}_\omega M_{\nu\lambda} - (\dot{\nabla}_\omega M_{\nu\mu} - \dot{\nabla}_\omega M_{\mu\nu}) A^{\varkappa}_\lambda.$$

Substituting (9.19), (9.22) and (9.24), we find

$$(9.32)\qquad \tfrac{1}{2}\dot{\nabla}_\omega P_{\nu\mu\lambda}^{\cdot\cdot\cdot\varkappa} = \nabla_{[\nu} Q_{|\omega|\mu]\lambda}^{\cdot\;\cdot\;\cdot\;\varkappa} + A^{\varkappa}_{[\nu} Q_{|\omega|\mu]\lambda} - Q_{\omega[\nu\mu]} A^{\varkappa}_\lambda.$$

Contracting this equation with respect to $\varkappa$ and ν and taking account of (9.27), we obtain

$$(9.33)\qquad Q_{\nu\mu\lambda} = -\frac{1}{n-1} \nabla_\rho Q_{\nu\mu\lambda}^{\cdot\cdot\cdot\rho},$$

from which we see that

$$(9.34)\qquad Q_{\nu\mu\lambda} = Q_{(\nu\mu\lambda)}$$

and consequently from (9.32)

$$(9.35)\qquad \tfrac{1}{2}\dot{\nabla}_\omega P_{\nu\mu\lambda}^{\cdot\cdot\cdot\varkappa} = \nabla_{[\nu} Q_{|\omega|\mu]\lambda}^{\cdot\;\cdot\;\cdot\;\varkappa} + A^{\varkappa}_{[\nu} Q_{|\omega|\mu]\lambda}.$$

If we contract this equation with respect to $\varkappa$ and ω, then we find

$$(9.36)\qquad \tfrac{1}{2}\dot{\nabla}_\rho P_{\nu\mu\lambda}^{\cdot\cdot\cdot\rho} = 0.$$

We next substitute (9.23) and (9.24) in (9.18) and take account of $Q_{\nu\mu\lambda}^{\cdot\cdot\cdot\varkappa}\xi^\nu = 0$. Then we obtain

$$(9.37)\qquad \nabla_{[\omega} P_{\nu\mu]\lambda}^{\cdot\cdot\;\cdot\varkappa} + A^{\varkappa}_{[\omega} P_{\nu\mu]\lambda} + P_{[\omega\nu\mu]} A^{\varkappa}_\lambda + P_{[\omega\nu|\sigma}^{\cdot\;\cdot\;\cdot\rho} \xi^\sigma Q_{\rho\ \mu]\lambda}^{\cdot\;\cdot\;\cdot\varkappa} = 0$$

by virtue of (9.21).

Contracting this equation with respect to $\varkappa$ and λ and taking account of (9.27), we find

(9.38) $$P_{[\omega\nu\mu]} = 0$$

and consequently (9.37) becomes

(9.39) $$\nabla_{[\omega} P_{\nu\mu]\lambda}^{\cdot\cdot\cdot\varkappa} + A_{[\omega}^{\varkappa} P_{\nu\mu]\lambda} + P_{[\omega\nu|\sigma}^{\cdot\cdot\cdot\rho} \dot{\xi}^{\sigma} Q_{\rho|\mu]\lambda}^{\cdot\cdot\cdot\varkappa} = 0.$$

If we contract this equatin wih respect to $\varkappa$ and ω, then we obtain

(9.40) $$\nabla_{\rho} P_{\nu\mu\lambda}^{\cdot\cdot\cdot\rho} + (n-2)P_{\nu\mu\lambda} + 2P_{\tau[\nu|\sigma}^{\cdot\cdot\cdot\rho} \dot{\xi}^{\sigma} Q_{\rho|\mu]\lambda}^{\cdot\cdot\cdot\tau} = 0.$$

If, by a suitable projective change, we can transform the equations of geodesics into the equations of geodesics in an E_n, we say that the general projective space of geodesics is *projectively Euclidean.*

A necessary condition for a general projective space to be projectively Euclidean is that

(9.41) $$P_{\nu\mu\lambda}^{\cdot\cdot\cdot\varkappa} = 0, \quad Q_{\nu\mu\lambda}^{\cdot\cdot\cdot\varkappa} = 0.$$

Conversely, if (9.41) holds then, as we can see from (9.28), by a suitable projective change the functions $\Gamma_{\mu\lambda}^{\varkappa}$ become independent of the direction element $\dot{\xi}^{\varkappa}$, and

$$C_{\mu\lambda} = 0,$$

$$M_{\mu\lambda} = -\frac{1}{n^2-1}(nR_{\mu\lambda} + R_{\lambda\mu}).$$

Hence $P_{\nu\mu\lambda}^{\cdot\cdot\cdot\varkappa}$ coincides with the projective curvature tensor of Weyl. Thus, for $n > 2$, $P_{\nu\mu\lambda}^{\cdot\cdot\cdot\varkappa} = 0$ implies that the space is projectively Euclidean. Hence

THEOREM 9.1. *In order that an n-dimensional general projective space of geodesics, $n > 2$, be projectively Euclidean, it is necessary and sufficient that $P_{\nu\mu\lambda}^{\cdot\cdot\cdot\varkappa} = 0$ and $Q_{\nu\mu\lambda}^{\cdot\cdot\cdot\varkappa} = 0$.*

§ 10. Projective motions in a general projective space of geodesics.

If a system of equations (9.1) is given, we can construct the functions $\Gamma_{\mu\lambda}^{\varkappa} = \frac{1}{2}\dot{\partial}_{\mu}\dot{\partial}_{\lambda}\Gamma^{\varkappa}$ and the normal projective connexion $\Pi_{\mu\lambda}$, $C_{\mu\lambda}$ and $\Pi_{\mu\lambda}^{\varkappa}$

in such a way that the system of geodesics of the projective connexion coincides with the given system of curves given by (9.1).

If we consider an extended point transformation

$$'\xi^{\varkappa} = f^{\varkappa}(\xi^{\nu}), \; '\dot{\xi}^{\varkappa} = (\partial_{\lambda} f^{\varkappa})\dot{\xi}^{\lambda}, \tag{10.1}$$

we get the deformed linear connexion $'\Gamma^{\varkappa}_{\mu\lambda}$ and from this we can construct the deformed normal projective connexion.

If the original normal projective connexion and the deformed one are the same, that is, if there exist functions p_{λ} such that we have (9.29), we call the transformation a projective motion. Since, for a normal projective connexion, the first and the second equation of (9.29) follow from the third, we have

THEOREM 10.1. *In order that an extended point transformation* (10.1) *be a projective motion in a general projective space of geodesics, it is necessary and sufficient that*

$$'\Pi^{\varkappa}_{\mu\lambda} = \Pi^{\varkappa}_{\mu\lambda} + p_{\mu} A^{\varkappa}_{\lambda} + p_{\lambda} A^{\varkappa}_{\mu}. \tag{10.2}$$

Considering an infinitesimal extended point transformation

$$'\xi^{\varkappa} = \xi^{\varkappa} + v^{\varkappa}(\xi)dt, \; '\dot{\xi}^{\varkappa} = \dot{\xi}^{\varkappa} + (\partial_{\lambda} v^{\varkappa})\dot{\xi}^{\lambda} dt, \tag{10.3}$$

we get

THEOREM 10.2. *In order that* (10.3) *be a projective motion, it is necessary and sufficient that the Lie derivative* $\underset{v}{\pounds}\Pi^{\varkappa}_{\mu\lambda}$ *of* $\Pi^{\varkappa}_{\mu\lambda}$ *have the form*

$$\underset{v}{\pounds}\Pi^{\varkappa}_{\mu\lambda} = p_{\mu} A^{\varkappa}_{\lambda} + p_{\lambda} A^{\varkappa}_{\mu}. \tag{10.4}$$

If we eliminate the p_{λ} from (10.4), we find

$$\underset{v}{\pounds}\overset{p}{\Pi}{}^{\varkappa}_{\mu\lambda} = 0; \; \overset{p}{\Pi}{}^{\varkappa}_{\mu\lambda} = \Pi^{\varkappa}_{\mu\lambda} - \frac{1}{n+1}(\Pi^{\rho}_{\mu\rho} A^{\varkappa}_{\lambda} + \Pi^{\rho}_{\lambda\rho} A^{\varkappa}_{\mu}). \tag{10.5}$$

Conversely, if we have (10.5), then $\underset{v}{\pounds}\Pi^{\varkappa}_{\mu\lambda}$ must have the form (10.4). Hence

THEOREM 10.3. *In order that* (10.3) *be a projective motion, it is necessary and sufficient that the Lie derivative of* $\overset{p}{\Pi}{}^{\varkappa}_{\mu\lambda}$ *vanish.*

The formulae on Lie derivatives (6.13), (6.14), (6.15) (6.16), (6.17)

and (6.18) become in the present case

$$(10.6)\qquad (\underset{v}{\pounds}\nabla_\mu - \nabla_\mu\underset{v}{\pounds})T_\lambda^{\cdot\kappa} = (\underset{v}{\pounds}\Pi^\kappa_{\mu\rho})T_\lambda^{\cdot\rho} - (\underset{v}{\pounds}\Pi^\rho_{\mu\lambda})T_\rho^{\cdot\kappa} - (\underset{v}{\pounds}\Pi^\rho_{\mu\sigma})\xi^\sigma\dot\nabla_\rho T_\lambda^{\cdot\kappa},$$

$$(10.7)\qquad (\underset{v}{\pounds}\dot\nabla_\mu - \dot\nabla_\mu\underset{v}{\pounds})T_\lambda^{\cdot\kappa} = 0,$$

$$(10.8)\qquad \nabla_\nu\underset{v}{\pounds}\Pi^\kappa_{\mu\lambda} - \nabla_\mu\underset{v}{\pounds}\Pi^\kappa_{\nu\lambda} = \underset{v}{\pounds}N_{\nu\mu\lambda}^{\cdots\kappa} + (\underset{v}{\pounds}\Pi^\rho_{\nu\sigma})\xi^\sigma U_{\rho\mu\lambda}^{\cdots\kappa} - (\underset{v}{\pounds}\Pi^\rho_{\mu\sigma})\xi^\sigma U_{\rho\nu\lambda}^{\cdots\kappa},$$

$$(10.9)\qquad \dot\nabla_\nu\underset{v}{\pounds}\Pi^\kappa_{\mu\lambda} = \underset{v}{\pounds}U_{\nu\mu\lambda}^{\cdots\kappa},$$

$$(10.10)\qquad (\underset{c}{\pounds}\underset{b}{\pounds})\Pi^\kappa_{\mu\lambda} = \underset{cb}{\pounds}\Pi^\kappa_{\mu\lambda},$$

$$(10.11)\qquad (\underset{c}{\pounds}\underset{b}{\pounds})\Pi^\kappa_{\mu\lambda} = c^a_{cb}\pounds\Pi^\kappa_{\mu\lambda}$$

respectively.

From (10.10) and (10.11), we find

$$(10.12)\qquad \overset{p}{(\underset{c}{\pounds}\underset{b}{\pounds})}\Pi^\kappa_{\mu\lambda} = \overset{p}{\underset{cb}{\pounds}}\Pi^\kappa_{\mu\lambda},$$

$$(10.13)\qquad \overset{p}{(\underset{c}{\pounds}\underset{b}{\pounds})}\Pi^\kappa_{\mu\lambda} = c^a_{cb}\overset{p}{\underset{a}{\pounds}}\Pi^\kappa_{\mu\lambda}.$$

Making use of these equations, we can prove Theorems corresponding to Theorems 2.1, 2.2, 2.3, 2.4, 2.5, 2.6 of Ch. III and to Theorem 1.3 of Ch. V.

§ 11. Integrability conditions of $\underset{v}{\pounds}\Pi^\kappa_{\mu\lambda} = p_\mu A^\kappa_\lambda + p_\lambda A^\kappa_\mu$.

In this section, we examine the conditions that

$$(11.1)\qquad \underset{v}{\pounds}\Pi^\kappa_{\mu\lambda} = \nabla_\mu\nabla_\lambda v^\kappa + U_{\nu\mu\lambda}^{\cdots\kappa}\xi^\rho\nabla_\rho v^\nu + N_{\nu\mu\lambda}^{\cdots\kappa}v^\nu = p_\mu A^\kappa_\lambda + p_\lambda A^\kappa_\mu$$

admits solutions v^κ and p_λ, the v^κ being functions of ξ^κ only and the p_λ being homogeneous functions of degree zero with respect to ξ^κ.

Substituting (11.1) in (10.8), we find

$$(11.2)\qquad \underset{v}{\pounds}N_{\nu\mu\lambda}^{\cdots\kappa} + A^\kappa_\nu\nabla_\mu p_\lambda - A^\kappa_\mu\nabla_\nu p_\lambda - (\nabla_\nu p_\mu - \nabla_\mu p_\nu)A^\kappa_\lambda + (U_{\nu\mu\lambda}^{\cdots\kappa} - U_{\mu\nu\lambda}^{\cdots\kappa})\xi^\rho p_\rho = 0,$$

from which, by contraction with respect to κ and ν,

$$(11.3)\qquad \nabla_\mu p_\lambda = \underset{v}{\pounds}M_{\mu\lambda} + C_{\mu\lambda}\xi^\rho p_\rho.$$

Substituting (11.1) in (10.9), we get

$$\underset{v}{\pounds} U_{\nu\mu\lambda}^{\cdots\varkappa} - \dot{\nabla}_\nu p_\mu A_\lambda^\varkappa - \dot{\nabla}_\nu p_\lambda A_\mu^\varkappa = 0, \tag{11.4}$$

from which, by contraction with respect to $\varkappa$ and λ

$$\dot{\nabla}_\mu p_\lambda = \underset{v}{\pounds} C_{\mu\lambda}. \tag{11.5}$$

Thus we are led to consider the following system of partial differential equations

$$\left\{\begin{array}{ll} \text{(i)} \quad \nabla_\lambda v^\varkappa = v_\lambda^{\cdot\varkappa}, & \text{(ii)} \quad \dot{\nabla}_\lambda v^\varkappa = 0, \\ \text{(iii)} \quad \nabla_\mu v_\lambda^{\cdot\varkappa} = - U_{\nu\mu\lambda}^{\cdots\varkappa} \dot{\xi}^\rho v_\rho^{\cdot\nu} - N_{\nu\mu\lambda}^{\cdots\varkappa} v^\nu + p_\mu A_\lambda^\varkappa + p_\lambda A_\mu^\varkappa, & \\ \text{(iv)} \quad \dot{\nabla}_\mu v_\lambda^{\cdot\varkappa} = U_{\mu\lambda\rho}^{\cdots\varkappa} v^\rho, & \\ \text{(v)} \quad \nabla_\mu p_\lambda = \underset{v}{\pounds} M_{\mu\lambda} + C_{\mu\lambda} \dot{\xi}^\rho p_\rho, & \text{(vi)} \quad \dot{\nabla}_\mu p_\lambda = \underset{v}{\pounds} C_{\mu\lambda}, \end{array}\right. \tag{11.6}$$

with the unknown functions $v^\varkappa$, $v_\lambda^{\cdot\varkappa}$ and p_λ. If the system admits the solutions $v^\varkappa$, $v_\lambda^{\cdot\varkappa}$ and p_λ, the $v^\varkappa$ do not contain $\dot{\xi}^\varkappa$, and the p_λ are homogeneous funtions of degree zero of $\dot{\xi}^\varkappa$, because

$$\dot{\xi}^\mu \dot{\nabla}_\mu p_\lambda = \dot{\xi}^\mu \underset{v}{\pounds} C_{\mu\lambda} = \underset{v}{\pounds} \dot{\xi}^\mu C_{\mu\lambda} = 0.$$

Moreover (11.6, vi) shows that there exists a homogeneous function p of degree one of $\dot{\xi}^\varkappa$ such that

$$p_\lambda = \dot{\nabla}_\lambda p. \tag{11.7}$$

Substituting (11.3) in (11.2) and (11.5) in (11.4), we find

$$\underset{v}{\pounds} P_{\nu\mu\lambda}^{\cdots\varkappa} = 0 \tag{11.8}$$

and

$$\underset{v}{\pounds} Q_{\nu\mu\lambda}^{\cdots\varkappa} = 0 \tag{11.9}$$

respectively.

Substituting (11.3) and (11.5) in the Ricci formula

$$(\dot{\nabla}_\nu \nabla_\mu - \nabla_\mu \dot{\nabla}_\nu) p_\lambda = - U_{\nu\mu\lambda}^{\cdots\varkappa} p_\varkappa - U_{\nu\mu\sigma}^{\cdots\rho} \dot{\xi}^\sigma \dot{\nabla}_\rho p_\lambda,$$

we find

$$\dot{\nabla}_\nu \underset{v}{\pounds} M_{\mu\lambda} - (\dot{\nabla}_\nu C_{\mu\lambda}) \dot{\xi}^\rho p_\rho - p_\nu C_{\mu\lambda} - \nabla_\mu \underset{v}{\pounds} C_{\nu\lambda} = - U_{\nu\mu\lambda}^{\cdots\varkappa} p_\varkappa - U_{\nu\mu\sigma}^{\cdots\rho} \dot{\xi}^\sigma \underset{v}{\pounds} C_{\rho\lambda}. \tag{11.10}$$

On the other hand, applying the formula (10.6) to $C_{\nu\lambda}$, we get

$$\underset{v}{\pounds}\nabla_\mu C_{\nu\lambda} - \nabla_\mu \underset{v}{\pounds} C_{\nu\lambda} = -(\underset{v}{\pounds}\Pi^{\rho}_{\mu\nu})C_{\rho\lambda} - (\underset{v}{\pounds}\Pi^{\rho}_{\mu\lambda})C_{\nu\rho} - (\underset{v}{\pounds}\Pi^{\rho}_{\mu\sigma})\xi^\sigma \dot{\nabla}_\rho C_{\nu\lambda},$$

from which

$$(11.11)\quad \nabla_\mu \underset{v}{\pounds} C_{\nu\lambda} = \underset{v}{\pounds}\nabla_\mu C_{\nu\lambda} + p_\mu C_{\nu\lambda} + p_\nu C_{\mu\lambda} + C_{\mu\nu} p_\lambda + (\dot{\nabla}_\mu C_{\nu\lambda})\xi^\rho p_\rho.$$

Taking account of $\dot{\nabla}_\nu C_{\mu\lambda} = \dot{\nabla}_\mu C_{\nu\lambda}$ and (11.4), we obtain from (11.10) and (11.11),

$$(11.12)\qquad \underset{v}{\pounds} Q_{\nu\mu\lambda} + Q_{\nu\mu\lambda}{}^{\cdot\cdot\cdot\varkappa} p_\varkappa = 0.$$

Applying the formula (10.6) to the tensor $M_{\mu\lambda}$, we obtain

$$\underset{v}{\pounds}\nabla_\nu M_{\mu\lambda} - \nabla_\nu \underset{v}{\pounds} M_{\mu\lambda} = -(\underset{v}{\pounds}\Pi^{\rho}_{\nu\mu})M_{\rho\lambda} - (\underset{v}{\pounds}\Pi^{\rho}_{\nu\lambda})M_{\mu\rho} - (\underset{v}{\pounds}\Pi^{\rho}_{\nu\sigma})\xi^\sigma \dot{\nabla}_\rho M_{\mu\lambda}$$

from which

$$\underset{v}{\pounds}\nabla_\nu M_{\mu\lambda} - \nabla_\nu\nabla_\mu p_\lambda + (\nabla_\nu C_{\mu\lambda})\xi^\rho p_\rho + C_{\mu\lambda}\xi^\rho \underset{v}{\pounds} M_{\nu\rho}$$
$$= -p_\nu M_{\mu\lambda} - p_\mu M_{\nu\lambda} - p_\nu M_{\mu\lambda} - p_\lambda M_{\mu\nu} - (\dot{\nabla}_\nu M_{\mu\lambda})\xi^\rho p_\rho.$$

Taking the alternating part of this equation with respect to ν and μ, we find

$$\underset{v}{\pounds}(\nabla_\nu M_{\mu\lambda} - \nabla_\mu M_{\nu\lambda}) + N_{\nu\mu\sigma}{}^{\cdot\cdot\cdot\rho}\xi^\sigma \dot{\nabla}_\rho p_\lambda$$
$$= -P_{\nu\mu\lambda}{}^{\cdot\cdot\cdot\varkappa} p_\varkappa - (\dot{\nabla}_\nu M_{\mu\lambda} - \dot{\nabla}_\mu M_{\nu\lambda} + \nabla_\nu C_{\mu\lambda} - \nabla_\mu C_{\nu\lambda})\xi^\rho p_\rho$$
$$+ (\underset{v}{\pounds} M_{\mu\rho})C_{\nu\lambda}\xi^\rho - (\underset{v}{\pounds} M_{\nu\rho})C_{\mu\lambda}\xi^\rho$$

or

$$\underset{v}{\pounds}(\nabla_\nu M_{\mu\lambda} - \nabla_\mu M_{\nu\lambda}) + (\underset{v}{\pounds} N_{\nu\mu\sigma}{}^{\cdot\cdot\cdot\rho})\xi^\sigma C_{\rho\lambda} + N_{\nu\mu\sigma}{}^{\cdot\cdot\cdot\rho}\xi^\sigma \underset{v}{\pounds} C_{\rho\lambda}$$
$$= -P_{\nu\mu\lambda}{}^{\cdot\cdot\cdot\varkappa} p_\varkappa - 2(\dot{\nabla}_{[\nu} M_{\mu]\lambda} - \nabla_{[\mu} C_{\nu]\lambda} + U_{[\nu\mu]\sigma}{}^{\cdot\cdot\cdot\rho}\xi^\sigma C_{\rho\lambda})\xi^\tau p_\tau$$
$$+ \underset{v}{\pounds}(N_{\nu\mu\sigma}{}^{\cdot\cdot\cdot\rho} + 2A^{\rho}_{[\nu} M_{\mu]\sigma} + 2M_{[\nu\mu]}A^{\rho}_{\sigma})\xi^\sigma C_{\rho\lambda}$$

or

$$(11.13)\qquad \underset{v}{\pounds} P_{\nu\mu\lambda} + p_{\nu\mu\lambda}{}^{\cdot\cdot\cdot\varkappa} p_\varkappa = 0,$$

by virtue of

$$0 = Q_{[\nu\mu]\lambda}^{\cdot\cdot\cdot\varkappa} = U_{[\nu\mu]\lambda}^{\cdot\cdot\cdot\varkappa} + A_{[\nu}^{\varkappa} C_{\mu]\lambda} = 0,$$

$$\xi^\rho C_{\rho\lambda} = 0 \text{ and } \underset{v}{\pounds} P_{\nu\mu\lambda}^{\cdot\cdot\cdot\varkappa} = 0.$$

Thus the integrability conditions of the system (11.6) are given by (11.8), (11.9), (11.12), (11.13) and the equations obtained from these by successive covariant differentiations and by eliminations of $\nabla_\lambda v^\varkappa$, $\dot{\nabla}_\lambda v^\varkappa$, $\nabla_\mu v_\lambda^{\cdot\varkappa}$, $\dot{\nabla}_\mu v_\lambda^{\cdot\varkappa}$, $\nabla_\mu p_\lambda$ and $\dot{\nabla}_\mu p_\lambda$ by the use of (11.6).

First we show that the conditions (11.12) and (11.13) are consequences of (11.8), (11.9) and their successive covariant derivatives.

Applying the oprator $\underset{v}{\pounds}$ to (9.33), we get

$$\underset{v}{\pounds} Q_{\nu\mu\lambda} = -\frac{1}{n-1} \underset{v}{\pounds} \nabla_\rho Q_{\nu\mu\lambda}^{\cdot\cdot\cdot\rho}.$$

On the other hand, applying the formulae (10.6) to $Q_{\nu\mu\lambda}^{\cdot\cdot\cdot\varkappa}$, we obtain

$$\begin{aligned}\underset{v}{\pounds}\nabla_\rho Q_{\nu\mu\lambda}^{\cdot\cdot\cdot\rho} - \nabla_\rho \underset{v}{\pounds} Q_{\nu\mu\lambda}^{\cdot\cdot\cdot\rho} &= (\underset{v}{\pounds}\Pi_{\rho\sigma}^{\rho}) Q_{\nu\mu\lambda}^{\cdot\cdot\cdot\sigma} - (\underset{v}{\pounds}\Pi_{\rho\nu}^{\sigma}) Q_{\sigma\mu\lambda}^{\cdot\cdot\cdot\rho} \\ &\quad - (\underset{v}{\pounds}\Pi_{\rho\mu}^{\sigma}) Q_{\nu\sigma\lambda}^{\cdot\cdot\cdot\rho} - (\underset{v}{\pounds}\Pi_{\rho\lambda}^{\sigma}) Q_{\nu\mu\sigma}^{\cdot\cdot\cdot\varkappa} - (\underset{v}{\pounds}\Pi_{\rho\tau}^{\sigma}) \xi^\tau \dot{\nabla}_\sigma Q_{\nu\mu\lambda}^{\cdot\cdot\cdot\rho} \\ &= (n-1) Q_{\nu\mu\lambda}^{\cdot\cdot\cdot\varkappa} p_\varkappa\end{aligned}$$

by virtue of $\dot{\nabla}_\rho Q_{\nu\mu\lambda}^{\cdot\cdot\cdot\rho} = \dot{\nabla}_\nu Q_{\rho\mu\lambda}^{\cdot\cdot\cdot\rho} = 0$.

From the above two equations, we find

$$\underset{v}{\pounds} Q_{\nu\mu\lambda} + Q_{\nu\mu\lambda}^{\cdot\cdot\cdot\varkappa} p_\varkappa = -\frac{1}{n-1} \nabla_\rho (\underset{v}{\pounds} Q_{\nu\mu\lambda}^{\cdot\cdot\cdot\rho}), \tag{11.14}$$

which shows that (11.12) is obtained from (11.9) and its covariant derivaties.

Next, applying the operator $\underset{v}{\pounds}$ to (9.40), we find

$$\underset{v}{\pounds}\nabla_\rho P_{\nu\mu\lambda}^{\cdot\cdot\cdot\rho} + (n-2)\underset{v}{\pounds} P_{\nu\mu\lambda} + 2(\underset{v}{\pounds} P_{\tau[\nu|\sigma}^{\cdot\cdot\cdot\rho}) \xi^\sigma Q_{\rho|\mu]\lambda}^{\cdot\cdot\cdot\tau} + 2P_{\tau[\nu|\sigma}^{\cdot\cdot\cdot\rho} \xi^\sigma (\underset{v}{\pounds} Q_{\rho|\mu]\lambda}^{\cdot\cdot\cdot\tau}) = 0.$$

On the other hand, applying the formula (10.6) to $P_{\nu\mu\lambda}^{\cdot\cdot\cdot\varkappa}$, we obtain

$$\begin{aligned}\underset{v}{\pounds}\nabla_\rho P_{\nu\mu\lambda}^{\cdot\cdot\cdot\rho} - \nabla_\rho \underset{v}{\pounds} P_{\nu\mu\lambda}^{\cdot\cdot\cdot\rho} &= (n-2) P_{\nu\mu\lambda}^{\cdot\cdot\cdot\varkappa} p_\varkappa - (\dot{\nabla}_\rho P_{\nu\mu\lambda}^{\cdot\cdot\cdot\rho}) \xi^\sigma p_\sigma \\ &= (n-2) P_{\nu\mu\lambda}^{\cdot\cdot\cdot\varkappa} p_\varkappa\end{aligned}$$

by virtue of (9.36). From these two equations, we get

$$\nabla_\rho \underset{v}{\pounds} P_{\nu\mu\lambda}^{\cdot\cdot\cdot\rho} + (n-2)\underset{v}{\pounds} P_{\nu\mu\lambda} + 2(\underset{v}{\pounds} P_{\tau[\nu|\sigma}^{\cdot\cdot\cdot\rho})\xi^\sigma Q_{\rho|\mu]\lambda}^{\cdot\cdot\cdot\tau}$$
$$+ 2P_{\tau[\nu|\sigma}^{\cdot\cdot\cdot\rho}\xi^\sigma(\underset{v}{\pounds} Q_{\rho|\mu]\lambda}^{\cdot\cdot\cdot\tau}) + (n-2)P_{\nu\mu\lambda}^{\cdot\cdot\cdot\varkappa}p_\varkappa = 0$$

or

$$(11.15)\quad (n-2)[\underset{v}{\pounds} P_{\nu\mu\lambda} + P_{\nu\mu\lambda}^{\cdot\cdot\cdot\varkappa}p_\varkappa]$$
$$= -\nabla_\rho \underset{v}{\pounds} P_{\nu\mu\lambda}^{\cdot\cdot\cdot\rho} - 2(\underset{v}{\pounds} P_{\tau[\nu|\sigma}^{\cdot\cdot\cdot\rho})\xi^\sigma Q_{\rho|\mu]\lambda}^{\cdot\cdot\cdot\tau} - 2P_{\tau[\nu|\sigma}^{\cdot\cdot\cdot\rho}\xi^\sigma(\underset{v}{\pounds} Q_{\rho|\mu]\lambda}^{\cdot\cdot\cdot\tau}) = 0,$$

which shows that the condition (11.13) is obtained from (11.8), (11.9) and their covariant derivatives.

Thus, as integrability conditions of (11.6), we have only to consider (11.8), (11.9) and their successive covariant derivatives.

We first consider the successive covariant derivatives of (11.8) and we show that the equation

$$(11.16)\qquad \dot{\nabla}_\omega \underset{v}{\pounds} P_{\nu\mu\lambda}^{\cdot\cdot\cdot\varkappa} = 0$$

obtained from (11.8) by covariant differentiation with respect to ξ^ω does not give a new condition. Indeed, applying the operator $\underset{v}{\pounds}$ to (9.35), we find

$$(11.17)\qquad \underset{v}{\pounds}\dot{\nabla}_\omega P_{\nu\mu\lambda}^{\cdot\cdot\cdot\varkappa} = 2\underset{v}{\pounds}\nabla_{[\nu} Q_{|\omega|\mu]\lambda}^{\cdot\cdot\cdot\varkappa} + 2A^{\varkappa}_{[\nu}\underset{v}{\pounds} Q_{|\omega|\mu]\lambda}.$$

On the other hand, applying the formula (10.6) to $Q_{\omega\mu\lambda}^{\cdot\cdot\cdot\varkappa}$, we get

$$\underset{v}{\pounds}\nabla_\nu Q_{\omega\mu\lambda}^{\cdot\cdot\cdot\varkappa} - \nabla_\nu \underset{v}{\pounds} Q_{\omega\mu\lambda}^{\cdot\cdot\cdot\varkappa} = (\underset{v}{\pounds}\Pi^{\varkappa}_{\nu\rho})Q_{\omega\mu\lambda}^{\cdot\cdot\cdot\rho} - (\underset{v}{\pounds}\Pi^{\rho}_{\nu\omega})Q_{\rho\mu\lambda}^{\cdot\cdot\cdot\varkappa}$$
$$- (\underset{v}{\pounds}\Pi^{\rho}_{\nu\mu})Q_{\omega\rho\lambda}^{\cdot\cdot\cdot\varkappa} - (\underset{v}{\pounds}\Pi^{\rho}_{\nu\lambda})Q_{\omega\mu\rho}^{\cdot\cdot\cdot\varkappa} - (\underset{v}{\pounds}\Pi^{\rho}_{\nu\sigma})\xi^\sigma\dot{\nabla}_\rho Q_{\omega\mu\lambda}^{\cdot\cdot\cdot\varkappa}$$
$$= A^{\varkappa}_{\nu}Q_{\omega\mu\lambda}^{\cdot\cdot\cdot\rho}p_\rho - p_\nu Q_{\omega\mu\lambda}^{\cdot\cdot\cdot\varkappa} - p_\omega Q_{\nu\mu\lambda}^{\cdot\cdot\cdot\varkappa} - p_\mu Q_{\omega\nu\lambda}^{\cdot\cdot\cdot\varkappa} - p_\lambda Q_{\omega\mu\nu}^{\cdot\cdot\cdot\varkappa}$$
$$- (\dot{\nabla}_\nu Q_{\omega\mu\lambda}^{\cdot\cdot\cdot\varkappa})\xi^\rho p_\rho,$$

from which

$$(11.18)\qquad \underset{v}{\pounds}\nabla_{[\nu} Q_{|\omega|\mu]\lambda}^{\cdot\cdot\cdot\varkappa} = \nabla_{[\nu}\underset{v}{\pounds} Q_{|\omega|\mu]\lambda}^{\cdot\cdot\cdot\varkappa} + A^{\varkappa}_{[\nu} Q_{|\omega|\mu]\lambda}^{\cdot\cdot\cdot\rho}p_\rho.$$

From (11.17) and (11.18), we obtain

$$(11.19)\quad \dot{\nabla}_\omega \underset{v}{\pounds} P_{\nu\mu\lambda}^{\cdot\cdot\cdot\varkappa} = 2\nabla_{[\nu}\underset{v}{\pounds} Q_{|\omega|\mu]\lambda}^{\cdot\cdot\cdot\varkappa} + 2A^{\varkappa}_{[\nu}(\underset{v}{\pounds} Q_{|\omega|\mu]\lambda} + Q_{|\omega|\mu]\lambda}^{\cdot\cdot\cdot\rho}p_\rho),$$

which shows that (11.16) is obtained from (11.9), (11.12) and the covariant derivative of (11.9). Thus (11.16) is obtained from (11.8), (11.9) and their covariant derivatives.

From (11.8), we get

$$\nabla_{\omega_1}(\underset{v}{\pounds} P_{\nu\mu\lambda}^{\cdot\cdot\cdot\kappa}) = 0.$$

We can show by a similar method that the covariant derivative of this equation with respect to $\dot{\xi}^\omega$ does not give a new condition. Thus from the above equation, we get

$$\nabla_{\omega_2\omega_1}(\underset{v}{\pounds} P_{\nu\mu\lambda}^{\cdot\cdot\cdot k}) = 0.$$

We can show that the covariant derivative of this equation with respect to $\dot{\xi}^\omega$ does not give a new condition.

Repeating this process, we obtain

$$\nabla_{\omega_r\cdots\omega_2\omega_1}\underset{v}{\pounds} P_{\nu\mu\lambda}^{\cdot\cdot\cdot\kappa} = 0, \qquad r = 1, 2, \ldots. \tag{11.20}$$

We next consider the successive covariant derivatives of (11.9). The equation (9.17) shows that the conditions obtained from (11.9) applying first the covariant differentiation with respect to ξ^κ and next the covariant differentiation with respect to $\dot{\xi}^\kappa$ and the conditions obtained from (11.9) applying the covariant differentiations in the reverse way are equivalent.

Thus we consider first the conditions obtained from (11.9) applying successively only the covariant differentiation with respect to $\dot{\xi}^\omega$:

$$\dot{\nabla}_{\omega_s\cdots\omega_2\omega_1}(\underset{v}{\pounds} Q_{\nu\mu\lambda}^{\cdot\cdot\cdot\kappa}) = 0; \qquad s = 1, 2, \ldots. \tag{11.21}$$

But by virtue of the homogeneity property of $Q_{\nu\mu\lambda}^{\cdot\cdot\cdot\kappa}$ with respect to $\dot{\xi}^\kappa$, any equation of (11.21) contains the preceding equations, and consequently, the equation (11.21) can be written as

$$\dot{\nabla}_{\omega_s\cdots\omega_2\omega_1}(\underset{v}{\pounds} Q_{\nu\mu\lambda}^{\cdot\cdot\cdot\kappa}) = 0; \qquad \text{(for some } s \text{ fixed)} \tag{11.22}$$

Thus the conditions obtained from (11.9) by successive covariant differentiations are

$$\nabla_{\pi_t\cdots\pi_2\pi_1}\dot{\nabla}_{\omega_s\cdots\omega_2\omega_1}\underset{v}{\pounds} Q_{\nu\mu\lambda}^{\cdot\cdot\cdot\kappa} = 0. \qquad t = 1, 2, \ldots \tag{11.23}$$

Hence we obtain

THEOREM 11.1. *In order that a general projective space of geodesics admit a group of projective motions, it is necessary and sufficient that, for a certain value of s, there exist a positive integer N such that the first N sets of the equations*

$$\nabla_{\omega_r \cdots \omega_2 \omega_1} \underset{v}{\pounds} P_{\nu\mu\lambda}^{\cdot\cdot\cdot\varkappa} = 0, \tag{11.24}$$
$$\nabla_{\pi_t \cdots \pi_2 \pi_1} \dot{\nabla}_{\omega_s \cdots \omega_2 \omega_1} \underset{v}{\pounds} Q_{\nu\mu\lambda}^{\cdot\cdot\cdot\varkappa} = 0, \qquad r, t = 0, 1, 2, \ldots$$

in which the derivatives of $v^\varkappa$, $v_\lambda^{\cdot\varkappa}$ and p_λ are eliminated by the use of (11.6), be algebraically consistent in $v^\varkappa$, $v_\lambda^{\cdot\varkappa}$ and p_λ and that all their solutions satisfy the $(N+1)$st set of equations.

If there exist $n^2 + 2n - r$ linearly independent equations in the first N sets, the space admits an r-parameter complete group of projective motions.

If (11.6) is completely integrable, then (11.8) and (11.9) must be identities in $v^\varkappa$, $v_\lambda^{\cdot\varkappa}$ and p_λ and consequently we must have $P_{\nu\mu\lambda}^{\cdot\cdot\cdot\varkappa} = 0$ and $Q_{\nu\mu\lambda}^{\cdot\cdot\cdot\varkappa} = 0$. Hence we obtain

THEOREM 11.2. *In order that an n-dimensional general projective space of geodesics admit a group of projective motions of the maximum order $n^2 + 2n$, it is necessary and sufficient that the space be projectively Euclidean.*

§ 12. Affine spaces of k-spreads.[1]

Consider an n-dimensional space in which a system of k-dimensional subspaces $\xi^\varkappa = \xi^\varkappa(\eta^h)$, $h, i, j, \ldots = 1, 2, \ldots, k$, is given by a completely integrable system of partial differential equations

$$\partial_j \dot{\xi}_i^\varkappa + \Gamma_{ji}^\varkappa(\xi, \dot{\xi}) = 0, \quad \dot{\xi}_i^\varkappa = \partial_i \xi^\varkappa, \quad \partial_i = \partial/\partial\eta^i, \tag{12.1}$$

where the functions $\Gamma_{ji}^\varkappa(\xi, \dot{\xi})$ are symmetric in j and i and form a so-called *homogeneous function system*[2] of $\dot{\xi}_i^\varkappa$ with respect to the lower indices. This means that they satisfy the generalized Euler relations:

$$\dot{\xi}_k^\lambda \dot{\partial}_\lambda^h \Gamma_{ji}^\varkappa = \delta_j^h \Gamma_{ki}^\varkappa + \delta_i^h \Gamma_{jk}^\varkappa, \quad \dot{\partial}_\lambda^h = \partial/\dot{\xi}_h^\lambda. \tag{12.2}$$

We assume that the left-hand of (12.1) transforms like a contravariant vector with respect to the upper index $\varkappa$ under the coordinate transfor-

---

[1] DOUGLAS [2].
[2] DOUGLAS [2].

mation

$$\xi^{\varkappa'} = \xi^{\varkappa'}(\xi^{\nu}), \quad \dot{\xi}_i^{\varkappa'} = A_{\varkappa}^{\varkappa'} \dot{\xi}_i^{\varkappa} \tag{12.3}$$

and like a covariant tensor with respect to the lower indices j and i under the affine parameter transformation

$$\eta^{h'} = A_h^{h'} \eta^h + B^{h'}, \tag{12.4}$$

where $A_h^{h'}$ and $B^{h'}$ are constants and $\det(A_h^{h'}) \neq 0$.

Now, under the coordinate transformation (12.3), the functions $\Gamma_{ji}^{\varkappa}(\xi, \dot{\xi})$ are transformed into

$$\Gamma_{ji}^{\varkappa'} = A_{\varkappa}^{\varkappa'} \Gamma_{ji}^{\varkappa} - (\partial_{\mu} A_{\lambda}^{\varkappa'}) \dot{\xi}_j^{\mu} \dot{\xi}_i^{\lambda}, \tag{12.5}$$

from which

$$\dot{\partial}_{\lambda'}^{i} \Gamma_{ji}^{\varkappa'} = A_{\varkappa \lambda'}^{\varkappa' \lambda} \dot{\partial}_{\lambda}^{i} \Gamma_{ji}^{\varkappa} - (k+1) A_{\lambda'}^{\lambda} (\partial_{\mu} A_{\lambda}^{\varkappa'}) \dot{\xi}_j^{\mu}, \tag{12.6}$$

$$\dot{\partial}_{\mu'}^{j} \dot{\partial}_{\lambda'}^{i} \Gamma_{ji}^{\varkappa'} = A_{\varkappa \mu' \lambda'}^{\varkappa' \mu \lambda} \dot{\partial}_{\mu}^{j} \dot{\partial}_{\lambda}^{i} \Gamma_{ji}^{\varkappa} - k(k+1) A_{\mu' \lambda'}^{\mu \lambda} (\partial_{\mu} A_{\lambda}^{\varkappa'}). \tag{12.7}$$

The last equation shows that the functions

$$\Gamma_{\mu\lambda}^{\varkappa} \stackrel{\text{def}}{=} \frac{1}{k(k+1)} \dot{\partial}_{\mu}^{j} \dot{\partial}_{\lambda}^{i} \Gamma_{ji}^{\varkappa} \tag{12.8}$$

have the transformation law

$$\Gamma_{\mu'\lambda'}^{\varkappa'} = A_{\varkappa \mu' \lambda'}^{\varkappa' \mu \lambda} \Gamma_{\mu\lambda}^{\varkappa} - A_{\mu'\lambda'}^{\mu\lambda} (\partial_{\mu} A_{\lambda}^{\varkappa'}) \tag{12.9}$$

or

$$\Gamma_{\mu'\lambda'}^{\varkappa'} = A_{\varkappa}^{\varkappa'} (A_{\mu'\lambda'}^{\mu\lambda} \Gamma_{\mu\lambda}^{\varkappa} + \partial_{\mu'} A_{\lambda'}^{\varkappa}). \tag{12.10}$$

Thus the $\Gamma_{\mu\lambda}^{\varkappa}$ defined by (12.8) are components of a linear connexion.

Because of the homogeneity property (12.2) of $\Gamma_{ji}^{\varkappa}$, we have

$$\Gamma_{ji}^{\varkappa} = \Gamma_{\mu\lambda}^{\varkappa} \dot{\xi}_j^{\mu} \dot{\xi}_i^{\lambda} \tag{12.11}$$

and consequently, the equations of k-spreads are also written as

$$\partial_j \dot{\xi}_i^{\varkappa} + \Gamma_{\mu\lambda}^{\varkappa} \dot{\xi}_j^{\mu} \dot{\xi}_i^{\lambda} = 0, \quad \dot{\xi}_i^{\varkappa} = \partial_i \xi^{\varkappa}. \tag{12.12}$$

We define the covariant differential of $\dot{\xi}_i^{\varkappa}$ by

$$\delta \dot{\xi}_i^{\varkappa} = d\dot{\xi}_i^{\varkappa} + \Gamma_{\mu\lambda}^{\varkappa} d\xi^{\mu} \dot{\xi}_i^{\lambda} \tag{12.13}$$

and the covariant differential of a contravariant vector $v^{\varkappa}$ by

$$\delta v^{\varkappa} = dv^{\varkappa} + \Gamma_{\mu\lambda}^{\varkappa} d\xi^{\mu} v^{\lambda}. \tag{12.14}$$

Then the covariant derivatives of $v^{\varkappa}$ are given by

(12.15) $$\nabla_{\mu} v^{\varkappa} = \partial_{\mu} v^{\varkappa} - \Gamma^{\rho}_{\mu\sigma} \dot{\xi}^{\sigma}_i \dot{\partial}^i_{\rho} v^{\varkappa} + \Gamma^{\varkappa}_{\mu\lambda} v^{\lambda},$$

(12.16) $$\dot{\nabla}^j_{\mu} v^{\varkappa} = \dot{\partial}^j_{\mu} v^{\varkappa}.$$

For the covariant derivatives of $\dot{\xi}^{\varkappa}_i$, we have

(12.17) $$\nabla_{\mu} \dot{\xi}^{\varkappa}_i = 0, \quad \dot{\nabla}^j_{\mu} \dot{\xi}^{\varkappa}_i = A^j_i A^{\varkappa}_{\mu},$$

Now, as generalizations of the Ricci identities, we find

(12.18) $$(\nabla_{\nu} \nabla_{\mu} - \nabla_{\mu} \nabla_{\nu}) v^{\varkappa} = R_{\nu\mu\lambda}^{\cdots\varkappa} v^{\lambda} - R_{\nu\mu\sigma}^{\cdots\rho} \dot{\xi}^{\sigma}_i \dot{\nabla}^i_{\rho} v^{\varkappa},$$

(12.19) $$(\dot{\nabla}^k_{\nu} \nabla_{\mu} - \nabla_{\mu} \dot{\nabla}^k_{\nu}) v^{\varkappa} = T_{\nu\mu\lambda}^{k\cdots\varkappa} v^{\lambda},$$

(12.20) $$(\dot{\nabla}^k_{\nu} \dot{\nabla}^j_{\mu} - \dot{\nabla}^j_{\mu} \dot{\nabla}^k_{\nu}) v^{\varkappa} = 0,$$

where

(12.21) $$R_{\nu\mu\lambda}^{\cdots\varkappa} \stackrel{\bullet}{=} (\partial_{\nu} \Gamma^{\varkappa}_{\mu\lambda} - \Gamma^{\rho}_{\nu\sigma} \dot{\xi}^{\sigma}_i \dot{\partial}^i_{\rho} \Gamma^{\varkappa}_{\mu\lambda}) - (\partial_{\mu} \Gamma^{\varkappa}_{\nu\lambda} - \Gamma^{\rho}_{\mu\sigma} \dot{\xi}^{\sigma}_i \dot{\partial}^i_{\rho} \Gamma^{\varkappa}_{\nu\lambda})$$
$$+ \Gamma^{\varkappa}_{\nu\rho} \Gamma^{\rho}_{\mu\lambda} - \Gamma^{\varkappa}_{\mu\rho} \Gamma^{\rho}_{\nu\lambda},$$

(12.22) $$T_{\nu\mu\lambda}^{k\cdots\varkappa} = \dot{\partial}^k_{\nu} \Gamma^{\varkappa}_{\mu\lambda}$$

are curvature tensors of the space of k-spreads. The Bianchi identities for the curvature tensors take the form

(12.23) $$R_{[\nu\mu\lambda]}^{\cdots\varkappa} = 0,$$

(12.24) $$\nabla_{[\omega} R_{\nu\mu]\lambda}^{\cdots\varkappa} + R_{[\omega\nu|\sigma}^{\cdots\rho} \dot{\xi}^{\sigma}_i T_{\rho|\mu]\lambda}^{i\cdots\varkappa} = 0.$$

Moreover, we have the following identities

(12.25) $$\dot{\nabla}^l_{\omega} R_{\nu\mu\lambda}^{\cdots\varkappa} = 2\nabla_{[\nu} T_{\mu]\omega\lambda}^{l\cdots\varkappa} - 2T_{\omega[\nu|\sigma}^{l\cdots\rho} \dot{\xi}^{\sigma}_i T_{\rho|\mu]\lambda}^{i\cdots\varkappa},$$

(12.26) $$\dot{\nabla}^l_{\omega} T_{\nu\mu\lambda}^{k\cdots\varkappa} = \dot{\nabla}^k_{\nu} T_{\omega\mu\lambda}^{l\cdots\varkappa}.$$

From (12.25) we have

(12.27) $$\dot{\nabla}^l_{[\omega} R_{\nu\mu]\lambda}^{\cdots\varkappa} = 0,$$

from which

(12.28) $$\dot{\nabla}^l_{[\omega} V_{\nu\mu]} = 0.$$

The Lie derivatives in an affine space of k-spreads with respect to an infinitesimal extended point transformation

$$'\xi^\kappa = \xi^\kappa + v^\kappa(\xi)dt,\ '\dot\xi^\kappa_i = \dot\xi^\kappa_i + (\partial_\lambda v^\kappa)\dot\xi^\lambda_i dt \tag{12.29}$$

are defined in exactly the same way as was used in § 6.

The Lie derivatives of a contravariant vector u^κ, a covariant vector w_λ and a mixed tensor $T_\lambda^{\cdot\kappa}$ are respectively given by

$$\underset{v}{\pounds} u^\kappa = v^\mu \nabla_\mu u^\kappa + (\dot\xi^\rho_i \nabla_\rho v^\mu)\dot\nabla^i_\mu u^\kappa - u^\rho \nabla_\rho v^\kappa, \tag{12.30}$$

$$\underset{v}{\pounds} w_\lambda = v^\mu \nabla_\mu w_\lambda + (\dot\xi^\rho_i \nabla_\rho v^\mu)\dot\nabla^i_\mu w_\lambda + w_\rho \nabla_\lambda v^\rho, \tag{12.31}$$

$$\underset{v}{\pounds} T_\lambda^{\cdot\kappa} = v^\mu \nabla_\mu T_\lambda^{\cdot\kappa} + (\dot\xi^\rho_i \nabla_\rho v^\mu)\dot\nabla^i_\mu T_\lambda^{\cdot\kappa} - T_\lambda^{\cdot\rho}\nabla_\rho v^\kappa + T_\rho^{\cdot\kappa}\nabla_\lambda v^\rho. \tag{12.32}$$

We verify easily that

$$\underset{v}{\pounds} \dot\xi^\kappa_i = 0. \tag{12.33}$$

As to the Lie derivative of the linear connexion we find

$$\underset{v}{\pounds} \Gamma^\kappa_{\mu\lambda} = \partial_\mu \partial_\lambda v^\kappa + v^\nu \partial_\nu \Gamma^\kappa_{\mu\lambda} + \dot\xi^\nu_i(\partial_\nu v^\rho)\dot\partial^i_\rho \Gamma^\kappa_{\mu\lambda} - \Gamma^\rho_{\mu\lambda}\partial_\rho v^\kappa + \Gamma^\kappa_{\rho\lambda}\partial_\mu v^\rho + \Gamma^\kappa_{\mu\rho}\partial_\lambda v^\rho \tag{12.34}$$

and

$$\underset{v}{\pounds} \Gamma^\kappa_{\mu\lambda} = \nabla_\mu \nabla_\lambda v^\kappa + R_{\nu\mu\lambda}^{\cdot\cdot\cdot\kappa} v^\nu + T^{i\cdot\cdot\kappa}_{\nu\mu\lambda}\dot\xi^\rho_i \nabla_\rho v^\nu. \tag{12.35}$$

We can also verify the following identities:

$$(\underset{v}{\pounds}\nabla_\mu - \nabla_\mu \underset{v}{\pounds})T_\lambda^{\cdot\kappa} = (\underset{v}{\pounds}\Gamma^\kappa_{\mu\rho})T_\lambda^{\cdot\rho} - (\underset{v}{\pounds}\Gamma^\rho_{\mu\lambda})T_\rho^{\cdot\kappa} - (\underset{v}{\pounds}\Gamma^\rho_{\mu\sigma})\dot\xi^\sigma_i \dot\nabla^i_\rho T_\lambda^{\cdot\kappa}, \tag{12.36}$$

$$(\underset{v}{\pounds}\dot\nabla^i_\mu - \dot\nabla^i_\mu \underset{v}{\pounds})T_\lambda^{\cdot\kappa} = 0, \tag{12.37}$$

$$\nabla_\nu(\underset{v}{\pounds}\Gamma^\kappa_{\mu\lambda}) - \nabla_\mu(\underset{v}{\pounds}\Gamma^\kappa_{\nu\lambda}) = \underset{v}{\pounds} R_{\nu\mu\lambda}^{\cdot\cdot\cdot\kappa} + (\underset{v}{\pounds}\Gamma^\rho_{\nu\sigma})\dot\xi^\sigma_i T^{i\cdot\cdot\kappa}_{\rho\mu\lambda} - (\underset{v}{\pounds}\Gamma^\rho_{\mu\sigma})\dot\xi^\sigma_i T^{i\cdot\cdot\kappa}_{\rho\nu\lambda}, \tag{12.38}$$

$$\dot\nabla^i_\nu(\underset{v}{\pounds}\Gamma^\kappa_{\mu\lambda}) = \underset{v}{\pounds} T^{i\cdot\cdot\kappa}_{\nu\mu\lambda}, \tag{12.39}$$

$$(\underset{c}{\pounds}\underset{b}{\pounds})\Gamma^\kappa_{\mu\lambda} = \underset{cb}{\pounds}\Gamma^\kappa_{\mu\lambda}, \tag{12.40}$$

$$(\underset{c}{\pounds}\underset{b}{\pounds})\Gamma^\kappa_{\mu\lambda} = c^a_{cb}\underset{a}{\pounds}\Gamma^\kappa_{\mu\lambda}. \tag{12.41}$$

The last equation holds if $\underset{b}{v}^{\varkappa}$ generate an r-parameter group with structural constants c_{cb}^{a}.

Now, if the extended point transformation (12.29) changes every k-spread into a k-spread and every set of affine parameters on a k-spread into a set of affine parameters of the deformed k-spread, then the transformation is called an affine motion in the affine space of k-spreads.

As in § 6, we can state

THEOREM 12.1. *In order that* (12.29) *be an affine motion in an affine space of k-spreads, it is necessary and sufficient that the Lie derivative of the linear connexion with respect to* (12.29) *vanish.*

The remarks following Theorem 7.2 hold also for affine motions in an affine space of k-spreads. [1]

Examining the integrability conditions of $\underset{v}{\pounds}\Gamma_{\mu\lambda}^{\varkappa} = 0$, we get a theorem corresponding to Theorem 8.1, the equation (8.14) being replaced by

$$(12.42)\qquad \begin{cases} \underset{v}{\pounds}(\nabla_{\omega_r \cdots \omega_2\omega_1} R_{\nu\mu\lambda}^{\cdot\cdot\cdot\varkappa}) = 0, \\ \underset{v}{\pounds}(\nabla_{\pi_t \cdots \pi_2\pi_1} \dot{\nabla}_{\omega_s \cdots \omega_2\omega_1}^{l_s \cdots l_2 l_1} T_{\nu\mu\lambda}^{i\cdot\cdot\varkappa}) = 0. \end{cases}$$

A theorem corresponding to Theorem 8.2 is also valid, if we replace (8.18) by

$$(12.43)\qquad \frac{\partial^2 \xi^{\varkappa}}{\partial\eta^j \partial\eta^i} + \Gamma_{\mu\lambda}^{\varkappa}(\xi)\frac{\partial\xi^{\mu}}{\partial\eta^j}\frac{\partial\xi^{\lambda}}{\partial\eta^i} = 0.$$

§ 13. Projective spaces of *k*-spreads.

Let us consider an n-dimensional space of k-spreads referred to a coordinate system $(\varkappa)$, the k-spreads being given by a completely integrable system of partial differential equations

$$(13.1)\qquad \partial_j \dot{\xi}_i^{\varkappa} + \Gamma_{ji}^{\varkappa}(\xi, \dot{\xi}) = 0,\ \dot{\xi}_i^{\varkappa} = \partial_i \xi^{\varkappa}.$$

If the functions $\Gamma_{ji}^{\varkappa}(\xi, \dot{\xi})$ are such that (13.1) is completely integrable, then a system of k-spreads is uniquely determined. But, when a system of k-spreads is given, a system of the functions $\Gamma_{ji}^{\varkappa}(\xi, \dot{\xi})$ is not uniquely determined. J Douglas [2] has shown that if $'\Gamma_{ji}^{\varkappa}(\xi, \dot{\xi})$ and $\Gamma_{ji}^{\varkappa}(\xi, \dot{\xi})$ give the same system of k-spreads, then they should be related by the equa-

---

[1] Theorem 4.2 of Ch. I in an affine space of k-spreads was proved by SU [3].

[2] DOUGLAS [2].

tions of the form

$$(13.2) \qquad {}'\Gamma^{\kappa}_{ji} = \Gamma^{\kappa}_{ji} + \dot{\xi}^{\kappa}_{h} p^{h}_{ji},$$

and consequently $'\Gamma^{\kappa}_{\mu\lambda}$ and $\Gamma^{\kappa}_{\mu\lambda}$ by

$$(13.3) \qquad {}'\Gamma^{\kappa}_{\mu\lambda} = \Gamma^{\kappa}_{\mu\lambda} + p_{\mu} A^{\kappa}_{\lambda} + p_{\lambda} A^{\kappa}_{\mu} + p^{h}_{\mu\lambda} \dot{\xi}^{\kappa}_{h},$$

where

$$(13.4) \qquad p_{\lambda} = \frac{1}{k(k+1)} \dot{\partial}^{j}_{\lambda} p^{h}_{jh}, \quad p^{h}_{\mu\lambda} = \frac{1}{k(k+1)} \dot{\partial}^{j}_{\mu} \dot{\partial}^{i}_{\lambda} p^{h}_{ji}.$$

The equations (13.2) and (13.3) give the so-called projective change of Γ^{κ}_{ji} and $\Gamma^{\kappa}_{\mu\lambda}$ respectively. The study of the properties of the spaces of k-spreads which are invariant under a projective change of $\Gamma^{\kappa}_{\mu\lambda}$ is called the projective geometry of k-spreads.

It is known[1] that the projective geometry of k-spreads is equivalent to the theory of the space of elements $(\xi^{\kappa}, \dot{\xi}^{\kappa}_{i})$ with a normal projective connexion whose family of k-dimensional geodesic subspaces is given by (13.1). The components $\Pi_{\mu\lambda}$, $C^{h}_{\mu\lambda}$ and $\Pi^{\kappa}_{\mu\lambda}$ of this normal projective connexion referred to a semi-natural frame of reference are given by

$$(13.5) \qquad \begin{cases} \Pi_{\mu\lambda} = \Pi^{\rho}_{\mu\sigma} \dot{\xi}^{\sigma}_{i} C^{i}_{\rho\lambda} - \dfrac{1}{n^2-1}(nN_{\mu\lambda} + N_{\lambda\mu}), \\[2ex] C^{i}_{\mu\lambda} = \dfrac{1}{n+1} \dot{\partial}^{i}_{\mu} \Pi^{\rho}_{\lambda\rho} = \dfrac{1}{n+1} \dot{\partial}^{i}_{\mu} \Gamma^{\rho}_{\lambda\rho}, \\[2ex] \Pi^{\kappa}_{\mu\lambda} = \Gamma^{\kappa}_{\mu\lambda} - \dfrac{1}{n-k}\left[\dot{\partial}^{i}_{\rho} \Gamma^{\rho}_{\mu\lambda} - \dfrac{1}{n+1}(\dot{\partial}^{i}_{\mu} \Gamma^{\rho}_{\lambda\rho} + \dot{\partial}^{i}_{\lambda} \Gamma^{\rho}_{\mu\rho})\right] \dot{\xi}^{\kappa}_{i}, \end{cases}$$

where

$$(13.6) \qquad N_{\nu\mu\lambda}{}^{\kappa} = (\partial_{\nu} \Pi^{\kappa}_{\mu\lambda} - \Pi^{\rho}_{\nu\sigma} \dot{\xi}^{\sigma}_{i} \dot{\partial}^{i}_{\rho} \Pi^{\kappa}_{\mu\lambda}) - (\partial_{\mu} \Pi^{\kappa}_{\nu\lambda} - \Pi^{\rho}_{\mu\sigma} \dot{\xi}^{\sigma}_{i} \dot{\partial}^{i}_{\rho} \Pi^{\kappa}_{\nu\lambda}) + \Pi^{\kappa}_{\nu\rho} \Pi^{\rho}_{\mu\lambda} - \Pi^{\kappa}_{\mu\rho} \Pi^{\rho}_{\nu\lambda},$$

$$(13.7) \qquad N_{\mu\lambda} = N_{\rho\mu\lambda}{}^{\rho}.$$

The $\Pi_{\mu\lambda}$ and $\Pi^{\kappa}_{\mu\lambda}$ are homogeneous functions of degree zero and $C^{i}_{\mu\lambda}$ is a homogeneous function system with respect to $\dot{\xi}^{\kappa}_{i}$. It is easily verified that the $C_{\mu\lambda}$ are components of a tensor and that the $\Pi^{\kappa}_{\mu\lambda}$ are components of a linear connexion.

[1] YANO and HIRAMATU [2].

We define the covariant derivatives of a tensor, say $T_{\lambda}^{\cdot\varkappa}$, by

$$(13.8)\quad \begin{cases} \nabla_{\mu} T_{\lambda}^{\cdot\varkappa} = \partial_{\mu} T_{\lambda}^{\cdot\varkappa} - \Pi_{\mu\sigma}^{\rho} \xi_i^{\sigma} \dot{\partial}_{\rho}^{i} T_{\lambda}^{\cdot\varkappa} + \Pi_{\mu\rho}^{\varkappa} T_{\lambda}^{\cdot\rho} - \Pi_{\mu\lambda}^{\rho} T_{\rho}^{\cdot\varkappa}, \\ \dot{\nabla}_{\mu}^{i} T_{\lambda}^{\cdot\varkappa} = \dot{\partial}_{\mu}^{i} T_{\lambda}^{\cdot\varkappa}. \end{cases}$$

The curvature tensors of the normal projective connexion are given by

$$(13.9)\qquad P_{\nu\mu\lambda} \overset{\text{def}}{=} \nabla_{\nu} M_{\mu\lambda} - \nabla_{\mu} M_{\nu\lambda} + N_{\nu\mu\sigma}^{\cdot\cdot\cdot\rho} \xi_i^{\sigma} C_{\rho\lambda}^{i},$$

$$(13.10)\qquad Q_{\nu\mu\lambda}^{i} \overset{\text{def}}{=} \dot{\nabla}_{\nu}^{i} M_{\mu\lambda} - \nabla_{\mu} C_{\nu\lambda}^{i} + U_{\nu\mu\sigma}^{i\cdot\cdot\rho} \xi_j^{\sigma} C_{\rho\lambda}^{j},$$

$$(13.11)\qquad P_{\nu\mu\lambda}^{\cdot\cdot\cdot\varkappa} \overset{\text{def}}{=} R_{\nu\mu\lambda}^{\cdot\cdot\cdot\varkappa} + A_{\nu}^{\varkappa} M_{\mu\lambda} - A_{\mu}^{\varkappa} M_{\nu\lambda} - (M_{\nu\mu} - M_{\mu\nu}) A_{\lambda}^{\varkappa},$$

$$(13.12)\qquad Q_{\nu\mu\lambda}^{i\cdot\cdot\varkappa} \overset{\text{def}}{=} U_{\nu\mu\lambda}^{i\cdot\cdot\varkappa} - C_{\nu\mu}^{i} A_{\lambda}^{\varkappa} - C_{\nu\lambda}^{i} A_{\mu}^{\varkappa},$$

where

$$(13.13)\qquad M_{\mu\lambda} \overset{\text{def}}{=} \Pi_{\mu\lambda} - \Pi_{\mu\sigma}^{\rho} \xi_i^{\sigma} C_{\rho\lambda}^{i} = -\frac{1}{n^2-1}(nN_{\mu\lambda} + N_{\lambda\mu})$$

and

$$(13.14)\qquad U_{\nu\mu\lambda}^{i\cdot\cdot\varkappa} \overset{\text{def}}{=} \dot{\partial}_{\nu}^{i} \Pi_{\mu\lambda}^{\varkappa}$$

are both tensors.

Theorem 9.1 holds also in a projective space of k-spreads.

The projective motions in a projective space of k-spreads are defined in exactly the same way as used in § 10, and all the theorems in § 10 hold also in a projective space of k-spreads. The discussions on the integrability conditions of $\underset{v}{\pounds}\Pi_{\mu\lambda}^{\varkappa} = p_{\mu} A_{\lambda}^{\varkappa} + p_{\lambda} A_{\mu}^{\varkappa}$ can also be carried out as in § 11 and Theorems 11.1 and 11.2 hold also in a projective space of k-spreads provided that the equations (11.24) are replaced by[1]

$$(13.15)\qquad \begin{aligned} &\nabla_{\omega_r \cdots \omega_2 \omega_1} (\underset{v}{\pounds} P_{\nu\mu\lambda}^{\cdot\cdot\cdot\varkappa} + 2 Q_{[\nu\mu]\lambda}^{i\cdot\cdot\varkappa} \xi_i^{\rho} p_{\rho}) = 0, \\ &\nabla_{\pi_t \cdots \pi_2 \pi_1} \dot{\nabla}_{\omega_s \cdots \omega_2 \omega_1}^{l_s \cdots l_2 l_1} (\underset{v}{\pounds} Q_{\nu\mu\lambda}^{i\cdot\cdot\varkappa}) = 0 \quad r, t = 0, 1, 2, \ldots . \end{aligned}$$

[1] Yano and Hiramatu [3].

CHAPTER IX

LIE DERIVATIVES IN A COMPACT ORIENTABLE RIEMANNIAN SPACE

§ 1. Theorem of Green.

Let us consider an n-dimensional space of class C^r $(r \geq 1)$ which is covered by a system of coordinate neighbourhoods $(\varkappa)$. If, from any covering of the space by a set of coordinate neighbourhoods we can choose a covering by a set of finite numbers of coordinate neighbourhoods, the space is said to be *compact.* If we can find a covering of the space by a set of coordinate neighbourhoods such that, in the overlapping domain of any two coordinate neighbourhoods U with $(\varkappa)$ and U' with $(\varkappa')$, we have always

$$\Delta = \det(A_{\varkappa}^{\varkappa'}) > 0, \tag{1.1}$$

the space is said to be *orientable.*

In this chapter, we consider an n-dimensional compact orientable Riemannian space of class C^3 with positive definite metric $ds^2 = g_{\lambda\varkappa}(\xi) d\xi^\lambda d\xi^\varkappa$.

We state first the following theorem of Green:

THEOREM 1.1.[1] *In a compact orientable V_n, we have*

$$\int_{V_n} \nabla_\mu v^\mu \, d\tau = 0, \tag{1.2}$$

for an arbitrary vector field $v^\varkappa$, where

$$d\tau \overset{\text{def}}{=} \sqrt{g}\, d\xi^1 d\xi^2 \ldots d\xi^n > 0 \tag{1.3}$$

is the volume element of the space.

Take a scalar f and consider

$$\Delta f \overset{\text{def}}{=} g^{\mu\lambda} \nabla_\mu \nabla_\lambda f. \tag{1.4}$$

Since this is also written as

$$\Delta f = \nabla_\mu (g^{\mu\lambda} \nabla_\lambda f), \tag{1.5}$$

[1] For the proof, see for instance BOCHNER [1]; YANO and BOCHNER [1].

applying Theorem 1.1, we get

THEOREM 1.2. *In a compact orientable V_n, we have*

$$\int_{V_n} \Delta f d\tau = 0. \tag{1.6}$$

Now consider the square of f and apply the operator Δ to it, then we obtain

$$\Delta f^2 = 2f\Delta f + 2g^{\mu\lambda}(\nabla_\mu f)(\nabla_\lambda f), \tag{1.7}$$

and consequently applying Theorem 1.2 to f^2, we obtain

$$\int_{V_n} [f\Delta f + g^{\mu\lambda}(\nabla_\mu f)(\nabla_\lambda f)]d\tau = 0. \tag{1.8}$$

Hence, if we have $\Delta f \geq 0$ everywhere in the V_n, then as we see from Theorem 1.2, we must have $\Delta f = 0$. Substituting this in (1.8), we find

$$g^{\mu\lambda}(\nabla_\mu f)(\nabla_\lambda f) = 0,$$

from which

$$\nabla_\lambda f = 0,$$

that is, f must be a constant. Thus we have

THEOREM 1.3. *If, in a compact orientable V_n, we have $\Delta f \geq 0$ everywhere, then $\Delta f = 0$ and f is a constant.*

§ 2. Harmonic tensors.

For an arbitrary alternating tensor field $w_{\lambda_p \cdots \lambda_1}$, the rotation and the divergence are respectively defined by[1]

$$\begin{cases} \text{Rot } w: & (p+1)\nabla_{[\mu} w_{\lambda_p \cdots \lambda_1]} \\ \text{Div } w: & \nabla_\mu w^{\mu\lambda_{p-1} \cdots \lambda_1} \end{cases} \tag{2.1}$$

If w is an alternating tensor of valence p, Rot w is alternating and of valence $p+1$ and Div w is also alternating and of valence $p-1$.

For two alternating tensors u and v of the same valence p, we define the global inner product (u, v) by

$$(u, v) = \int_{V_n} u_{\lambda_p \cdots \lambda_1} v^{\lambda_p \cdots \lambda_1} d\tau. \tag{2.2}$$

Since the metric is positive definite, we have always $(u, u) \geq 0$, and the equality holds if and only if $u_{\lambda_p \cdots \lambda_1} = 0$.

---

[1] SCHOUTEN [8], p. 83.

Now take two alternating tensors $u_{\lambda_p \cdots \lambda_1}$ of valence p and $v_{\lambda_{p+1} \cdots \lambda_1}$ of valence $p + 1$ and consider the vector

$$u_{\lambda_p \cdots \lambda_1} v^{\mu\lambda_p \cdots \lambda_1}.$$

Applying Theorem 1.1 to this vector, we obtain

$$\begin{aligned} 0 &= \int_{V_n} \nabla_\mu (u_{\lambda_p \cdots \lambda_1} v^{\mu\lambda_p \cdots \lambda_1}) d\tau \\ &= \int_{V_n} (\nabla_{[\mu} u_{\lambda_p \cdots \lambda_1]}) v^{\mu\lambda_p \cdots \lambda_1} d\tau + \int_{V_n} u_{\lambda_p \cdots \lambda_1} (\nabla_\mu v^{\mu\lambda_p \cdots \lambda_1}) d\tau, \end{aligned}$$

that is

$$(2.3) \qquad (\text{Rot } u, v) + (p + 1)(u, \text{Div } v) = 0.$$

An alternating tensor $w_{\lambda_p \cdots \lambda_1}$ is called a harmonic tensor if it satisfies

$$(2.4) \qquad \text{Rot } w = 0, \ \text{Div } w = 0.$$

It is evident that, for a harmonic tensor w, we have

$$(2.5) \qquad \Delta w \overset{\text{def}}{=} \text{Div Rot } w + \text{Rot Div } w = 0.$$

Conversely, take an alternating tensor $w_{\lambda_p \cdots \lambda_1}$ which satisfies (2.5). Putting $u = w$, $v = \text{Rot } w$ in (2.3), we obtain

$$(2.6) \qquad (\text{Rot } w, \text{Rot } w) + (p + 1)(w, \text{Div Rot } w) = 0.$$

Putting next $u = \text{Div } w$, $v = w$ in (2.3), we get

$$(2.7) \qquad (\text{Rot Div } w, w) + p(\text{Div } w, \text{Div } w) = 0.$$

From (2.6) and (2.7), we find

$$\begin{aligned} 0 &= (w, \text{Div Rot } w + \text{Rot Div } w) \\ &= -\frac{1}{p+1} (\text{Rot } w, \text{Rot } w) - p(\text{Div } w, \text{Div } w), \end{aligned}$$

from which

$$\text{Rot } w = 0, \ \text{Div } w = 0.$$

THEOREM 2.1.[1] *In order that an alternating tensor $w_{\lambda_p \cdots \lambda_1}$ in a V_n be harmonic, it is necessary and sufficient that*

$$\Delta w = \text{Div Rot } w + \text{Rot Div } w = 0.$$

---

[1] DE RHAM and KODAIRA [1].

By a straightforward calculation, we find[1]

$$(2.8)\qquad \Delta w_{\lambda_p\cdots\lambda_1} = g^{\nu\mu}\nabla_\nu\nabla_\mu w_{\lambda_p\cdots\lambda_1} - K_{[\lambda_p}{}^{\cdot\rho} w_{|\rho|\lambda_{p-1}\cdots\lambda_1]} - \tfrac{1}{2}(p-1)K_{[\lambda_p\lambda_{p-1}}{}^{\sigma\rho} w_{|\sigma\rho|\lambda_{p-2}\cdots\lambda_1]}.$$

Now suppose that an alternating tensor $w_{\lambda_p\cdots\lambda_1}$ is harmonic and is equal to a rotation of another alternating tensor $u_{\lambda_{p-1}\cdots\lambda_1}$:

$$w = \text{Rot}\, u.$$

Then we have, by the definition of a harmonic tensor,

$$\text{Rot Rot}\, u = 0,\ \text{Div Rot}\, u = 0.$$

Putting $v = \text{Rot}\, u$ in (2.3), we have

$$(\text{Rot}\, u, \text{Rot}\, u) + p(u, \text{Div Rot}\, u) = 0,$$

from which

$$\text{Rot}\, u = 0.$$

Thus we have

THEOREM 2.2. *A harmonic tensor which is the rotation of an alternating tensor is identically zero.*

§ 3. Lie derivative of a harmonic tensor.

Suppose that the V_n admits a one-parameter group of motions generated by an infinitesimal transformation

$$(3.1)\qquad '\xi^\varkappa = \xi^\varkappa + v^\varkappa(\xi)dt,$$

then we have $\underset{v}{\pounds} g_{\lambda\varkappa} = 0$ and the operators ∇_μ and $\underset{v}{\pounds}$ are commutative.

Suppose furthermore that there exists in the V_n a harmonic tensor $w_{\lambda_p\cdots\lambda_1}$, then we have

$$\nabla_{[\mu} w_{\lambda_p\cdots\lambda_1]} = 0,\ g^{\nu\mu}\nabla_\nu w_{\mu\lambda_{p-1}\cdots\lambda_1} = 0,$$

from which

$$\nabla_{[\mu} \underset{v}{\pounds} w_{\lambda_p\cdots\lambda_1]} = 0,\ g^{\nu\mu}\nabla_\nu \underset{v}{\pounds} w_{\mu\lambda_{p-1}\cdots\lambda_1} = 0,$$

which show that the Lie derivative $\underset{v}{\pounds} w_{\lambda_p\cdots\lambda_1}$ of a harmonic tensor $w_{\lambda_p\cdots\lambda_1}$

[1] SCHOUTEN [6] p. 109. (1α)

is also harmonic. But on the other hand we have

$$\underset{v}{\pounds} w_{\lambda_p \ldots \lambda_1} = v^\mu \nabla_\mu w_{\lambda_p \ldots \lambda_1} + w_{\mu\lambda_{p-1} \ldots \lambda_1} \nabla_{\lambda_p} v^\mu + \ldots + w_{\lambda_p \ldots \lambda_2 \mu} \nabla_{\lambda_1} v^\mu$$
$$= v^\mu (\nabla_{\lambda_p} w_{\mu\lambda_{p-1} \ldots \lambda_1} + \ldots + \nabla_{\lambda_1} w_{\lambda_p \ldots \lambda_2 \mu})$$
$$+ w_{\mu\lambda_{p-1} \ldots \lambda_1} \nabla_{\lambda_p} v^\mu + \ldots + w_{\lambda_p \ldots \lambda_2 \mu} \nabla_{\lambda_1} v^\mu$$
$$= p \nabla_{[\lambda_p} (v^\mu w_{|\mu|\lambda_{p-1} \ldots \lambda_1]}),$$

that is,

$$\underset{v}{\pounds} w_{\lambda_p \ldots \lambda_1} = \text{Rot}\, v^\mu w_{\mu\lambda_{p-1} \ldots \lambda_1}.$$

Thus according to Theorem 2.2, we have

$$\underset{v}{\pounds} w_{\lambda_p \ldots \lambda_1} = 0.$$

THEOREM 3.1.[1] *If a compact orientable V_n admits an infinitesimal motion, the Lie derivative of a harmonic tensor with respect to this motion vanishes identically.*

Suppose that w_λ is a harmonic vector and $v^\varkappa$ is a Killing vector, then, by the above theorem, we have

$$0 = \underset{v}{\pounds} w_\lambda = v^\mu \nabla_\mu w_\lambda + w_\mu \nabla_\lambda v^\mu$$
$$= v^\mu \nabla_\lambda w_\mu + w_\mu \nabla_\lambda v^\mu$$
$$= \nabla_\lambda (w_\mu v^\mu),$$

from which we get

THEOREM 3.2.[2] *In a compact orientable V_n, the inner product of a harmonic vector and a Killing vector is constant.*

§ 4. Motions in a compact orientable V_n.

Take an arbitrary vector field $v^\varkappa$ and calculate the divergence of $v^\lambda \nabla_\lambda v^\varkappa$:

$$\nabla_\mu (v^\lambda \nabla_\lambda v^\mu) = (\nabla_\mu v^\lambda)(\nabla_\lambda v^\mu) + v^\lambda \nabla_\mu \nabla_\lambda v^\mu$$
$$= (\nabla_\mu v^\lambda)(\nabla_\lambda v^\mu) + v^\lambda (\nabla_\lambda \nabla_\mu v^\mu + K_{\mu\lambda\varkappa}{}^{\mu} v^\varkappa)$$
$$= (\nabla^\mu v^\lambda)(\nabla_\lambda v_\mu) + v^\lambda \nabla_\lambda \nabla_\mu v^\mu + K_{\lambda\varkappa} v^\lambda v^\varkappa,$$

[1] YANO [18]; YANO and BOCHNER [1].
[2] BOCHNER [4]; YANO and BOCHNER [1].

where $\nabla^\mu = g^{\mu\lambda}\nabla_\lambda$. On the other hand, calculate the divergence of $v^\lambda \nabla_\mu v^\mu$;

$$\nabla_\lambda(v^\lambda \nabla_\mu v^\mu) = (\nabla_\lambda v^\lambda)(\nabla_\mu v^\mu) + v^\lambda \nabla_\lambda \nabla_\mu v^\mu.$$

From these two equations, we get

$$\nabla_\mu(v^\lambda \nabla_\lambda v^\mu) - \nabla_\lambda(v^\lambda \nabla_\mu v^\mu) = (\nabla^\mu v^\lambda)(\nabla_\lambda v_\mu) - (\nabla_\mu v^\mu)(\nabla_\lambda v^\lambda) + K_{\mu\lambda} v^\mu v^\lambda.$$

Since

$$\int_{V_n} [\nabla_\mu(v^\lambda \nabla_\lambda v^\mu) - \nabla_\lambda(v^\lambda \nabla_\mu v^\mu)] d\tau = 0,$$

we have

$$\int_{V_n} [(\nabla^\mu v^\lambda)(\nabla_\lambda v_\mu) - (\nabla_\mu v^\mu)(\nabla_\lambda v^\lambda) + K_{\mu\lambda} v^\mu v^\lambda] d\tau = 0. \tag{4.1}$$

Now suppose that a vector field $v^\varkappa$ generates a one-parameter group of motions in a V_n, then we have

$$\underset{v}{\pounds} g_{\mu\lambda} = \nabla_\mu v_\lambda + \nabla_\lambda v_\mu = 0, \ \nabla_\lambda v^\lambda = 0.$$

Substituting these equations in (4.1), we find

$$\int_{V_n} [(\nabla^\mu v^\lambda)(\nabla_\mu v_\lambda) - K_{\mu\lambda} v^\mu v^\lambda] d\tau = 0.$$

Thus, if the Ricci tensor $K_{\mu\lambda}$ is negative semi-definite everywhere in the V_n, we must have

$$\nabla_\mu v_\lambda = 0, \ K_{\mu\lambda} v^\mu v^\lambda = 0,$$

that is, the vector $v^\varkappa$ must be a covariant constant field.

If the Ricci tensor $K_{\mu\lambda}$ is negative definite everywhere in the V_n, we must have $v^\varkappa = 0$. Thus we have

THEOREM 4.1.[1] *In a compact orientable V_n whose Ricci tensor is negative semi-definite, vector a generating a one-parameter group of motions is a covariant constant field. In a V_n whose Ricci tensor is negative definite, there does not exist a continuous group of motions.*

Suppose that a V_n with $K_{\mu\lambda} = 0$ admits a transitive group of motions, then by Theorem 4.1 all the vectors generating the transitive group of motions are covariant constant. This means that the V_n admits more than n linearly independent covariant constant vector fields. Thus the V_n is locally Euclidean.

---

[1] BOCHNER [2]; YANO and BOCHNER [1].

THEOREM 4.2.[1] *A compact orientable V_n with $K_{\mu\lambda} = 0$ admitting a transitive group of motions is locally Euclidean.*

Suppose next that a vector $v^\varkappa$ generates a one-parameter group of conformal motions, then we have

$$\underset{v}{\pounds} g_{\mu\lambda} = \nabla_\mu v_\lambda + \nabla_\lambda v_\mu = 2\phi g_{\mu\lambda}, \ \nabla_\lambda v^\lambda = n\phi.$$

Substituting these equations in (4.1), we find

$$\int_{V_n} [(\nabla^\mu v^\lambda)(\nabla_\mu v_\lambda) + n(n-2)\phi^2 - K_{\mu\lambda} v^\mu v^\lambda] d\tau = 0.$$

Thus, if the Ricci tensor $K_{\mu\lambda}$ is negative semi-definite, we must have

$$\nabla_\mu v_\lambda = 0, \ \phi = 0, \ K_{\mu\lambda} v^\mu v^\lambda = 0,$$

that is the vector field v_λ must be covariant constant.

If the Ricci tensor $K_{\mu\lambda}$ is negative definite, we must have $v^\varkappa = 0$. Thus we have

THEOREM 4.3.[2] *In a compact orientable V_n whose Ricci tensor is negative semi-definite, a vector generating a one-parameter group of conformal motions is a covariant constant field. In a compact orientable V_n whose Ricci tensor is negative definite, there does not exist a one-parameter group of conformal motions.*

Now consider an arbitrary vector field $v^\varkappa$ and form

$$\tfrac{1}{2}\Delta(v_\varkappa v^\varkappa) = \tfrac{1}{2} g^{\mu\lambda}\nabla_\mu \nabla_\lambda (v_\varkappa v^\varkappa) = v_\varkappa g^{\mu\lambda}\nabla_\mu\nabla_\lambda v^\varkappa + (\nabla^\mu v^\lambda)(\nabla_\mu v_\lambda).$$

Since

$$\int_{V_n} \Delta(v_\varkappa v^\varkappa) d\tau = 0,$$

we get

$$\int_{V_n} [v_\varkappa g^{\mu\lambda}\nabla_\mu\nabla_\lambda v^\varkappa + (\nabla^\mu v^\lambda)(\nabla_\mu v_\lambda)] d\tau = 0. \tag{4.2}$$

Adding the equations (4.1) and (4.2), we obtain

$$\int_{V_n} [v_\varkappa (g^{\mu\lambda}\nabla_\mu\nabla_\lambda v^\varkappa + K_\lambda^{\cdot\varkappa} v^\lambda) + 2(\nabla^{(\mu} v^{\lambda)})(\nabla_{(\mu} v_{\lambda)}) - (\nabla_\mu v^\mu)(\nabla_\lambda v^\lambda)] d\tau = 0. \tag{4.3}$$

Now suppose that a vector $v^\varkappa$ generates a one-parameter group of

---

[1] LICHNEROWICZ [1].

[2] YANO [18]; YANO and BOCHNER [1].

motions in a V_n, then from

$$\underset{v}{\pounds} g_{\mu\lambda} = \nabla_\mu v_\lambda + \nabla_\lambda v_\mu = 0,$$

$$\underset{v}{\pounds} \{^{\ \varkappa}_{\mu\lambda}\} = \nabla_\mu \nabla_\lambda v^\varkappa + K_{\nu\mu\lambda}{}^{\varkappa} v^\nu = 0,$$

we find

$$g^{\mu\lambda} \nabla_\mu \nabla_\lambda v^\varkappa + K_\nu{}^{\varkappa} v^\nu = 0, \quad \nabla_\mu v^\mu = 0. \tag{4.4}$$

Conversely, suppose that a vector field $v^\varkappa$ in a V_n satisfies (4.4). Then substituting (4.4) in (4.3), we find

$$\int_{V_n} (\nabla^{(\mu} v^{\lambda)})(\nabla_{(\mu} v_{\lambda)}) d\tau = 0,$$

from which

$$\underset{v}{\pounds} g_{\mu\lambda} = 2\nabla_{(\mu} v_{\lambda)} = 0,$$

that is, the vector $v^\varkappa$ generates a one-parameter group of motions. Thus we have

THEOREM 4.4.[1] *In order that a vector $v^\varkappa$ generate a one-parameter group of motions in a compact orientable V_n, it is necessary and sufficient that $v^\varkappa$ satisfy* (4.4).

§ 5. Affine motions in a compact orientable V_n.

Suppose that a V_n admits a one-parameter group of affine motions generated by a vector field $v^\varkappa$:

$$\underset{v}{\pounds} \{^{\ \varkappa}_{\mu\lambda}\} = \nabla_\mu \nabla_\lambda v^\varkappa + K_{\nu\mu\lambda}{}^{\varkappa} v^\nu = 0, \tag{5.1}$$

from which

$$g^{\mu\lambda} \nabla_\mu \nabla_\lambda v^\varkappa + K_\nu{}^{\varkappa} v^\nu = 0, \tag{5.2}$$

and

$$\nabla_\mu \nabla_\lambda v^\lambda = 0. \tag{5.3}$$

From (5.3), we see that $\nabla_\lambda v^\lambda$ is a constant. But we have on the other hand

$$\int_{V_n} \nabla_\lambda v^\lambda d\tau = 0,$$

[1] YANO [18]; YANO and BOCHNER [1].

which shows that

$$\nabla_\lambda v^\lambda = 0. \tag{5.4}$$

Thus from the equations (5.2) and (5.4), we obtain, on account of Theorem 4.4,

THEOREM 5.1.[1] *A one-parameter group of affine motions in a compact orientable V_n is a group of motions.*

§ 6. Symmetric V_n.

A V_n, symmetric in the sense of Cartan [2] is characterized by the equation

$$\nabla_\omega K_{\nu\mu\lambda}^{\cdot\cdot\cdot\varkappa} = 0. \tag{6.1}$$

From this equation, we get

$$2\nabla_{[\pi}\nabla_{\omega]} K_{\nu\mu\lambda}^{\cdot\cdot\cdot\varkappa} = 0,$$

and consequently, for a symmetric space, we have

$$\begin{aligned} H_{\pi\omega\nu\mu\lambda}^{\cdot\cdot\cdot\cdot\cdot\varkappa} \overset{\text{def}}{=} K_{\pi\omega\rho}^{\cdot\cdot\cdot\varkappa} K_{\nu\mu\lambda}^{\cdot\cdot\cdot\rho} - K_{\pi\omega\nu}^{\cdot\cdot\cdot\rho} K_{\rho\mu\lambda}^{\cdot\cdot\cdot\varkappa} \\ - K_{\pi\omega\mu}^{\cdot\cdot\cdot\rho} K_{\nu\rho\lambda}^{\cdot\cdot\cdot\varkappa} - K_{\pi\omega\lambda}^{\cdot\cdot\cdot\rho} K_{\nu\mu\rho}^{\cdot\cdot\cdot\varkappa} = 0. \end{aligned} \tag{6.2}$$

On the other hand, we have from (6.1)

$$\nabla_\omega K_{\mu\lambda} = 0. \tag{6.3}$$

Thus, for a symmetric V_n, we have (6.2) and (6.3). We shall prove the converse of this:

THEOREM 6.1.[3] *A compact orientable V_n satisfying* (6.2) *and* (6.3) *is symmetric in the sense of E. Cartan.*

Using the identities

$$K_{\nu\mu\lambda\varkappa} = K_{\lambda\varkappa\nu\mu},$$

$$\nabla_{[\omega} K_{\nu\mu]\lambda\varkappa} = 0,$$

$$2\nabla_{[\pi}\nabla_{\omega]} K_{\nu\mu\lambda}^{\cdot\cdot\cdot\varkappa} = H_{\pi\omega\nu\mu\lambda}^{\cdot\cdot\cdot\cdot\cdot\varkappa},$$

---

[1] YANO [18]; YANO and BOCHNER [1].
[2] CARTAN [1, 2, 6, 8, 11].
[3] LICHNEROWICZ [1].

we get a general formula

(6.4) $\frac{1}{2}\Delta(K_{\nu\mu\lambda\kappa}K^{\nu\mu\lambda\kappa})$

$$= 4(\nabla_\mu\nabla_\lambda K_{\nu\kappa})K^{\nu\mu\lambda\kappa} + 2H_{\pi\nu\lambda\kappa\mu}^{\cdot\cdot\cdot\cdot\cdot\pi}K^{\nu\mu\lambda\kappa} + (\nabla_\omega K_{\nu\mu\lambda\kappa})(\nabla^\omega K^{\nu\mu\lambda\kappa}).$$

Consequently, if we assume (6.2) and (6.3), we have

$$\tfrac{1}{2}\Delta(K_{\nu\mu\lambda\kappa}K^{\nu\mu\lambda\kappa}) = (\nabla_\omega K_{\nu\mu\lambda\kappa})(\nabla^\omega K^{\nu\mu\lambda\kappa}),$$

which is positive definite. Thus by Theorem 1.3, we conclude

$$\nabla_\omega K_{\nu\mu\lambda\kappa} = 0,$$

which proves Theorem 6.1.

§ 7. Isotropy groups and holonomy groups.

We know that a symmetric V_n admits a transitive group G of motions and that the linear isotropy group $\tilde{G}(P)$ at a point P contains the homogeneous holonomy group $\sigma(P)$ at P of the space as a subgroup.

Conversely, we assume that an irreducible V_n[1] admits a transitive group G of motions and that the linear isotropy group $\tilde{G}(P)$ at P contains the homogeneous holonomy group $\sigma(P)$ at P of the space as a subgroup for every point of the space.

Denoting by $\underset{v}{\pounds}$ the infinitesimal operator corresponding to one of the generators of the group $\tilde{G}(P)$, we obtain

(7.1) $$\underset{v}{\pounds}K_{\nu\mu\lambda}^{\cdot\cdot\cdot\kappa} = -K_{\nu\mu\lambda}^{\cdot\cdot\cdot\rho}\nabla_\rho v^\kappa + K_{\rho\mu\lambda}^{\cdot\cdot\cdot\kappa}\nabla_\nu v^\rho + K_{\nu\rho\lambda}^{\cdot\cdot\cdot\kappa}\nabla_\mu v^\rho + K_{\nu\mu\rho}^{\cdot\cdot\cdot\kappa}\nabla_\lambda v^\rho = 0.$$

But we assumed that $\tilde{G}(P)$ contains $\sigma(P)$ and the $\overset{\kappa}{\lambda}$-domain of $K_{\nu\mu\lambda}^{\cdot\cdot\cdot\kappa}$ is contained in the $\overset{\kappa}{\lambda}$-domain of $\nabla_\lambda \underset{a}{v}^\kappa$ formed from all generators $\underset{a}{v}^\kappa$ of the group $\tilde{G}(P)$. Thus from (7.1) we get

(7.2) $$H_{\pi\omega\nu\mu\lambda}^{\cdot\cdot\cdot\cdot\cdot\kappa} = 0.$$

On the other hand, from (7.1), we find

(7.3) $$\underset{v}{\pounds}K_{\mu\lambda} = 0.$$

But we have assumed that $\sigma(P)$ is irreducible and consequently $\tilde{G}(P)$

---

[1] When the holonomy group σ of a V_n is irreducible, the space V_n is said to be irreducible.

is also irreducible. Thus we get from (7.3)

$$K_{\mu\lambda} = \frac{K}{n} g_{\mu\lambda},$$

from which

$$\nabla_\omega K_{\mu\lambda} = 0. \tag{7.4}$$

The equations (6.4), (7.2) and (7.4) show that

$$\tfrac{1}{2}\Delta(K_{\nu\mu\lambda\kappa}K^{\nu\mu\lambda\kappa}) = (\nabla_\omega K_{\nu\mu\lambda\kappa})(\nabla^\omega K^{\nu\mu\lambda\kappa}), \tag{7.5}$$

The group G of motions is transitive and consequently from

$$\underset{a}{\pounds}(K_{\nu\mu\lambda\kappa}K^{\nu\mu\lambda\kappa}) = 0,$$

we can conclude that

$$K_{\nu\mu\lambda\kappa}K^{\nu\mu\lambda\kappa} = \text{constant}$$

hence

$$\Delta(K_{\nu\mu\lambda\kappa}K^{\nu\mu\lambda\kappa}) = 0. \tag{7.6}$$

From (7.5) and (7.6) we get

$$\nabla_\omega K_{\nu\mu\lambda\kappa} = 0 \tag{7.7}$$

which proves the following theorem.

THEOREM 7.1.[1] *If an irreducible V_n (not necessarily compact and orientable) admits a transitive group of motions whose linear isotropy group at any point contains the homogeneous holonomy group at that point, the V_n is symmetric in the sense of E. Cartan.*

---

[1] NOMIZU [4, 6].

Chapter X

LIE DERIVATIVES IN AN ALMOST COMPLEX SPACE

§ 1. Almost complex spaces.

Consider a $2n$-dimensional real space X_{2n} covered by a set of neighbourhoods with real coordinates $(\eta^\varkappa, \zeta^\varkappa)$; $\varkappa, \lambda, \mu, \ldots = 1, 2, \ldots, n$. The complex numbers

$$\xi^\varkappa = \eta^\varkappa + i\zeta^\varkappa, \quad \xi^{\bar\varkappa} = \eta^\varkappa - i\zeta^\varkappa \tag{1.1}$$

$$\bar\varkappa, \bar\lambda, \bar\mu, \ldots = \bar 1, \bar 2, \ldots, \bar n,$$

can be regarded as complex coordinates of a point in the X_{2n} whose real coordinates are $(\eta^\varkappa, \zeta^\varkappa)$. If it is possible to choose a set of coordinate neighbourhoods in such a way that, in the domain of intersection of two coordinate neighbourhoods $U(\eta^{\varkappa'}, \zeta^{\varkappa'})$ and $U(\eta^\varkappa, \zeta^\varkappa)$, we have

$$\xi^{\varkappa'} = f^{\varkappa'}(\xi^\varkappa), \quad \xi^{\bar\varkappa'} = f^{\bar\varkappa'}(\xi^{\bar\varkappa}), \quad \det\left(\frac{\partial f^{\varkappa'}}{\partial \xi^\varkappa}\right) \neq 0. \tag{1.2}$$

where $f^{\bar\varkappa'}$ are complex conjugate functions of $f^{\varkappa'}$, we say that the space admits a *complex analytic structure* or simply a *complex structure* and we call such a space an n-dimensional *complex space*. Since (1.2) can be written as

$$\eta^{\varkappa'} = g^{\varkappa'}(\eta, \zeta), \quad \zeta^{\varkappa'} = h^{\varkappa'}(\eta, \zeta), \tag{1.3}$$

and since the functions $g^{\varkappa'}$ and $h^{\varkappa'}$ are real analytic, a complex space is of class C^ω. If we write (1.2) as

$$\xi^{\alpha'} = f^{\alpha'}(\xi^\alpha), \tag{1.4}$$

$$\alpha, \beta, \gamma, \ldots = 1, 2, \ldots, n; \bar 1, \bar 2, \ldots, \bar n,$$

then the Jacobian Δ of the transformation is given by

$$\Delta = \det\left(\frac{\partial f^{\varkappa'}}{\partial \xi^\varkappa}\right) \det\left(\overline{\frac{\partial f^{\varkappa'}}{\partial \xi^\varkappa}}\right) > 0, \tag{1.5}$$

where the bar denotes the complex conjugate. Thus the Jacobian of

(1.3) is also positive and consequently a complex space is orientable.

A mixed tensor of valence 2, is defined as a geometric object which has $(2n)^2$ components $T_{\beta}^{\cdot\alpha}$ in every complex coordinate system $(\xi^{\kappa}, \xi^{\bar\kappa})$, and whose transformation law under a coordinate transformation (1.4) is

$$T_{\beta'}^{\cdot\alpha'} = A_{\beta'\alpha}^{\beta\alpha'} T_{\beta}^{\cdot\alpha}. \tag{1.6}$$

In a complex space, there exists a mixed tensor field $F_{\beta}^{\cdot\alpha}$ which has the numerical components

$$F_{\lambda}^{\cdot\kappa} = +i\delta_{\lambda}^{\kappa},\quad F_{\bar\lambda}^{\cdot\kappa} = 0,\quad F_{\lambda}^{\cdot\bar\kappa} = 0,\quad F_{\bar\lambda}^{\cdot\bar\kappa} = -i\delta_{\bar\lambda}^{\bar\kappa} \tag{1.7}$$

in all complex coordinate systems and which satisfies

$$F_{\gamma}^{\cdot\beta} F_{\beta}^{\cdot\alpha} = -A_{\gamma}^{\alpha}. \tag{1.8}$$

In such a space, the differential equations

$$\text{(a)}\ \tfrac{1}{2}(A_{\beta}^{\alpha} - iF_{\beta}^{\cdot\alpha})d\xi^{\beta} = 0, \quad \text{(b)}\ \tfrac{1}{2}(A_{\beta}^{\alpha} + iF_{\beta}^{\cdot\alpha})d\xi^{\beta} = 0 \tag{1.9}$$

are both completely integrable. In fact, (a) admits the solutions $\xi^{\kappa} =$ const. and (b) admits the solution $\xi^{\bar\kappa} =$ const.

When, in a $2n$-dimensional real space X_{2n} of class $C^r (r \geq 2)$, there is given a mixed tensor field $F_i^{\cdot h}$; $h, i, j, \ldots = 1, 2, \ldots, 2n$, satisfying

$$F_i^{\cdot l} F_l^{\cdot h} = -A_i^h, \tag{1.10}$$

we say that the space admits an *almost complex structure* and we call such a space an *almost complex space.* [1] In such a space, we can choose at each point $2n$ linearly independent vectors

$$\underset{1}{e^h},\ F_i^{\cdot h}\underset{1}{e^i},\ \underset{2}{e^h},\ F_i^{\cdot h}\underset{2}{e^i},\ \ldots,\ \underset{n}{e^h},\ F_i^{\cdot h}\underset{n}{e^i}.$$

Since the corresponding orientation of the space depends only on the tensor $F_i^{\cdot h}$, an almost complex structure determines a unique orientation of the space.

If there exists a complex coordinate system with respect to which the tensor $F_i^{\cdot h}$ has the components (1.7), then, in a domain in which two such coordinate systems ξ^{α} and $\xi^{\alpha'}$ are valid, we have

$$\frac{\partial \xi^{\alpha}}{\partial \xi^{\alpha'}} F_{\beta'}^{\cdot\alpha'} = \frac{\partial \xi^{\beta}}{\partial \xi^{\beta'}} F_{\beta}^{\cdot\alpha}, \tag{1.11}$$

---

[1] EHRESMANN [2].

from which it follows that ξ^{κ} are functions of $\xi^{\kappa'}$ only and $\xi^{\bar{\kappa}}$ are functions of $\xi^{\bar{\kappa}'}$ only. Thus the space is a complex space. In this case, we say that the almost complex structure is *induced* by a complex structure.

If an almost complex structure $F_i^{\cdot h}$ is induced by a complex structure, then, in a complex coordinate system, we have

$$N_{\gamma\beta}^{\cdot\cdot\alpha} \stackrel{\text{def}}{=} 2F_{[\gamma}^{\cdot\,\varepsilon}(\partial_{|\varepsilon|}F_{\beta]}^{\cdot\alpha} - \partial_{\beta]}F_{\varepsilon}^{\cdot\alpha}) = 0.$$

Since $N_{\gamma\beta}^{\cdot\cdot\alpha}$ is a tensor, we have[1]

$$N_{ji}^{\cdot\cdot h} = 2F_{[j}^{\cdot\,l}(\partial_{|l|}F_{i]}^{\cdot h} - \partial_{i]}F_{l}^{\cdot h}) = 0 \tag{1.12}$$

with respect to an arbitrary coordinate system (h).

Conversely, suppose that an almost complex structure $F_i^{\cdot h}$ of class C^{ω} satisfies (1.12). Then the differential equations

$$\text{(a)}\ B_i^h d\xi^i = 0, \qquad \text{(b)}\ C_i^h d\xi^i = 0 \tag{1.13}$$

are both completely integrable, where

$$B_i^h \stackrel{\text{def}}{=} \tfrac{1}{2}(A_i^h - iF_i^{\cdot h}),\quad C_i^h \stackrel{\text{def}}{=} \tfrac{1}{2}(A_i^h + iF_i^{\cdot h}), \tag{1.14}$$

and consequently

$$A_i^h = B_i^h + C_i^h,\quad F_i^{\cdot h} = i(B_i^h - C_i^h). \tag{1.15}$$

Indeed, the integrability conditions of (a) and (b) are identically satisfied:

$$\begin{cases} \text{(a)}\ C_j^l C_i^k \partial_{[l} B_{k]}^h = \tfrac{1}{8}(N_{ji}^{\cdot\cdot h} - iN_{ji}^{\cdot\cdot l}F_l^{\cdot h}) = 0, \\ \text{(b)}\ B_j^l B_i^k \partial_{[l} C_{k]}^h = \tfrac{1}{8}(N_{ji}^{\cdot\cdot h} + iN_{ji}^{\cdot\cdot l}F_l^{\cdot h}) = 0. \end{cases} \tag{1.16}$$

Denoting the solutions of (1.13a) and (1.13b) by $\xi^{\kappa'} = \xi^{\kappa'}(\xi^i) = \text{const.}$ and $\xi^{\bar{\kappa}'} = \xi^{\bar{\kappa}'}(\xi^i) = \text{const.}$ respectively, we get

$$\frac{\partial\xi^h}{\partial\xi^{\bar{\kappa}'}} = iF_i^{\cdot h}\frac{\partial\xi^i}{\partial\xi^{\kappa'}}, \qquad \frac{\partial\xi^h}{\partial\xi^{\kappa'}} = -\,iF_i^{\cdot h}\frac{\partial\xi^i}{\partial\xi^{\kappa'}}, \tag{1.17}$$

which shows that $F_i^{\cdot h}$ has the components (1.7) with respect to the

[1] The tensor $N_{ji}^{\cdot\cdot h}$ was found by NIJENHUIS [1] for a more general case. We call $N_{ji}^{\cdot\cdot h}$ defined here the *Nijenhuis tensor* of $F_i^{\cdot h}$. Cf. SCHOUTEN [8], p. 248. It is also called the torsion tensor but we prefer to use this expression for the tensor $S_{ji}^{\cdot\cdot h} = \Gamma_{[ji]}^h$ of the connexion Γ_{ji}^h.

coordinate system $(\xi^{\varkappa'}, \xi^{\bar{\varkappa}'})$. Thus we have [1]

THEOREM 1.1. *If an almost complex structure* $F_i^{\cdot h}$ *of class* $C^r\ (r \geq 2)$ *is induced by a complex structure, we have* $N_{ji}^{\cdot\cdot h} = 0$. *Conversely, if an almost complex structure* $F_i^{\cdot h}$ *of class* C^ω *satisfies* $N_{ji}^{\cdot\cdot h} = 0$, *then it is induced by a complex structure.*

The Nijenhuis tensor satisfies the following identities:

$$N_{(ji)}^{\cdot\cdot h} = 0, \quad N_{ji}^{\cdot\cdot i} = 0, \tag{1.18}$$

$$N_{ji}^{\cdot\cdot h} F_i^{\cdot l} = - N_{ji}^{\cdot\cdot l} F_l^{\cdot h} = - N_{il}^{\cdot\cdot h} F_j^{\cdot l} \text{ [2]} \tag{1.19}$$

$$\text{(1.20)} \quad \begin{aligned} &\text{a)}\ N_{ji}^{\cdot\cdot h} + F_j^{\cdot l} F_i^{\cdot k} N_{lk}^{\cdot\cdot h} = 0, \\ &\text{b)}\ N_{ji}^{\cdot\cdot h} - F_i^{\cdot l} F_k^{\cdot h} N_{jl}^{\cdot\cdot k} = 0. \end{aligned}$$

An almost complex structure which need not be of class C^ω is called a *pseudo-complex structure* if $N_{ji}^{\cdot\cdot h} = 0$. A space with a pseudo-complex structure is called a *pseudo-complex space.*

§ 2. Linear connexions in an almost complex space.

It is always possible to introduce in an almost complex manifold a linear connexion Γ_{ji}^h such that $\nabla_j F_i^{\cdot h} = 0$. If $\overset{*}{\Gamma}{}_{ji}^h$ is an arbitrary symmetric connexion and

$$T_{ji}^{\cdot\cdot h} \overset{\text{def}}{=} \Gamma_{ji}^h - \overset{*}{\Gamma}{}_{ji}^h, \tag{2.1}$$

we have

$$0 = \nabla_j F_i^{\cdot h} = \overset{*}{\nabla}_j F_i^{\cdot h} + T_{jm}^{\cdot\cdot h} F_i^{\cdot m} - T_{ji}^{\cdot\cdot l} F_l^{\cdot h},$$

from which

$$\tfrac{1}{2}(\overset{*}{\nabla}_j F_i^{\cdot l}) F_l^{\cdot h} = - \tfrac{1}{2} T_{ji}^{\cdot\cdot h} - \tfrac{1}{2} T_{jm}^{\cdot\cdot l} F_i^{\cdot m} F_l^{\cdot h} = - \tfrac{1}{2}\,(A_i^m A_l^h + F_i^{\cdot m} F_l^{\cdot h}) T_{jm}^{\cdot\cdot l}.$$

The operators

$$\text{(2.2)} \quad \begin{cases} O_{i\,l}^{mh} \overset{\text{def}}{=} \tfrac{1}{2}(A_i^m A_l^h - F_i^{\cdot m} F_l^{\cdot h}), \\ \overset{*}{O}{}_{i\,l}^{mh} \overset{\text{def}}{=} \tfrac{1}{2}(A_i^m A_l^h + F_i^{\cdot m} F_l^{\cdot h}) \end{cases}$$

---

[1] ECKMANN and FRÖLICHER [1]; CALABI and SPENCER [1]; YANO [22].
[2] ECKMANN [1, 2].

are idempotent but not reversible and from this it follows that there are more solutions and that

$$T_{ji}^{\cdot\cdot h} = -\tfrac{1}{2}(\overset{*}{\nabla}_j F_i^{\cdot l})F_l^{\cdot h} \tag{2.3}$$

is one of them. A tensor is called *pure* (*hybrid*) in two indices if it is annihilated by transvection of $\overset{*}{O}(O)$ on these indices. So $N_{ji}^{\cdot\cdot h}$ is pure in j, i and hybrid in $\overset{h}{i}$.

To this solution every term can be added that is made zero by the operator $\overset{*}{O}$, for instance, $+\tfrac{1}{2}(\overset{*}{\nabla}_{[j}F_{i]}^{\cdot l})F_l^{\cdot h} - \tfrac{1}{2}(\overset{*}{\nabla}_{[j}F_{l]}^{\cdot h})F_i^{\cdot l}$. Then we get the solution

$$T_{ji}^{\cdot\cdot h} = -\tfrac{1}{2}(\overset{*}{\nabla}_{(j}F_{i)}^{\cdot l})F_l^{\cdot h} - \tfrac{1}{2}(\overset{*}{\nabla}_{[j}F_{l]}^{\cdot h})F_i^{\cdot l} \tag{2.4}$$ [1]

On the other hand, the Nijenhuis tensor $N_{ji}^{\cdot\cdot h}$ can be written also in the form

$$\begin{aligned} N_{ji}^{\cdot\cdot h} = {} & 2F_{[j}^{\cdot l}(\nabla_{|l|}F_{i]}^{\cdot h} - \nabla_{i]}F_l^{\cdot h}) \\ & + 2(S_{ji}^{\cdot\cdot h} - F_j^{\cdot l}F_i^{\cdot k}S_{lk}^{\cdot\cdot h} + F_j^{\cdot l}F_k^{\cdot h}S_{li}^{\cdot\cdot k} - F_i^{\cdot l}F_k^{\cdot h}S_{lj}^{\cdot\cdot k}), \end{aligned} \tag{2.5}$$

where ∇_j denotes the covariant differentiation with respect to an arbitrary linear connexion Γ_{ji}^h and $S_{ji}^{\cdot\cdot h}$ its torsion tensor.

Thus, if the space is pseudo-complex and if we introduce a linear connexion such that $\nabla_j F_i^{\cdot h} = 0$, then the torsion tensor satisfies

$$S_{ji}^{\cdot\cdot h} - F_j^{\cdot l}F_i^{\cdot k}S_{lk}^{\cdot\cdot h} + F_j^{\cdot l}F_k^{\cdot h}S_{li}^{\cdot\cdot k} - F_i^{\cdot l}F_k^{\cdot h}S_{lj}^{\cdot\cdot k} = 0. \tag{2.6}$$

Conversely, if we can introduce, in an almost complex space, a linear connexion such that $\nabla_j F_i^{\cdot h} = 0$ and (2.6) holds, then the space is pseudo-complex. Thus we have [2]

THEOREM 2.1. *In order that an almost complex space be a pseudo-complex space, it is necessary and sufficient that we can introduce in it a linear connexion such that* $\nabla_j F_i^{\cdot h} = 0$ *and that* (2.6) *holds.*

Furthermore, if the space is pseudo-complex, we can introduce a

---

[1] ECKMANN [1]; FRÖLICHER [1].

[2] YANO and MOGI [2].

symmetric linear connexion such that $\nabla_j F_i^{\ h} = 0$, because the linear connexion $\Gamma_{ji}^h = \overset{*}{\Gamma}{}_{ji}^h + T_{ji}^{\cdot\cdot h}$ given by (2.4) satisfies

$$S_{ji}^{\cdot\cdot h} = -\tfrac{1}{8} N_{ji}^{\cdot\cdot h} = 0. \tag{2.7}$$

Conversely, if we can introduce, in an almost complex space, a symmetric linear connexion such that $\nabla_j F_i^{\ h} = 0$, then $N_{ji}^{\cdot\cdot h} = 0$, and the space is pseudo-complex. Thus we get [1]

THEOREM 2.2. *In order that an almost complex space be a pseudo-complex space, it is necessary and sufficient that we can introduce in it a symmetric linear connexion such that* $\nabla_j F_i^{\ h} = 0$.

§ 3. Almost complex metric spaces.

If an almost (pseudo-) complex space has a positive definite Riemannian metric $ds^2 = g_{ji} d\xi^j d\xi^i$ which satisfis

$$F_j^{\cdot l} F_i^{\cdot k} g_{lk} = g_{ji}, \tag{3.1}$$

then the space is called an *almost (pseudo-) Hermitian space.* In this case the tensor $F_{ih} \overset{\text{def}}{=} F_i^{\cdot l} g_{lh}$ is antisymmetric in i and h. Note that $F_i^{\cdot h}$ is pure but that F_{ih} and g_{ih} are hybrid. A. Lichnerowicz [2] has proved

THEOREM 3.1. *In an almost complex space, it is always possible to define a Hermitian metric.*

In fact, let a_{ji} be a tensor which defines a positive definite Riemannian metric in an almost complex space and let

$$g_{ji} \overset{\text{def}}{=} \tfrac{1}{2}(a_{ji} + F_j^{\cdot l} F_i^{\cdot k} a_{lk}), \tag{3.2}$$

(g_{ji} is the hybrid part of a_{ji}), then g_{ji} defines another positive definite Riemannian metric and satisfies (3.1).

The equation (3.1) and the antisymmetry of the tensor F_{ih} show that the transformation $v^h \to F_i^{\cdot h} v^i$ changes a vector v^h into a vector orthogonal to it and does not change its length.

---

[1] ECKMANN [1]; HODGE [1]; PATTERSON [1].

[2] LICHNEROWICZ [2, 5].

Moreover we can easily see that

$$F_{jih} \overset{\text{def}}{=} 3\partial_{[j}F_{ih]} \tag{3.3}$$

are components of an antisymmetric tensor.

If an almost (pseudo-) Hermitian space satisfies $F_{jih} = 0$, the space is called an almost (pseudo-) *Kählerian space*. It can be proved that the bivector F_{ih} is harmonic in an almost Kählerian space.[1]

The relations between these spaces may be seen in the diagram:

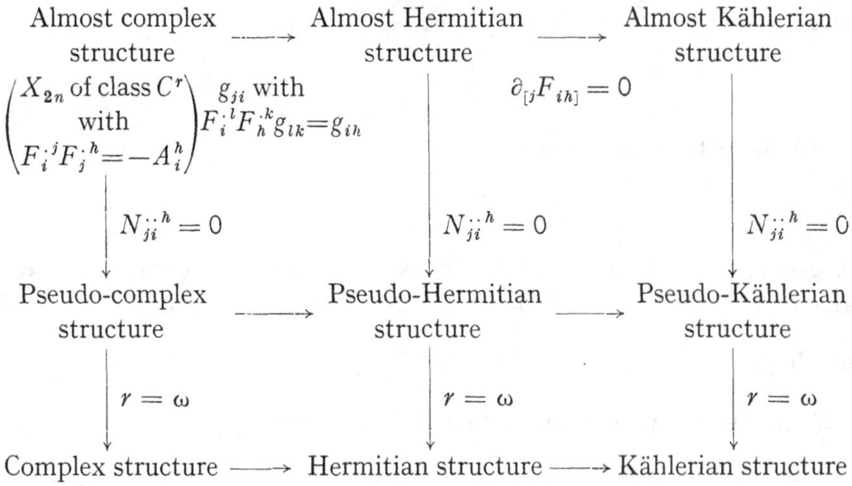

In an almost Hermitian space, we denote by $\overset{0}{\nabla}_j$ the covariant differentiation with respect to the Christoffel symbols $\{{}^{\,h}_{ji}\}$. If $\overset{0}{\nabla}_j F_{ih}$ vanishes, then the tensors $N_{ji}^{\cdot\cdot h}$ and F_{jih} vanish too, and consequently the space is pseudo-Kählerian.

Conversely, since the Nijenhuis tensor can be written also in the form

$$N_{jih} = -2(F_{[j}^{\cdot\, l}F_{i]hl} - F_j^{\cdot\, l}\overset{0}{\nabla}_h F_{il}), \tag{3.4}$$

$\overset{0}{\nabla}_j F_{ih}$ vanishes if the tensors $N_{ji}^{\cdot\cdot h}$ and F_{jih} vanish. Thus we obtain

THEOREM 3.2.[2] *In order that an almost Hermitian space be pseudo-Kählerian, it is necessary and sufficient that $\overset{0}{\nabla}_j F_{ih}$ vanish.*

---

[1] SCHOUTEN and YANO [3].

[2] ECKMANN [2]; YANO [22]; YANO and MOGI [1].

In an almost Hermitian space, the four following connexions occur in literature:

(3.5) (I)[1] $$\overset{1}{\Gamma}{}^h_{ji} = \left\{{h \atop ji}\right\} - \tfrac{1}{2}(\overset{0}{\nabla}_j F_{il})F^{lh},$$

(3.6) (II)[2] $$\overset{2}{\Gamma}{}^h_{ji} = \left\{{h \atop ji}\right\} - \tfrac{1}{2}(\overset{0}{\nabla}_j F_{il} + \overset{0}{\nabla}_i F_{jl} + \overset{0}{\nabla}_l F_{ji})F^{lh},$$

(3.7) (III)[2] $$\overset{3}{\Gamma}{}^h_{ji} = \left\{{h \atop ji}\right\} - \tfrac{1}{2}(\overset{0}{\nabla}_j F_{il} - \overset{0}{\nabla}_i F_{jl} - \overset{0}{\nabla}_l F_{ji})F^{lh},$$

(3.8) (IV)[3] $$\overset{4}{\Gamma}{}^h_{ji} = \left\{{h \atop ji}\right\} - \tfrac{1}{2}(\overset{0}{\nabla}_j F_{il} + \overset{0}{\nabla}_i F_{jl})F^{lh} + \tfrac{1}{2}F_{jl}(\overset{0}{\nabla}_k F_i^{\cdot l})g^{kh} - \tfrac{1}{4}N_{hij}.$$

All these four connexions satisfy

(3.9) $$\nabla_j F_{ih} = 0.$$

A geometrical characterization for each of these connexions has been given by Schouten and Yano.[4] In the following, we put

(3.10) $$\Gamma^h_{ji} = \left\{{h \atop ji}\right\} + T_{ji}^{\cdot\cdot h}.$$

With respect to the first connexion, we have

(3.11) $$\overset{1}{\nabla}_j g_{ih} = 0,$$

(3.12) $$\overset{1}{T}_{jih} = \tfrac{1}{4}N_{hij} - \tfrac{1}{2}F_{[h}^{\cdot\, l}F_{i]jl}.$$

In the Hermitian case, this connexion reduces to the connexion of Lichnerowicz.[5]

With respect to the second connexion, we have

(3.13) $$\overset{2}{T}_{jih} = \tfrac{1}{2}N_{hji} - \tfrac{1}{2}F_j^{\cdot\, l}F_{ihl}.$$

Thus in a pseudo-Hermitian space, we have $\overset{2}{T}_{jih} = -\tfrac{1}{2}F_j^{\cdot\, l}F_{ihl}$ and

[1] LICHENROWICZ [5].
[2] YANO [23].
[3] SCHOUTEN and YANO [1].
[4] SCHOUTEN and YANO [1].
[5] LICHNEROWICZ [5].

consequently

$$\overset{2}{\nabla}_j g_{ih} = 0. \tag{3.14}$$

In a Hermitian space, this connexion reduces to the connexion found by Schouten and van Dantzig [1] and used by Chern [2] and Liebermann. [3]

With respect to the third connexion, we have

$$\overset{3}{T}_{jih} = \tfrac{1}{4}N_{jih} + \tfrac{1}{2}F_j^{\cdot m} F_i^{\cdot l} F_h^{\cdot k} F_{mlk}. \tag{3.15}$$

Thus in a pseudo-Hermitian space, we have $\overset{3}{T}_{jih} = \frac{1}{2}F_j^{\cdot m}F_i^{\cdot l}F_h^{\cdot k}F_{mlk}$ and consequently

$$\overset{3}{\nabla}_j g_{ih} = 0. \tag{3.16}$$

With respect to the fourth connexion, we have

$$\overset{4}{\nabla}_j g_{ih} = 0, \ \overset{4}{\nabla}_j F_{ih} = 0 \tag{3.17}$$

and

$$S_{ji}^{\cdot\cdot h} = O_{ji}^{ml} S_{ml}^{\cdot\cdot h} \tag{3.18}$$

$$N_{ji}^{\cdot\cdot h} = 8\overset{*}{O}{}_{tl}^{mh} S_{jm}^{\cdot\cdot l}. \tag{3.19}$$

The equation (3.18) expresses that $S_{ji}^{\cdot\cdot h}$ is pure in j, i. This means geometrically that there exist infinitesimal parallelograms in every E_2 that is invariant for the linear transformation $F_i^{\cdot h}$. [4]

§ 4. The curvature in a pseudo-Kählerian space.

In a pseudo-Kählerian space, we have

$$\overset{0}{\nabla}_j g_{ih} = 0, \ \overset{0}{\nabla}_j F_{ih} = 0. \tag{4.1}$$

Applying the Ricci formula to $F_i^{\cdot h}$, we get

$$K_{kji}^{\cdot\cdot\cdot l} F_l^{\cdot h} = K_{kjl}^{\cdot\cdot\cdot h} F_i^{\cdot l}, \tag{4.2}$$

$$K_{kjil} F_h^{\cdot l} = K_{kjhl} F_i^{\cdot l}, \tag{4.3}$$

$$K_{kjih} = K_{kjml} F_i^{\cdot m} F_h^{\cdot l} \ \text{or} \ O_{ih}^{ml} K_{kjml} = 0. \tag{4.4}$$

Hence K_{kjih} is hybrid in the first two and in the last two indices.

---

[1] SCHOUTEN and VAN DANTZIG [1]. Cf. SCHOUTEN [8], p. 397.

[2] CHERN [1].

[3] LIEBERMANN [1].

[4] SCHOUTEN and YANO [1].

Transvecting (4.2) with g^{ji}, we find

$$K_{k}^{\cdot l} F_{l}^{\cdot h} = K_{kml}^{\cdot\cdot\cdot h} F^{ml} = \tfrac{1}{2}(K_{kml}^{\cdot\cdot\cdot h} - K_{klm}^{\cdot\cdot\cdot h}) F^{ml},$$

from which

$$K_{k}^{\cdot l} F_{l}^{\cdot h} = -\tfrac{1}{2} K_{mlk}^{\cdot\cdot\cdot h} F^{ml}. \tag{4.5}$$

Thus

$$K_{k}^{\cdot l} F_{lh} + K_{h}^{\cdot l} F_{lk} = 0,$$

from which

$$K_{j}^{\cdot i} = -K_{m}^{\cdot l} F_{j}^{\cdot m} F_{l}^{\cdot i}, \tag{4.6}$$

$$K_{ji} = K_{ml} F_{j}^{\cdot m} F_{i}^{\cdot l} \quad \text{or} \quad O_{j\,i}^{ml} K_{ml} = 0. \tag{4.7}$$

Hence K_{ji} is hybrid.

Using these relations, we can prove

THEOREM 4.1.[1] *If a pseudo-Kählerian space is of constant curvature, then it is of zero curvature.*

THEOREM 4.2.[2] *If a pseudo-Kählerian space is conformally Euclidean, it is of zero curvature.*

THEOREM 4.3.[3] *A projective correspondence between two pseudo-Kählerian spaces is necessarily affine.*

THEOREM 4.4.[4] *A conformal correspondence between two pseudo-Kählerian spaces is necessarily a trivial one.*

THEOREM 4.5.[5] *A necessary and sufficient condition that a 2n-dimensional pseudo-Hermitian space be conformal to a pseudo-Kählerian space is that, for* $2n > 4$,

$$C_{jih} \overset{\text{def}}{=} F_{jih} - \frac{1}{2(n-1)} (F_{ji} F_h + F_{ih} F_j + F_{hj} F_i) = 0 \tag{4.8}$$

and for $2n = 4$

$$C_{ji} \overset{\text{def}}{=} 2\partial_{[j} F_{i]} = 0, \tag{4.9}$$

where $F_j = F_{jih} F^{ih}$.

---

[1] BOCHNER [3].

[2] YANO and MOGI [2].

[3] BOCHNER [3]; WESTLAKE [1]; YANO [20].

[4] WESTLAKE [2].

[5] WESTLAKE [2]; YANO [21].

Now we put

$$H_{kj} \overset{\text{def}}{=} K_{kjih}F^{ih} = -2K_h^{\cdot l}F_{lj}. \tag{4.10}$$

We see that H_{kj} is zero if and only if $K_{ji} = 0$ and that

$$F^{kj}H_{kj} = -2K. \tag{4.11}$$

Moreover from the Bianchi identity, we have

$$\overset{0}{\nabla}_{[i}H_{kj]} = 0. \tag{4.12}$$

On the other hand, we have

$$\begin{aligned} g^{lk}\overset{0}{\nabla}_l H_{kj} &= g^{lk}\nabla_l(2K_{km}F_j^{\cdot m}) = 2\overset{0}{\nabla}_l K_m^{\cdot l}F_j^{\cdot m} \\ &= (\overset{0}{\nabla}_m K)F_j^{\cdot m}. \end{aligned} \tag{4.13}$$

Thus

THEOREM 4.6. *The tensor H_{kj} is harmonic if and only if $K =$ const, and it is effective (that is, $F^{kj}H_{kj} = 0$) if and only if $K = 0$.*

Applying a theorem of Hodge,[1] we get from this

THEOREM 4.7. *If, in a compact pseudo-Kählerian space, $K_{ji} \neq 0$, $K = 0$, then the second Betti number $B_2 \geq 2$.*

THEOREM 4.8. *If, in a compact pseudo-Kählerian space, $K_{ji} \neq 0$, $K =$ const. $\neq 0$ and $B_2 = 1$, then $K_{ji} = \frac{1}{2n}Kg_{ji}$.*

§ 5. Pseudo-analytic vectors.

In a pseudo-Kählerian space, we call a field v_i whose covariant derivative is pure

$$F_j^{\cdot l}\overset{0}{\nabla}_i v_l - F_i^{\cdot l}\overset{0}{\nabla}_l v_j = 0 \quad \text{or} \quad \overset{*}{O}{}_{ji}^{lk}\overset{0}{\nabla}_l w_k = 0 \tag{5.1}$$

a covariant *pseudo-analytic vector field.*[2] From (5.1), we can deduce

$$g^{ji}\overset{0}{\nabla}_j\overset{0}{\nabla}_i v_h - K_h^{\cdot l}v_l = 0. \tag{5.2}$$

---

[1] HODGE [1, 3].

[2] In a Kählerian space, the equations (5.1) can be written as $\partial_{\bar{\mu}}v_\lambda = 0$, $\partial_\mu v_{\bar{\lambda}} = 0$ with respect to a complex coordinate system. Hence $v_\lambda(v_{\bar{\lambda}})$ are complex analytic functions of $z^\mu(z^{\bar{\mu}})$.

This is a necessary and sufficient condition that a vector field v_h in a compact orientable Riemannian space be harmonic.[1]

Conversely, if v_h is harmonic, then we have (5.2), from which

$$g^{ji}\overset{0}{\nabla}_j\overset{0}{\nabla}_i F_m^{\cdot h}v_h - K_m^{\cdot h}F_h^{\cdot l}v_l = 0 \tag{5.3}$$

by virtue of $F_m^{\cdot h}K_h^{\cdot l} = K_m^{\cdot h}F_h^{\cdot l}$. From this, it follows that $F_m^{\cdot h}v_h$ is also harmonic.

Thus, from

$$\overset{0}{\nabla}_j v_l = \overset{0}{\nabla}_l v_j \text{ and } \overset{0}{\nabla}_i(F_j^{\cdot l}v_l) - \overset{0}{\nabla}_j(F_i^{\cdot l}v_l) = 0,$$

we get (5.1). Thus

THEOREM 5.1. *In order that a vector field in a compact pseudo-Kählerian space be covariant pseudo-analytic, it is necessary and sufficient that the vector be harmonic.*

Thus applying a theorem of Bochner,[2] we obtain

THEOREM 5.2. *If the Ricci curvature of a compact pseudo-Kählerian space is positive definite, there does not exist a covariant pseudo-analytic vector field.*

In a pseudo-Kählerian space, we call a vector field v^h whose covariant derivative is pure

$$\underset{v}{\pounds} F_i^{\cdot h} = -F_i^{\cdot l}\overset{0}{\nabla}_l v^h + F_l^{\cdot h}\overset{0}{\nabla}_i v^l = 0 \text{ or } \overset{*}{O}{}_{jk}^{lh}\overset{0}{\nabla}_l v^k = 0 \tag{5.4}$$

a contravariant *pseudo-analytic vector field*. From (5.4) we see that if v^h is contravariant pseudo-analytic, then $F_i^{\cdot h}v^i$ is also contravariant pseudo-analytic. Moreover, if u^h and v^h are both contravariant pseudo-analytic, then denoting the Lie derivations with respect to u^h and v^h by $\underset{u}{\pounds}$ and $\underset{v}{\pounds}$ respectively, we have

$$\underset{u}{\pounds} F_i^{\cdot h} = 0 \text{ and } \underset{v}{\pounds} F_i^{\cdot h} = 0,$$

from which $(\underset{v}{\pounds}\underset{u}{\pounds})F_i^{\cdot h} = 0$, where $(\underset{v}{\pounds}\underset{u}{\pounds})$ denotes the Lie derivation with respect to the vector $\underset{v}{\pounds} u^h$. Thus the vector $\underset{u}{\pounds} v^h$ is also contravariant pseudo-analytic. Thus we have

---

[1] DE RHAM and KODAIRA [1]; YANO and BOCHNER [1].

[2] BOCHNER [2]; YANO and BOCHNER [1].

THEOREM 5.3. *If u^h and v^h are both contravariant pseudo-analytic vector fields in a pseudo-Kählerian space, then*

$$F_i^{\ h}u^i,\ F_i^{\ h}v^i,\ \underset{v}{\pounds}u^h,\ \underset{Fv}{\pounds}u^h,\ \underset{v}{\pounds}F_i^{\ h}u^i,\ \underset{Fv}{\pounds}F_i^{\ h}u^i$$

are all contravariant pseudo-analytic vector fields.

In an almost complex space, we have the following identity:[1]

$$\underset{v}{\pounds}u^h + F_i^{\ h}\underset{v}{\pounds}F_j^{\ i}u^j + F_i^{\ h}\underset{Fv}{\pounds}u^i - \underset{Fv}{\pounds}F_i^{\ h}u^i = N_{ji}^{\ \ h}u^jv^i, \tag{5.5}$$

from which

THEOREM 5.4. *In order that an almost complex space be pseudo-complex, it is necessary and sufficient that the left-hand side of* (5.5) *vanish for any vectors u^h and v^h.*

Now, from (5.4), we obtain

$$g^{ji}\nabla_j\nabla_i v^h + K_i^{\ h}v^i = 0, \tag{5.6}$$

from which we get the following theorems which hold in a compact pseudo-Kählerian space.

First, from Theorem 4.4 of Ch. IX and the equation (5.6), we get

THEOREM 5.5. *A contravariant pseudo-analytic vector field v^h satisfying $\nabla_i v^i = 0$ is a Killing vector.*

For a contravariant pseudo-analytic vector field v^h, we have

$$\begin{aligned}\Delta(v_h v^h) &= 2[v_h g^{ji}\nabla_j\nabla_i v^h + (\nabla_i v_h)(\nabla^i v^h)]\\ &= 2[-K_{ih}v^iv^h + (\nabla_i v_h)(\nabla^i v^h)].\end{aligned}$$

Thus, from Theorem 1.3 of Ch. IX, we obtain[2]

THEOREM 5.6. *If a compact pseudo-Kählerian space has a negative definite Ricci tensor, there does not exist a contravariant pseudo-analytic vector field other than the zero vector.*

If a vector v^h is contravariant pseudo-analytic, then $F_i^{\ h}v^i$ is also pseudo-analytic. Hence, if a contravariant pseudo-analytic vector v^h satisfies $F^{ih}\nabla_i v_h = 0$, then according to Theorem 5.5., $F_i^{\ h}v^i$ is a Killing vector. Thus

---

[1] ECKMANN [1, 2]; FRÖLICHER [1].

[2] BOCHNER [2]; YANO and BOCHNER [1].

THEOREM 5.7. *If a contravariant pseudo-analytic vector v^h satisfies $F^{ih}\nabla_i v_h = 0$, then $F_i{}^h v^i$ is a Killing vector.*

If a contravariant pseudo-analytic vector v^h satisfies $g^{ih}\nabla_i v_h = 0$ and $F^{ih}\nabla_i v_h = 0$, then according to Theorems 5.5 and 5.7, v^h and $F_i{}^h v^i$ are both Killing vectors. Hence

$$\nabla_i v_h + \nabla_h v_i = 0,$$

$$F_i{}^l \nabla_h v_l + F_h{}^l \nabla_i v_l = 0,$$

from which

$$- F_i{}^l \nabla_l v^h - F_l{}^h \nabla_i v^l = 0.$$

Comparing this equation with (5.4), we conclude $\nabla_i v^l = 0$. Hence

THEOREM 5.8. *If a contravariant pseudo-analytic vector field v^h satisfies $g^{ih}\nabla_i v_h = 0$ and $F^{ih}\nabla_i v_h = 0$, it is a covariant constant field.*

If a vector is at the same time covariant and contravariant pseudo-analytic, then the vector is harmonic and the equations (5.1) and (5.4) can respectively be written as

$$F_j{}^l \nabla_l v_i - F_i{}^l \nabla_l v_j = 0$$

$$F_j{}^l \nabla_l v_i + F_i{}^l \nabla_l v_j = 0,$$

from which $\nabla_l v_j = 0$. Thus

THEOREM 5.9. *If a vector is at the same time covariant and contravariant pseudo-analytic, then it is covariant constant.*

§ 6. Pseudo-Kählerian spaces of constant holomorphic curvature.

We call a sectional curvature

$$k = - \frac{K_{mjlh} F_k{}^m u^k u^j F_i{}^l u^i u^h}{g_{kj} u^k u^j g_{ih} u^i u^h} \tag{6.1}$$

determined by two orthogonal vectors u^h and $F_i{}^h u^i$ the *holomorphic sectional curvature* with respect to the vector u^h. If the holomorphic sectional curvature is always constant with respect to any vector at every point of the space, then we call the space a space of constant holomorphic curvature.[1]

---

[1] BOCHNER [3]; HAWLEY [1]; SCHOUTEN and VAN DANTZIG [2]; YANO and MOGI [1, 2].

Now, if this is the case, then (6.1) or

$$K_{mjlh}F_q^{\cdot m}F_p^{\cdot l}u^q u^j u^p u^h = -kg_{qj}g_{ph}u^q u^j u^p u^h$$

should be satisfied for any u^h, from which we obtain

$$K_{mjlh}F_q^{\cdot m}F_p^{\cdot l} + K_{mplh}F_j^{\cdot m}F_q^{\cdot l} + K_{mqlh}F_p^{\cdot m}F_j^{\cdot l}$$
$$= -k(g_{qj}g_{ph} + g_{jp}g_{qh} + g_{pq}g_{jh}),$$

by virtue of the symmetry of $K_{mjlh}F_q^{\cdot m}F_p^{\cdot l}$ with respect to q, j and to p, h. Multiplying the above equation by $F_k^{\cdot q}F_i^{\cdot p}$ and contracting, we find

$$K_{kjih} - K_{jikh} - K_{iqlh}F_k^{\cdot q}F_j^{\cdot l} = -k(F_{kj}F_{ih} - F_{ji}F_{kh} + g_{ki}g_{jh}).$$

Taking the antisymmetric part of this equation with respect to k and j and taking account of

$$K_{iqlh}F_k^{\cdot q}F_j^{\cdot l} - K_{iqlh}F_j^{\cdot q}F_k^{\cdot l}$$
$$= (K_{iqlh} - K_{ilqh})F_k^{\cdot q}F_j^{\cdot l}$$
$$= -K_{qlih}F_k^{\cdot q}F_j^{\cdot l} = -K_{kjih},$$

we obtain

$$2K_{kjih} - K_{jikh} + K_{kijh} + K_{kjih}$$
$$= -k[2F_{kj}F_{ih} + (g_{ki}g_{jh} - g_{ji}g_{kh}) - (F_{ji}F_{kh} - F_{ki}F_{jh})]$$

or

$$(6.2)\quad K_{kjih} = \frac{k}{4}[(g_{kh}g_{ji} - g_{jh}g_{ki}) + (F_{kh}F_{ji} - F_{jh}F_{ki}) - 2F_{kj}F_{ih}].$$

It is easily to be seen from the Bianchi identity that, if the curvature tensor has the form (6.2), the scalar curvature k is an absolute constant. Hence we have proved[1]

THEOREM 6.1. *If a pseudo-Kählerian space has a constant holomorphic sectional curvature at every point, then the curvature tensor of the space is of the form* (6.2), *where k is a constant.*

Using the formula (6.2), we can easily prove[1]

THEOREM 6.2. *In a pseudo-Kählerian space of constant holomorphic curvature, the general sectional curvature K determined by two orthogonal*

[1] YANO and MOGI [1, 2].

unit vectors u^h *and* v^h *is given by*

$$K = \frac{k}{4}(1 + 3a^2) \tag{6.3}$$

where $a \overset{\text{def}}{=} F_{ih}v^i u^h$ *is the cosine of the angle between two units vectors* $F_i^{\cdot h}v^i$ *and* u^h *and consequently* $a^2 \leq 1$. *Thus*

$$\frac{k}{4} \leq K \leq k \quad \text{for } k > 0, \tag{6.4}$$

$$k \leq K \leq \frac{k}{4} \quad \text{for } k < 0.$$

We now assume that, when there is given a holomorphic plane element, that is, a plane element determined by the vectors u^h and $F_i^{\cdot h}u^i$ at a point of the space, we can always draw a 2-dimensional totally geodesic surface passing through this point and being tangent to the given holomorphic plane element. If this is the case, we say that the space satisfies the *axiom of holomorphic planes*.

If we represent such a surface by the parametric equation

$$\xi^h = \xi^h(\eta^a) \qquad a, b, c, d = 1, 2, \tag{6.5}$$

then the fact that the surface is totally geodesic is represented by the equation

$$\partial_c B_b^h + B_{cb}^{ji}\left\{{h \atop ji}\right\} - B_a^h\,'\left\{{a \atop cb}\right\} = 0, \tag{6.6}$$

where $B_b^h = \partial_b \xi^h$ and where $'\left\{{a \atop cb}\right\}$ is the Christoffel symbol formed with the fundamental tensor $'g_{cb} \overset{\text{def}}{=} B_{cb}^{ji} g_{ji}$ of the surface.

The integrability conditions of (6.6) are

$$B_{dcb}^{kji} K_{kji}^{\cdot\cdot\cdot h} = B_a^h\,'K_{dcb}^{\cdot\cdot\cdot a}, \tag{6.7}$$

where $'K_{dcb}^{\cdot\cdot\cdot a}$ is the curvature tensor of the surface.

If we put

$$B_1^h = u^h, \; B_2^h = F_i^{\cdot h}u^i,$$

equation (6.7) must be satisfied by any unit vector u^h. Thus we must have

$$\begin{cases} F_s^{\cdot m}u^s u^j u^i K_{mji}^{\cdot\cdot\cdot h} = \alpha u^h + \beta F_p^{\cdot h}u^p, \\ F_s^{\cdot m}u^s u^j F_q^{\cdot l}u^q K_{mjl}^{\cdot\cdot\cdot h} = \lambda u^h + \mu F_p^{\cdot h}u^p. \end{cases} \tag{6.8}$$

From the first equation of (6.8), we obtain

$$(F_s^{\cdot m} K_{mji}^{\cdot\cdot\cdot h} - \alpha g_{sj} A_i^h - \beta g_{sj} F_i^{\cdot h}) u^s u^j u^i = 0,$$

from which

$$F_s^{\cdot m} K_{mji}^{\cdot\cdot\cdot h} + F_j^{\cdot m} K_{mis}^{\cdot\cdot\cdot h} + F_i^{\cdot m} K_{msj}^{\cdot\cdot\cdot h}$$
$$= \alpha(g_{sj} A_i^h + g_{ji} A_s^h + g_{is} A_j^h) + \beta(g_{sj} F_i^{\cdot h} + g_{ji} F_s^{\cdot h} + g_{is} F_j^{\cdot h}).$$

Transvecting this with $F_k^{\cdot s}$, we obtain

$$- K_{kji}^{\cdot\cdot\cdot h} + F_j^{\cdot m} F_k^{\cdot s} K_{mis}^{\cdot\cdot\cdot h} + K_{ikj}^{\cdot\cdot\cdot h}$$
$$= \alpha(F_{kj} A_i^h + g_{ji} F_k^{\cdot h} + F_{ki} A_j^h) + \beta(F_{kj} F_i^{\cdot h} - g_{ji} A_k^h + F_{ki} F_j^{\cdot h}),$$

from which, taking the alternating part with respect to k and j and using the relation

$$F_j^{\cdot m} F_k^{\cdot s} K_{mis}^{\cdot\cdot\cdot k} - F_k^{\cdot m} F_j^{\cdot s} K_{mis}^{\cdot\cdot\cdot h} = - K_{kji}^{\cdot\cdot\cdot h},$$

we find

$$- 4K_{kji}^{\cdot\cdot\cdot h} = \alpha(2F_{kj} A_i^h + g_{ji} F_k^{\cdot h} - g_{ki} F_j^{\cdot h} + F_{ki} A_j^h - F_{ji} A_k^h)$$
$$+ \beta(2F_{kj} F_i^{\cdot h} - g_{ji} A_k^h + g_{ki} A_j^h + F_{ki} F_j^{\cdot h} - F_{ji} F_k^{\cdot h}).$$

Contracting this equation with respect to h and i, we find $\alpha = 0$, and consequently we obtain

$$K_{kji}^{\cdot\cdot\cdot h} = \frac{\beta}{4} [(g_{ji} A_k^h - g_{ki} A_j^h) + (F_{ji} F_k^{\cdot h} - F_{ki} F_j^{\cdot h}) - 2F_{kj} F_i^{\cdot h})],$$

which shows that the space is of constant holomorphic curvature. Thus we have proved[1]

THEOREM 6.3. *If a pseudo-Kählerian space admits the axiom of holomorphic planes, then the space is of constant holomorphic curvature.*

If a pseudo-Kählerian space admits a group of motions which carry any two vectors u^h and $F_i^{\cdot h} u^i$ at a point P to any two vectors $'u^h$ and $'F_i^{\cdot h}{'u^i}$ at any point $'P$, then we say that the space admits a *holomorphic free mobility.*

If we denote by

$$'\xi^h = \xi^h + v^h(\xi) dt \tag{6.9}$$

---

[1] YANO and MOGI [1, 2].

an infinitesimal transformation of the group, then the fact that this is a motion is represented by

$$\underset{v}{\pounds} g_{ji} = \nabla_j v_i + \nabla_i v_j = 0, \tag{6.10}$$

and the fact that this carries a pair of vectors u^h and $F_i^{\cdot h}u^i$ into a pair of vectors $'u^h$ and $'F_i^{\cdot h}{}'u^i$ is represented by

$$\underset{v}{\pounds} F_i^{\cdot h} = - F_i^{\cdot l}\nabla_l v^h + F_l^{\cdot h}\nabla_i v^l = 0. \tag{6.11}$$

From (6.10) we get

$$\underset{v}{\pounds}\left\{ {h \atop ji} \right\} = \nabla_j \nabla_i v^h + K_{kji}^{\cdots h} v^k = 0, \tag{6.12}$$

and the integrability conditions of these differential equations are given by

$$\begin{aligned}\underset{v}{\pounds} K_{kji}^{\cdots h} = v^l \nabla_l K_{kji}^{\cdots h} - K_{kji}^{\cdots l}\nabla_l v^h + K_{lji}^{\cdots h}\nabla_k v^l \\ + K_{kli}^{\cdots h}\nabla_j v^l + K_{kjl}^{\cdots h}\nabla_i v^l = 0.\end{aligned} \tag{6.13}$$

Now, at a fixed point P of the space, we consider two arbitrary holomorphic plane elements, then by hypothesis there exists always a motion which fixes this point and carries one of these holomorphic plane elements into the other. Since the point P is arbitrary, the space must be of constant holomorphic curvature and consequently the curvature tensor of the space has the form

$$K_{kji}^{\cdots h} = \frac{k}{4}[(g_{ji}A_k^h - g_{ki}A_j^h) + (F_{ji}F_k^{\cdot h} - F_{ki}F_j^{\cdot h}) - 2F_{kj}F_i^{\cdot h}].$$

Conversely, if the curvature tensor of the space has the above form, then it is easily to be seen that the integrability condition $\underset{v}{\pounds} K_{kji}^{\cdots h} = 0$ is always satisfied by any v^h for which $\underset{v}{\pounds} g_{ji} = 0$ and $\underset{v}{\pounds} F_i^{\cdot h} = 0$, and that the differential equations (6.12) have solutions. But equation (6.12) is equivalent to

$$\nabla_k(\underset{v}{\pounds} g_{ji}) = 0,$$

and consequently, if the equation $\underset{v}{\pounds} g_{ji} = 0$ is satisfied by some initial values of v^h and $\nabla_i v^h$, then it is satisfied by any solutions belonging to them.

On the other hand, if v^h satisfies (6.12), then we have

$$\begin{aligned}\nabla_j(\underset{v}{\pounds} F_i{}^h) &= \nabla_j(- F_i{}^l \nabla_l v^h + F_l{}^h \nabla_i v^l)\\ &= - F_i{}^l \nabla_j \nabla_l v^h + F_l{}^h \nabla_j \nabla_i v^l\\ &= (F_i{}^l K_{kjl}{}^h - F_l{}^h K_{kji}{}^l) v^k\\ &= 0,\end{aligned}$$

and consequently, if the equation $\underset{v}{\pounds} F_i{}^h = 0$ is satisfied by some initial values of v^h and $\nabla_i v^h$, then it is satisfied by any solutions belonging to them. Thus the space admits the holomorphic free mobility, and we have [1]

THEOREM 6.4. *The necessary and sufficient condition that a pseudo-Kählerian space admit a holomorphic free mobility is that the space be of constant holomorphic curvature.*

---

[1] YANO and MOGI [1, 2].

BIBLIOGRAPHY

Berwald, L.

[1] Untersuchung der Krümmung allgemeiner metrischer Räume auf Grund des in ihnen herrschenden Parallelismus. Math. Zeitschr., **25** (1926), 40–73.

[2] Una forma normale invariantiva della seconda variazione. Rend. della R. Accad. dei Lincei, **5** (1928), 301–306.

[3] On the projective geometry of paths. Ann. of Math., **37** (1936), 879–898.

Bianchi, L.

[1] Lezioni sulla teoria dei gruppi continui finiti di trasformazioni. Spoerri Pisa (1918).

Birkhoff, G.

[1] Extensions of Lie groups. Math. Zeitschr., **53** (1950), 226–235.

Bochner, S.

[1] Remarks on the theorem of Green. Duke Math. Journal, **3** (1937), 334–338.

[2] Vector fields and Ricci curvature. Bull. of the Amer. Math. Soc., **52** (1946), 776–797.

[3] Curvature in Hermitian manifolds. Bull. of the Amer. Math. Soc., **53** (1947), 179–195.

[4] Vector fields on complex and real manifolds. Ann. of Math., **52** (1950), 642–649.

[5] Complex spaces with transitive cummutative groups of transformations. Proc. Nat. Acad. Sci. U.S.A., **37** (1951), 356–359.

[6] Tensor fields and Ricci curvature in Hermitian metric. Proc. Nat. Acad. Sci., U.S.A., **37** (1951), 704–706.

Borel, A. and A. Lichnerowicz

[1] Groupes d'holonomie des variétés riemanniens. C. R. Acad. Sci. Paris, **234** (1952), 1835–1837.

[2] Espaces riemanniens et hermitiens symétriques. C. R. Acad. Sci. Paris, **234** (1952), 2332–2334.

Calabi, E. and D. C. Spencer

[1] Completely integrable almost complex manifolds. Bull. of Amer. Math. Soc., **57** (1951), 254–255.

CARTAN, E.

[1] Sur une classe remarquable d'espaces de Riemann. Bull. Soc. Math. France, **54** (1926), 214–264; **55** (1927). 114–134.

[2] Sur les espaces de Riemann dans lesquels le transport par parallélisme conserve la courbure. Rend. della R. Accad. dei Lincei, (6) **3** (1926), 544–547.

[3] La géométrie des groupes de transformations. Journ. de Math., **6** (9), (1927), 1–119.

[4] La géométrie des groupes simples. Ann. di Mat., **4** (1927), 209–256.

[5] Sur l'écart géodesique et quelques notions connexes. Rend. della R. Accad. dei Lincei, (6a), **5** I (1927), 609–613.

[6] Leçons sur la géométrie des espaces de Riemann. Gauthier-Villars, Paris (1928).

[7] La théorie des groupes finis et continues et l'analysis situs. Mém. Sci. Math., **42**, Gauthier-Villars, Paris, (1930).

[8] Les espaces riemanniens symétriques. Verh. Int. Math. Kongr. Zürich I, (1932), 152–161.

[9] Les espaces métriques fondés sur la notion d'aire. Actualités Sci. et Ind., Hermann et Cie, Paris (1933).

[10] Les espaces de Finsler. Actualités Sci. et Ind., Hermann et Cie, Paris (1934).

[11] Leçons sur la géométrie des espaces de Riemann. Second edition. Gauthier-Villars, Paris, (1951).

CARTAN, E. and J. A. SCHOUTEN

[1] On the geometry of the group manifold of simple and semi-simple groups. Proc. Kon. Ned. Akad. Amsterdam, **29** (1926), 803–818.

CHERN, S. S.

[1] Characteristic classes of Hermitian manifolds. Annals of Math., **47** (1946), 85–121.

CHEVALLEY, C.

[1] Theory of Lie groups. I. Princeton Math. Ser., Princeton Univ. Press., (1946).

CLARK, R. S.

[1] Projective collineations in a space of K-spreads. Proc. Cambridge Phil. Soc., **41** (1945), 210–223.

COBURN, N.

[1] A characterization of Schouten's and Hayden's deformation methods. Journal of the London Math. Soc., **15** (1940), 123–136.

[2] Unitary spaces with corresponding geodesics. Bull. Amer. Math. Soc., **47** (1941), 901–910.

[3] Conformal unitary spaces. Trans. Amer. Math. Soc., **50** (1941), 26–39.

DANTZIG, D. VAN

[1] Theorie des projektiven Zusammenhangs n-dimensionaler Räume. Math. Ann., **106** (1932), 400—454.

[2] Zur allgemeinen projektiven Differentialgeometrie I, II. Proc. Kon. Akad. Amsterdam, **35** (1932), 524—534; 535—542.

[3] On the projective differential geometry III. Proc. Kon. Akad. Amsterdam, **37** (1934), 150—155.

[4] Electromagnetism, independent of metrical geometry I, II, III, IV. Proc. Kon. Akad. Amsterdam, **37** (1934), 521—525; 526—531; 644—652; 825—836.

[5] On the thermo-hydrodynamics of perfectly perfect fluids. I. II. Proc. Kon. Akad. Amsterdam, **43** (1940), 387—402, 609—618.

DAVIES, E. T.

[1] On the infinitesimal deformations of a space. Annali di Mat., (4) **12** (1933), 145—151.

[2] On the deformation of a subspace. Journal of the London Math. Soc., **11** (1936), 295—301.

[3] Analogous of the Frenet formulae determined by deformation operators. Journal of the London Math. Soc., **13** (1938), 210—216.

[4] On the deformation of the tangent m-plane of a V_n^m. Proc. Edinburgh Math. Soc., (2) **5** (1938), 202—206.

[5] Lie derivation in generalized metric spaces. Annali di Mat., **18** (1939), 261—274.

[6] The first and second variations of the volume integral in Riemannian space. Quart. Journal of Math., Oxford Ser., **13** (1942), 58—64.

[7] On the isomorphic transformations of a space of K-spreads. Journ. of the London Math. Soc., (2) **18** (1943), 100—107.

[8] Motions in a metric space based on the notion of area. Quart. Journal of Math., **16** (1945), 22—30.

[9] Subspaces of a Finsler space. Proc. of the London Math. Soc., (2) **49** (1945), 19—39.

[10] The geometry of a multiple integral. Journal of the London Math. Soc., **20** (1945), 163—170.

[11] On metric spaces based on a vector density. Proc. of the London Math. Soc., (2) **49** (1947), 241—259.

[12] The theory of surfaces in a geometry based on the notion of area. Proc. Cambridge Phil. Soc., **43** 1947), 307—313.

[13] On the second variation of a simple integral with movable end point. Journal of the London Math. Soc., **24** (1949), 241—247.

DIENES, P.

[1] Sur la déformation des espaces à connexion linéaire générale. C. R. Acda. Sci. Paris, **197** (1933), 1084—1086.

[2] Sur la déformation des sous-espaces dans un espace à connexion linéaire générale. C. R. Acad. Sci. Paris, **197** (1933), 1167–1169.

[3] On the deformation of tensor manifolds. Proc. of the London Math. Soc., **37** (1934), 512–519.

DIENES, P. and E. T. DAVIES

[1] On the infinitesimal deformations of tensor submanifolds. Journal de Math., **24** (1937), 111–150.

DOUGLAS, J.

[1] The general geometry of paths. Annals of Math., (2) **29** (1928), 143–168.

[2] Systems of K-dimensional manifolds in an N-dimensional space. Math. Ann., **105** (1931), 707–733.

DUSCHEK, A. and W. MAYER

[1] Zur geometrischen Variationsrechnung: Zweite Mitteilung: Über die zweite Variation des eindimensionalen Problems. Monatshefte für Math. und Physik, **40** (1933), 294–308.

ECKMANN, B.

[1] Sur les structures complexes et presques complexes. Géom. Diff. Coll. Inter. de C.N.S.R. Strasbourg (1953), 151–159.

[2] Structures complexes et transformations infinitésimales. Convegno di Geom. Diff. (1954), 1–9.

ECKMANN, B. and A. FRÖLICHER

[1] Sur l'integrabilité de structures presque complexes. C. R. Acad. Sci. Paris, **232** (1951), 2284–2286.

EGOROV, I. P.

[1] On the order of the group of motions of spaces with affine connexion. Doklady Akad. Nauk SSSR (N.S.), **57** (1947), 867–870 (Russian).

[2] On collineations in spaces with projective connexion. Doklady Akad. Nauk, SSSR (N.S.), **61** (1948), 605–608 (Russian).

[3] On the groups of motions of spaces with asymmetric affine connexion. Doklady Akad. Nauk, SSSR (N.S.), **64** (1949), 621–624 (Russian).

[4] On a strengthening of Fubini's theorem on the order of the group of motions of a Riemannian space. Doklady Akad. Nauk SSSR (N.S.), **66** (1949), 793–796 (Russian).

[5] On groups of motions of spaces with general asymmetrical affine connexion. Doklady Akad. Nauk SSSR (N.S.), **73** (1950), 265–267 (Russian).

[6] Collineations of projectively connected spaces. Doklady Akad. Nauk SSSR (N.S.), **80** (1951), 709–712. (Russian).

[7] A tensor characterisation of A_n of nonzero curvature with maximum mobility. Doklady Akad. Nauk SSSR (N.S.) **84** (1952), 209–212 (Russian).

[8] Maximally mobile L_n with a semi-symmetric connexion. Doklady Akad. Nauk SSSR (N.S.), **84** (1952), 433–435 (Russian).

[9] Motions in spaces with affine connexion. Doklady Acad. Nauk, SSSR (N.S.), **87** (1952), 693–696 (Russian).

EHRESMANN, C.

[1] Les connexions infinitésimales dans un espace fibré differentiable. Coll. de Topologie, Bruxelles (1950), 29–55.

[2] Sur les variétés presque complexes. Proc. Int. Congr. Math., II (1952), 412–419.

EISENHART, L. P.

[1] Linear connections of a space which are determined by simply transitive groups. Proc. Nat. Acad. Sci. U.S.A., **11** (1925), 246–250.

[2] Riemannian geometry. Princeton Univ. Press (1926).

[3] Non Riemannian geometry. Amer. Math. Soc. Coll. Publ. VIII (1927).

[4] Continuous groups of transformations. Princeton Univ. Press (1933).

[5] Groups of motions and Ricci directions. Annals of Math., **36** (1935), 826–832.

[6] Simply transitive groups of motions. Monatshefte für Math. und Physik, **43** (1936), 448–462.

[7] Riemannian geometry. Second printing. Princeton Univ. Press. (1949).

EISENHART, L. P. and M. S. KNEBELMAN

[1] Displacements in a geometry of paths which carry paths into paths. Proc. Nat. Acad. Sci. U.S.A., **13** (1927), 38–42.

FIALKOW, A.

[1] The conformal theory of curves. Trans. Amer. Math. Soc. **51** (1942), 435–501.

FICKEN, F. A.

[1] The Riemannian and affine differential geometry of product-spaces. Annals of Math., **40** (1939), 892–913.

FINSLER, P.

[1] Über Kurven und Flächen in allgemeinen Räumen. Dissertation, Göttingen (1918).

FRÖLICHER, A.

[1] Zur Differentialgeometrie der komplexen Strukturen. Math. Ann. **129** (1955), 50–95.

FUBINI, G.

[1] Sugli spazi che amettono un gruppo continuo di movimenti. Annali di Mat., (3) **8** (1903), 39–81.

Gołąb, S.

[1] Über die Klassifikation der geometrischen Objekte. Math. Zeitschr., **44** (1938), 104–114.

Goursat, E.

[1] Leçons sur le problème de Pfaff. Hermann et Cie, Paris, (1922).

Guggenheimer, H.

[1] Formes et vecteurs pseudo-analytiques. Annali di Mat. Pura Appl., (4) **36** (1954), 223–246.

Haantjes, J. and G. Laman

[1] On the definition of geometric objects. Indagationes Math., Proc. Kon. Ned Akad. Amsterdam, **15** (1953), 208–222.

Hawley, N. S.

[1] Constant holomorphic curvature. Canadian Journal of Math., **5** (1953), 53–56.

Hayden, H. A.

[1] Deformations of a curve in a Riemannian n-space, which displace certain vectors parallelly at each point. Proc. of the London Math., Soc., (2) **32** (1931), 321–336.

[2] Infinitesimal deformations of subspaces in a general metrical space. Proc. of the London Math. Soc., (2) **37** (1934), 410–440.

[3] Infinitesimal deformations of an L_m in an L_n. Proc. of the London Math. Soc., (2) **41** (1936), 332–336.

Hermann, R.

[1] Sur les isométries infinitésimales et le groupe d'holonomie d'un espace de Riemann. C. R. Acad. Sci. Paris, **239** (1954), 1178–1180.

Hiramatu, H.

[1] On affine collineations in a space of hyperplanes. Kumamoto Journal of Science. Series A, **1** (1952), 1–7.

[2] On projective collineations in a space of hyperplanes. Tensor **2** (1952), 1–14.

[3) Groups of homothetic transformations in a Finsler space. Tensor **3** (1954), 131–143.

[4] On some properties of groups of homothetic transformations in Riemannian and Finslerian spaces. Tensor, **4** (1954), 28–39.

Hlavatý, V.

[1] Sur la déformation infinitesimale d'une courbe dans la variété métrique avec torsion. Bull. Soc. Math. de France, **56** (1929), 18–25.

[2] Deformation theory of subspaces in a Riemann space. Proc. Amer. Math. Soc., **16** (1950), 600–617.

[3] Intrinsic deformation theory of subspaces in a Riemann space. Journal of Rational Mech. and Anal. **1** (1952), 49–72.

HODGE, W. V. D.

[1] The theory and applications of harmonic integrals. Cambridge Univ. Press. (1941).

[2] Structure problems for complex manifolds. Rend. Mat., Ser. V, **11** (1952), 101–110.

[3] The theory and applications of harmonic integrals. Second edition. Cambridge Univ. Press. (1953).

HOMBU, H. and M. MIKAMI

[1] Conics in the projectively connected manifolds. Mem. Fac. Sci. Kyūsyū Imp. Univ., **2** (1942), 217–239.

HOPF, H.

[1] Zur Topologie der komplexen Mannigfaltigkeiten. Studies and Essays presented to R. Courant. New York (1948), 167–185.

ISHIHARA, S.

[1] Homogeneous Riemann spaces of four dimensions. Journ. of the Math. Soc. Japan, **7** (1955), 345–369.

IYANAGA, S. and M. ABE

[1] Über das Helmholtzsche Raumproblem, I, II. Proc. Imp. Acad. Tokyo, **19** (1943), 174–180; 540–543.

KAGAN, B.

[1] Über eine Erweiterung des Begriffes vom projektiven Raume und dem zugehörigen Absolut. Abh. Sem. Vektor und Tensor. Moskau, **1** (1933), 12–101.

KÄHLER, E.

[1] Über eine bemerkenswerte Hermitische Metrik. Abh. Math. Sem. Hamburg Univ., **9** (1933), 173–186.

KALUZA, TH.

[1] Zum Unitätsproblem der Physik. S.-B. preuss. Akad. Wiss., (1921), 966–976.

KATSURADA, Y.

[1] Specialization of the theory of a space of higher order. II. On the extended Lie derivative. Tensor (N.S.), **2** (1952), 15–26.

KILLING, W.

[1] Über die Grundlagen der Geometrie. Journ. für der reine und angew. Math., **109** (1892), 121–186.

KLEIN, O.

[1] Quantentheorie und fünfdimensionale Relativitätstheorie. Zeitschr. für Physik, **37** (1926), 895–906.

KNEBELMAN, M. S.

[1] Groups of collineations in a space of paths. Proc. Nat. Acad. Sci. U.S.A., **13** (1927), 396–400.

[2] Motions and collineations in general space. Proc. Nat. Acad. Sci. U.S.A., **13** (1927), 607–611.

[3] Collineations of projectively related affine connections. Annals of Math., **29** (1928), 389–394.

[4] Collineations and motions in generalized spaces. Amer. Journ. of Math., **51** (1929), 527–564.

[5] On groups of motions in related spaces. Amer. Journ. of Math., **52** (1930), 280–282.

[6] Content-preserving transformations. Proc. Nat. Acad. Sci. U.S.A., **16** (1930), 156–159.

[7] On the equations of motion in a Riemannian space. Bull. Amer. Math. Soc., **51** (1945), 682–685.

KOBAYASHI, S.

[1] La connexion des variétés fibrées I, II. C. R. Acad. Sci. Paris, **238** (1954), 318–319; 443–444.

[2] Groupes de transformations qui laissent invariante une connexion infinitésimales. C. R. Acad. Sci. Paris, **238** (1954), 644–645.

[3] Le groupe des transformations qui laissent invariant le paraléllisme. Colloque de Topologie de Strasbourg. (1954).

KOSAMBI, D. D.

[1] Collineations in path-space. Journ. of the Ind. Math. Soc., N.S. **1** (1934), 69–72.

[2] Lie rings in paths space. Proc. Nat. Acad. Sci. U.S.A. **35** (1949), 389–394.

[3] Path-spaces admitting collineations. Quart. Journ. of Math. Oxford Ser. (2) **3** (1952), 1–11.

[4] Path geometry and continuous groups. Quart. Journ. of Math. Oxford Ser. (2) **3** (1952), 307–320.

KUIPER, N. H. and K. YANO

[1] On geometric objects and Lie groups of transformations. Indagationes Mathematicae, **17** (1955), 411–420.

KURITA, M.

[1] On the isometry of a homogeneous Riemann space. Tensor, **3** (1954), 91–100.

Laptev, B.

[1] La dérivée de Lie des objects géométriques qui dépendent de point et de direction. Bull. Soc. Phys. Math. Kazan, (3) **10** (1938), 3—39 (Russian).

[2] Une forme invariante de la variation et la dérivée de S. Lie. Bull. Soc. Phys.-Math. Kazan, (3) **12** (1940), 3—8 (Russian).

Laptev, G. F.

[1] Differential geometry of imbedded manifolds. Group-theoretical method of differential geometric investigations. Trudy Moskov Mat. Obšč. **2** (1953), 275—382 (Russian).

Ledger,

[1] Harmonic homogeneous spaces of Lie groups. Journ. of the London Math. Soc., **29** (1954), 345—347.

Lee, E. H.

[1] On even-dimensional skew-symmetric spaces and their groups of transformations. Amer. Journ. of Math., **67** (1945), 321—328.

Legrand, G.

[1] Connexions affines sur une variété presque hermitique. C. R. Acad. Sci. Paris, **237** (1953), 1626—1627.

Lelong-Ferrand, J.

[1] Formes différentielles définies sur une variété admettant un groupe continu d'isométries. C. R. Acad. Sci. Paris, **240** (1955), 268—269.

Levi-Civita, T.

[1] Sur l'écart géodésique. Math. Ann., **97** (1926), 291—320.

Levine, J.

[1] On a class of metric spaces admitting simply transitive groups of motions. Annali di Mat., **16** (1936), 49—59.

[2] Groups of motions in conformally flat spaces, I. Bull. Amer. Math. Soc., **42** (1937), 418—432.

[3] Groups of motions in conformally flat spaces. II, Bull. Amer. Math., Soc. **45** (1939), 766—773.

[4] Classifications of collineations in projectively and affinely connected spaces of two dimensions. Annals of Math., **52** (1950), 465—477.

[5] Collineations in Weyl spaces of two dimensions. Proc. Amer. Math. Soc., **2** (1951), 264—269.

[6] Collineations in generalized spaces. Proc. Amer. Math. Soc., **2** (1951), 447—455.

[7] Motions in linearly connected two-dimensional spaces. Proc. Amer. Math. Soc., **2** (1951), 932—941.

Lichnerowicz, A.

[1] Courbure, nombres de Betti et espaces symétriques. Proc. Intern. Congr. of Math., **2** (1950), 216—223.

[2] Généralisations de la géométrie kählériennes globale. Coll. de géom. diff. Louvain (1951), 99—122.

[3] Espaces homogènes kählériennes. Géom. Diff. Collogues Internationaux du C. N. R. S. Strasbourg (1953), 171—184.

[4] Sur les groupes d'automorphismes de certaines variétés kählériennes. C. R. Acad. Sci. Paris, **239** (1954), 1344—1346.

[5] Un théorème sur les espaces homogènes complexes. Archiv der Mathematik, **5** (1954), 207—215.

Lie, S. and F. Engel

[1] Theorie der Transformationsgruppen (1888).

Liebermann, P.

[1] Sur le problème d'équivalence de certaines structures infinitésimales. Thèse (1953).

McConnel, A. J.

[1] Strain and torsion in Riemannian space. Annali di Mat. (4) **6** (1928), 207—231.

[2] The variation of curvatures in the deformation of a curve in Riemannian space. Proc. Roy. Irish Acad., **39**A (1929), 1—9.

[3] Applications of the absolute differential calculus. Blackie and Son, London (1931).

Michal, A. D.

[1] Global groups of motions of some infinitely dimensional Riemannian spaces. Bull. Soc. Math. Grèce, **23** (1938), 143—151.

Montgomery, D. and H. Samelson

[1] Transformation groups of spheres. Annals of Math., **44** (1943), 454—470.

Mutō, Y.

[1] On the affinely connected space admitting a group of affine motions. Proc. Japan Acad., **26** (1950), 107—110.

[2] Some properties of a Riemannian space admitting a simply transitive group of translations. Tôhoku Math. Journ., **2** (1951), 205—213.

[3] On a curved affinely connected space admitting a group of affine motions of maximum order. Sci. Rep. of Yokohama Nat. Univ. Sec. I, **3** (1954), 1—12.

[4] On n-dimensional projectively flat spaces admitting a group of affine motions G_r of order $r > n^2 - n$. Sci. Rep. Yokohama Nat. Univ. Sec. I, **4** (1955), 3—18.

[5] On the curvature affinor of an affinely connected manifold A_n, $n \geq 7$ admitting a group of affine motions G_r of order $r > n^2 - 2n$. Tensor, **5** (1955), 39—53.

[6] On some properties of a kind of affinely connected manifolds admitting a group of affine motions. I, II. Tensor, (N.S.), **5** (1955), 127–142.

MYERS, S. B. and N. E. STEENROD

[1] The group of isometries of a Riemannian manifold. Annals of Math., **40** (1939), 400–416.

NAKAE, T.

[1] Sur un groupe de transformations d'éléments linéaires qui laissent $ds^2 = g_{ij}dx^i dx^j$ invariant. Mém. Coll. Sci. Kyoto Imp. Univ. Ser. A, **22** (1939), 455–458.

NEUMANN, J. v.

[1] Über die analytischen Eigenschaften von Gruppen linearer Transformationen und ihre Anwendungen. Math. Zeitschr. **30** (1924), 5–42.

NIJENHUIS, A.

[1] X_{n-1}-forming sets of eigenvectors. Proc. Kon. Ned. Akad. Amsterdam, 54 = Indagationes Math., **13** (1951), 200–212.

[2] Theory of the geometric objects. Thesis. Univ. of Amsterdam, (1952).

[3] A theorem on sequences of local affine collineations and isometries. Nieuw Archief voor Wiskunde (3) **2** (1954), 118–125.

NOMIZU, K.

[1] On the group of affine transformations of an affinely connected manifold. Proc. Amer. Math. Soc., **4** (1953), 816–823.

[2] Sur les transformations affines d'une variété riemannienne. C. R. Acad. Sci. Paris, **237** (1953), 1308–1310.

[3] Application de l'étude des transformations affines aux espaces homogènes riemanniens. C. R. Acad. Sci. Paris, **237** (1953), 1386–1387.

[4] Invariant affine connections on homogeneous spaces. Amer. Journ. of Math., **76** (1954), 33–65.

[5] Sur l'algèbre d'holonomie d'un espace homogène. C. R. Acad. Sci. Paris, **238** (1954), 319–321.

[6] Remarques sur les groupes d'holonomie et d'isométrie. Colloque de Topologie de Strasbourg (1954).

OBATA, M.

[1] On n-dimensional homogeneous spaces of Lie groups of dimension greater than $\frac{1}{2}n(n-1)$. Journ. of the Math. Soc. of Japan, **7** (1955), 371–388.

OTSUKI, T. and Y. TASHIRO

[1] On curves in Kählerian spaces. Math. Journal of Okayama Univ., **4** (1954), 57–78.

PALAIS,

[1] A definition of the exterior derivative in terms of Lie derivatives. Proc. Amer. Math. Soc., **56** (1954), 902–908.

PALATINI, A.

[1] Deduzioni invariantiva delle equazione gravitazionali dal principio di Hamilton. Rend. Circolo Mat. Palermo, **43** (1919), 203–212.

PATTERSON, E. M.

[1] A characterization of Kähler manifolds in terms of parallel fields of planes. Journ. of the London Math. Soc., **28** (1953), 260–269.

PICKERT, G.

[1] Elementare Behandlung des Helmholtzschen Raumproblems. Math. Ann., **102** (1949), 492–501.

RACHEVSKY, P.

[1] Caractères tensoriels de l'espace sous-projectifs. Abh. des Seminars für Vektor- und Tensoranalysis. Moskau **1** (1933), 126–140.

RENAUDIE, J.

[1] Un théorème sur les espaces harmoniques. C. R. Acad. Sci. Paris, **238** (1954), 199–201.

RHAM, G. DE and K. KODAIRA

[1] Harmonic Integrals. Inst. for Advanced Study. Princeton, New Jersey. (1950).

RICCI, G.

[1] Sur les groupes continus de mouvements d'une variété quelconque à trois dimensions. C. R. Acad. Sci. Paris, **127** (1898), 344–346.

[2] Sur les groupes de mouvements d'une variété quelconque. C. R. Acad. Sci. Paris, **127** (1898), 360–361.

[3] Sui gruppi continui di movimenti in una varietà qualunque a tre dimensioni. Mem. d. Soc. Ital. (III) **12** (1902), 61–92.

ROBERTSON, H. P.

[1] Groups of motions in spaces admitting absolute parallelism. Annals of Math., **33** (1932), 496–520.

RUSE, H. S.

[1] On simply harmonic spaces. Journ. of the London Math. Soc., **21** (1946), 243–247.

[2] On simply harmonic "kappa spaces" of four dimensions. Proc. of the London Math. Soc., **50** (1949), 317–329.

[3] Three dimensional spaces of recurrent curvature. Proc. of the London Math. Soc., **50** (1949), 438–446.

SASAKI, S.

[1] Geometry of the conformal connexion. Sci. Rep. Tôhoku Imp. Univ., **29** (1940), 219–267.

SCHMIDT, H.

[1] Winkeltreue und Streckentreue bei konformer Abbildung Riemannshcer Räume Math. Zeitschr. **5** (**1**1949), 700–701.

SCHOUTEN, J. A.

[1] Vorlesungen über die Theorie der halbeinfachen kontinuierlichen Gruppen. Univ. of Leiden, (1927).

[2] On infinitesimal deformations of V_m in V_n. Proc. Kon. Ned. Akad. Amsterdam, **31** (1927), 208–218.

[3] Über unitäre Geometrie. Proc. Kon. Ned. Akad. Amsterdam, **32** (1929), 457–465.

[4] Zur Geometrie der kontinuierlichen Transformationsgruppen. Math. Ann., **102** (1929), 244–272.

[5] Lie's differential operator. Math. Centrum. Amsterdam, Report Z.W. (1949).

[6] Tensor analysis for physicists. Clarendon Press, Oxford, (1951).

[7] Sur les tenseurs de V_n aux directions principales V_{n-1}-normales. Coll. de Géom. Diff. Louvain, (1951), 67–70.

[8] Ricci Calculus, second edition, Springer (1954).

SCHOUTEN, J. A. and D. VAN DANTZIG

[1] Über unitäre Geometrie. Math. Ann., **103** (1930), 319–346.

[2] Über unitäre Geometrie konstanter Krümmung. Proc. Kon. Ned. Akad. Amsterdam, **34** (1931), 1293–1314.

SCHOUTEN, J. A. and J. HAANTJES

[1] Zur allgemeinen projektiven Differentialgeometrie. Compositio Mathematica, **3** (1936), 1–51.

[2] On the theory of the geometric object. Proc. of the London Math. Soc., (2) **42** (1937), 356–376.

SCHOUTEN, J. A. and E. R. VAN KAMPEN

[1] Beiträge zur Theorie der Deformation. Prace Matematyczno-Fizycznych, Warszawa, **41** (1933), 1–19.

SCHOUTEN, J. A. and W. VAN DER KULK

[1] Pfaff's problem and its generalizations. Clarendon Press, Oxford, (1949).

SCHOUTEN, J. A. and D. J. STRUIK

[1] Einführung in die neueren Methoden der Differentialgeometrie I, (1935).

[2] Einführung in die neueren Methoden der Differentialgeometrie II, (1938).

SCHOUTEN, J. A. and K. YANO

[1] On an intrinsic connexion in an X_{2n} with an almost Hermitian structure. Indagationes Mathematicae, **17** (1955), 1–9.

[2] On the geometric meaning of the vanishing of the Nijenhuis tensor in an X_{2n} with an almost complex structure. Indagationes Mathematicae, **17** (1955), 132–138.

[3] On invariant subspaces in the almost complex X_{2n}. Indagationes Mathematicae, **17** (1955), 261–269.

SHABBAR, M.

[1] One-parameter groups of deformations in Riemannian spaces. Journ. of the Indian Math. Soc., **6** (1942), 186–191.

SHANKS, E. B.

[1] Homothetic correspondence between Riemannian spaces. Duke Math. Journ., **17** (1950), 299–311.

SLEBODZINSKI, W.

[1] Sur les équations de Hamilton. Bull. Acad. Roy. de Belg. (5) **17** (1931), 864–870.

[2] Sur les transformations isomorphiques d'une variété à connexion affine. Prace Mat. Fiz. Warszawa, **39** (1932), 55–62.

SOÓS, GY.

[1] Über Gruppen von Affinitäten und Bewegungen in Finslerschen Räumen. Acta Mathematica Academiae Scientiarum Hungaricae, **5** (1954), 73–83.

SU, BUCHIN

[1] Descriptive collineations in spaces of K-spreads. Trans. Amer. Math. Soc., **61** (1947), 495–507.

[2] On the isomorphic transformations of K-spreads in a Douglas space. I, II. Science Records, Academia Sinica, **2** (1947), 11–19; **2** (1948), 139–146.

[3] A characteristic property of affine collineations in a space of K-spreads. Bull. Amer. Math. Soc., **54** (1948), 136–138.

[4] Geodesic deviation in generalized metric spaces. Science Records, Academia Sinica, **2** (1949), 220–226.

[5] Lie derivation in the geometry of conformal connexions. Science Records. Academia Sinica, **2** (1949), 331–339.

[6] A generalization of descriptive collineations in a space of K-spreads. Journ. of the London Math. Soc., **25** (1950), 236–238.

[7] Integrability conditions in a descriptive geometry of K-spreads. Revista Ci. Lima, **52** (1950), 49–58.

[8] Extremal deviation in a geometry based on the notion of area. Acta Math., **85** (1951), 99–116.

SUN, J. T.

[1] A note on the isometric correspondence of Riemannian spaces. Duke Math. Journ., **16** (1949), 571–573.

SYNGE, J. L.

[1] A generalization of the Riemannian line element. Trans. Amer. Math. Soc., **27** (1925), 61–67.

[2] The first and second variations of the length integral. Proc. of the London Math. Soc., **25** (1928), 247–264.

TAKANO, K.

[1] Homothetic transformations in Finsler spaces. Rep. Univ. of Electro- Commun., **4** (1952), 61–69.

TASHIRO, Y.

[1] Sur la dérivée de Lie de l'être géométrique et son groupe d'invariance. Tôhoku Math. Journ., **2** (1950), 166–181.

[2] Note sur la dérivée de Lie d'un être géométrique. Math. Journal of Okayama Univ., **1** (1952), 125–128.

TAUB, A. H.

[1] A characterisation of conformally flat spaces. Bull. Amer. Math. Soc., **55** (1949), 85–89.

TAYLOR, J. H.

[1] A generalization of Levi-Civita's parallelism and the Frenet-formulas. Trans. Amer. Math. Soc., **27** (1925), 246–264.

TELEMAN, C.

[1] Les groupes transitifs de mouvements des espaces de Riemann V_5. Studii şi Cercetări Matematice. **4** (1953), 503–526.

THOMAS, T. Y.

[1] Reducible Riemann spaces and their characterization. Mon. f. Math. u. Phys., **48** (1939), 228–292.

[2] The decomposition of Riemann spaces in the large. Mon. f. Math. u. Phys., **48** (1939), 388–418.

[3] The differential invariants of generalized spaces. Cambr. Univ. Press (1934).

ULANOVSKII, M. A.

[1] On stationary groups of motions of spaces with linear projective and affine connection. Doklady Akad. Nauk SSSR (N.S.), **71** (1950), 629–631. (Russian).

VEBLEN, O.

[1] Invariants of quadratic differential forms. Cambridge Tracts, No. 24 (1927).

[2] Generalized projective geometry. Journ. of the London Math. Soc., **4** (1929), 140–160.

[3] Projektive Relativitätstheorie. Erg. d. Math., (1933).

VRANCEANU, G.

[1] Leçons de géométrie différentielle, vol. 1. Bucarest (1947).

[2] Sur les espaces à connexion à groupe maximum des transformations en eux-mêmes. C. R. Acad. Sci. Paris, **229** (1949), 543–545.

[3] On spaces with non-Euclidean affine connexion with a maximal group of transformations into itself. Acad. Repub. Pop. Române. Bul. Sti. A. **1** (1949), 813–821. (Roumanian. Russian and French summaries).

[4] Groupes de mouvements des espaces à connexion. Studii şi Cercetări Matematice, **2** (1951), 387–444.

[5] Lectii de geometrie Diferentială. vol. 2. Bucarest (1951).

[6] Sur les groupes de mouvements d'un espace de Riemann à quatre dimensions. Studii şi Cercetări Matematice, **4** (1953), 121–153 (Roumanian).

VRIES, H. L. DE

[1] Über Riemannsche Räume, die infinitesimale konforme Transformationen gestatten. Math. Zeitschr., **60** (1954), 235–242.

WAGNER, V.

[1] The theory of geometric objects and the theory of finite and infinite continuous group of transformations. C. R. Acad. Sci. URSS. **46** (1945), 347–349 (Russian).

[2] Classification of simple differential-geometric objects. Doklady Acad. Nauk SSSR **69** (1949), 293–296 (Russian).

[3] On the theory of pseudogroups of transformations. Doklady Acad. Nauk SSSR **72** (1950), 453–456 (Russian).

WALKER, A. G.

[1] On Ruse's spaces of recurrent curvature. Proc. of the London Math. Soc., **52** (1950), 36–64.

WANG, H. C.

[1] On Finsler spaces with completely integrable equations of Killing. Journ. of the London Math. Soc., **22** (1947), 5–9.

WANG, H. C. and K. YANO

[1] A class of affinely connected spaces. Trans. Amer. Math. Soc., **80** (1955), 72–96.

WESTLAKE, W. J.

[1] Hermitian spaces in geodesic correspondence. Proc. Amer. Math. Soc., **5** (1954), 301–306.

[2] Conformally Kähler manifolds. Proc. Cambridge Philos. Soc., **50** (1954), 16–19.

WEYL, H.

[1] Zur Infinitesimalgeometrie: Einordnung der projektiven und konformen Auffassung. Göttinger Nachrichten (1921), 99–112.

[2] Theorie der Darstellung kontinuierlicher halb-einfacher Gruppen durch lineare Transformationen, I. Math. Zeitschr. **23** (1925), 271–309.

[3] Gruppentheorie und Quantenmechanik, (1929).

WEYL, H. and H. P. ROBERTSON

[1] On a problem in the theory of groups arising in the foundations of infinitesimal geometry. Bull. Amer. Math. Soc., **35** (1929), 686–690.

WHITEHEAD, J. H. C.

[1] The representation of projective spaces. Annals of Math., **32** (1931), 327–360.

YANO, K.

[1] Sur les circonférences généralisées dans les espaces à connexion conforme. Proc. Imp. Acad. Tokyo, **14** (1938), 329–332.

[2] Sur l'espace projectif de M. D. van Dantzig. C. R. Acad. Sci. Paris, **206** (1938), 1610–1612.

[3] La relativité non holonome et la théorie unitaire d'Einstein et Mayer. Mathematica, **14** (1938), 124–132.

[4] Concircular geometry I, II, III, IV, V. Proc. Imp. Acad. Tokyo, **16** (1940), 195–200, 354–360, 442–448, 505–511, **18** (1942), 446–451.

[5] Les espaces d'éléments linéaires à connexion projective normale et la géométrie projective générale des paths. Proc. Phys.-Math. Soc. Japan, **24** (1942), 7–24.

[6] Sur le parallélisme et la concourance dans l'espace de Riemann. Proc. Imp. Acad. Tokyo, **19** (1943), 189–197.

[7] Lie derivatives in general space of paths. Proc. Japan Acad., **21** (1945), 363–371.

[8] Bemerkungen über infinitesimale Deformationen eines Raumes. Proc. Japan Acad., **21** (1945), 171–178.

[9] Sur la déformation infinitésimale des sous-espaces dans un espace affine. Proc. Japan Acad., **21** (1945), 248–260.

[10] Sur la déformation infinitésimale tangentielle d'un sous-espace. Proc. Japan Acad., **21** (1945), 261–268.

[11] Quelques remarques sur un article de M. N. Coburn "A characterization of Schouten's and Hayden's deformation methods". Proc. Japan Acad., **21** (1945), 330–336.

[12] Quelques remarques sur les groupes de transformations dans les espaces à connexion linéaire. I, III, VI. Proc. Japan Acad., **22** (1946), 41–47, 167–172, **23** (1947), 143–146.

[13] Groups of transformations in generalized spaces. Akademeia Press, Tokyo, (1949).

[14] Sur la théorie des déformations infinitésimales. Journ. of Fac. of Sci., Univ. Tokyo, **6** (1949), 1—75.

[15] On groups of homothetic transformations in Riemannian spaces. Journ. of the Ind. Math. Soc., **15** (1951), 105—117.

[16] On harmonic and Killing vector fields. Annals of Math., **55** (1952), 38—45.

[17] Some remarks on tensor fields and curvature. Annals of Math., **55** (1952), 328—347.

[18] On Killing vector fields in a Kählerian space. Journ. of the Math. Soc. of Japan, **5** (1953), 6—12.

[19] On n-dimensional Riemannian spaces admitting a group of motions of order $\frac{1}{2}n(n-1)+1$. Trans. Amer. Math. Soc., **74** (1953), 260—279.

[20] Sur la correspondence projective entre deux espaces pseudo-hermitiens. C. R. Acad. Sci. Paris, **239** (1954), 1346—1348.

[21] Geometria conforme in varietà quasi hermitiane. Rend. Accad. Lincei. **16** (1954), 449—454.

[22] Quelques remarques sur les variétés à structure presque complexe Bull.. Soc. Math. France. **83** (1955), 57—80.

[23] On three remarkable affine connexions in almost Hermitian spaces. Indagationes Mathematicae, **17** (1955), 24—32.

YANO, K. and T. ADATI

[1] On certain spaces admitting concircular transformations. Proc. Japan Acad., **25** (1949), 188—195.

YANO, K. and S. BOCHNER

[1] Curvature and Betti numbers. Annals of Math. Studies, **32** (1953).

YANO, K. and H. HIRAMATU

[1] Affine and projective geometries of systems of hypersurfaces. Journal of the Math. Soc. of Japan, **3** (1951), 116—136.

[2] On projective geometry of K-spreads. Compositio Mathematica, **10** (1952), 286—296.

[3] On groups of projective collineations in a space of K-spreads. Journal of the Math. Soc. of Japan, **6** (1954), 131—150.

YANO, K. and T. IMAI

[1] Remarks on affine and projective collineations. Journal of the Math. Soc. of Japan, **1** (1950), 287—288.

YANO, K. and I. MOGI

[1] Sur les variétés pseudo-kählériennes à courbure holomorphique constante. C. R. Acad. Sci. Paris, **237** (1953), 962—964.

[2] On real representations of Kählerian manifolds. Annals of Math., **61** (1955), 170–189.

YANO, K. and K. TAKANO

[1] Sur les coniques dans les espaces à connexion affine ou projective, I, II. Proc. Imp. Acad. Tokyo, **20** (1944), 410–417; 418–424.

[2] Quelques remarques sur les groupes de transformations dans les espaces à connexion linéaire. II. Proc. Japan Acad., **22** (1946), 69–74.

[3] Conics in D. van Dantzig's projective space. Proc. Japan Acad., **21** (1949), 179–187.

YANO, K., K. TAKANO and Y. TOMONAGA

[1] On infinitesimal deformations of curves in spaces with linear connexion. Jap. Journ. of Math., **19** (1948), 433–477.

YANO, K. and Y. TASHIRO

[1] Some theorems on geometric objects and their applications. Nieuw Arch. Wiskunde, (3) **2** (1954), 134–142.

YANO, K. and Y. TOMONAGA

[1] Quelques remarques sur les groupes de transformations dans les espaces à connexion linéaire, IV, V. Proc. Japan Acad., **22** (1946), 173–183; 275–283.

APPENDIX [1]

§ 1. Groups of motions.

In § 10 of Chapter IV, we have studied an n-dimensional Riemannian manifold V_n which admits a group G_r of motions of the order $r = \frac{1}{2}n(n-1)+1$. But, the cases $n = 3$, $n = 4$ and $n = 8$ were exceptional cases in our results.

The V_3 with G_4 was studied by E. Cartan [11] and G. I. Kručkovič [1].

The case $n = 4$ was studied by S. Ishihara [1] as was mentioned in the text.

The case $n = 4$ was also studied by I. P. Egorov [10]. He proved the following two theorems:

THEOREM 1.1. *There exist two and only two different Riemannian spaces of four dimensions which are maximally mobile and are of non constant curvature and for which the line element is defined by the formula*

$$(1.1) \qquad ds^2 = \frac{(\varepsilon + \Sigma y^{i^2}) \Sigma dy^{i^2} - (\Sigma y^i dy^i)^2 - (y^1 dy^2 - y^2 dy^1 + y^3 dy^4 - y^4 dy^3)^2}{(\varepsilon + \Sigma y^{i^2})^2}$$

$$(\varepsilon = \pm 1)$$

The group G_8 of motions is compact for $\varepsilon = +1$ and non-compact for $\varepsilon = -1$.

If we put

$$z^1 = y^1 + iy^2, \quad z^2 = y^3 + iy^4,$$

then we have

$$(1.2) \qquad ds^2 = e^{-2v}[\varepsilon(dz^1 d\bar{z}^1 + dz^2 d\bar{z}^2) + (z^2 dz^1 - z^1 dz^2)(\bar{z}^2 d\bar{z}^1 - \bar{z}^1 d\bar{z}^2)]$$

where

$$e^v = z^1\bar{z}^1 + z^2\bar{z}^2 + \varepsilon.$$

THEOREM 1.2. *Maximally mobile V_4 's of non constant curvature are real representations of the space of point-line couples of a complex projective plane in which the points and lines are harmonic with respect to the Hermitian quadric*

$$z^1\bar{z}^1 + z^2\bar{z}^2 + \varepsilon = 0.$$

---

[1] Added Oct. 31st 1956.

The case $n = 4$ was also studied by G. Vranceanu [6], and the case $n = 5$ by C. Teleman [1].

C. Teleman [2] has proved

THEOREM 1.3. *Every subgroup of the complete group of rotations in n variables is a motion group of a space V_{n-1} of constant positive curvature which keeps two or more complementary systems of Pfaff invariant.*

THEOREM 1.4. *A group G_r of rotations in n variables, real and irreducible, with $r < \frac{1}{2}n(n-1)$ parameters, has at most p^2 parameters if $n = 2p$ or $n = 2p + 1$. Hence the space $V_{2p}(\lambda)$ of Vranceanu are those irreducible Riemannian V_{2p} 's of variable curvature which has a motion group of the maximum number of parameters $p^2 + 2p$.*

A $V_{2p}(\lambda)$ of Vranceanu is defined as a Riemannian V_{2p} with a transitive group of motions and the group

$$X_{2s-1,2r},\ X_{2s-1,2r-1} + X_{2s,2r},\ X_{2s-1,2r} - X_{2s,2r-1}\ (X_{ji} = x_j\partial_i f - x_i\partial_j f)$$

as group of stability.

THEOREM 1.5. *Such a $V_{2p}(\lambda)$ can be realized as a non homogeneous manifold on the spheres S_{2p+1} in a euclidean E_{2p+2}.*

THEOREM 1.6. *The stability group G_{p^2} of these V_{2p} admits a particular transformation $X_{12} + X_{34} + \ldots + X_{2p-1,2p}$ and is the largest orthogonal group of this property. The V_{2p} has thus the maximum group of motions among the spaces of which the group of stability has this particular transformation.*

THEOREM 1.7. *The group G_{p^2} is closed and is composed of this particular transformation and a simple G_{p^2-1}. Hence the group G_{2p+p^2} is simple and is closed.*

THEOREM 1.8. *The $V_{2p}(\lambda)$ has the metric*

$$(1.3) \qquad ds^2 = \frac{(d\xi^1)^2 + \ldots + (d\xi^n)^2}{u} - k\frac{(\Sigma\, \xi^h d\xi^h)^2 + w^2}{u^2},$$

$$u = 1 + k[(\xi^1)^2 + \ldots + (\xi^n)^2]$$

$$w = \xi^1 d\xi^2 - \xi^2 d\xi^1 + \ldots + \xi^{2p-1}d\xi^{2p} - \xi^{2p}d\xi^{2p-1},$$

$$(k = \text{constant}).$$

As it was mentioned in the text, I. P. Egorov proved that a Riemannian

V_n which is not an Einstein space has a maximum group of motions of $\frac{1}{2}n(n-1)+1$ parameters and that this maximum is reached.

G. Vranceanu [7] proved

THEOREM 1.9. *If the* V_n *is not conformally Euclidean this maximum is* $\frac{1}{2}(n-1)(n-2)+3$ *and this maximum is reached. Moreover if the Einstein* V_n *is not of constant curvature, the maximum is* $\frac{1}{2}(n-1)(n-2)+5$, *reached for* $n=4$ *and* $n=6$ *but not reached for* $n \geq 7$.

M. Obata [1] obtained the following theorem in which the case $n=8$ is not exceptional.

THEOREM 1.10. *If an n-dimensional connected Riemannian manifold* M *for* $n \geq 3$, $n \neq 4$ *admits a group* G_r *of motions of the order* r, $\frac{1}{2}n(n-1) < r < \frac{1}{2}n(n+1)$, *then* G_r *is of the order* $\frac{1}{2}n(n-1)+1$ *and* M *is one of the followings*:

as a Riemannian manifold	*as a topological space*
$S_1^0 \times S_{n-1}^+$	$E_1 \times S_{n-1}$, if it is simply connected,
$S_1^0 \times S_{n-1}^-$	E_n or $S_1 \times E_{n-1}$
S_n^0	E_n or $S_1 \times E_{n-1}$
S_n^-	E_n

where

S_n^+: *n-dimensional Riemannian manifold of positive constant curvature,*
S_n^-: *n-dimensional Riemannian manifold of negative constant curvature,*
S_n^0: *n-dimensional locally flat Riemannian manifold,*
E_n: *n-dimensional Euclidean space,*
S_n: *n-dimensional sphere.*

H. Wakakuwa [1] studied a similar problem.

In a Riemannian V_n with constant rotation coefficients, there exists a real simply transitive group of motions G_n and conversely. G. Vranceanu [10] proved the following

THEOREM 1.11. *A necessary condition that the Ricci tensor be positive definite is that* G_n *coincides with its derived group* G_n'.

Let M be a Riemannian manifold and let $A(M)$ and $I(M)$ the group of all affine motions and the group of all isometries of M onto itself respectively. We denote by $A_0(M)$ and $I_0(M)$ the connected components of the identity in $A(M)$ and $I(M)$ respectively.

In § 5 of Chapter IX, we have proved the

THEOREM 1.12. *In a compact Riemannian manifold M an infinitesimal affine motion is an isometry. Therefore $A_0(M)$ coincides with $I_0(M)$.*

There appeared recently several generalizations of this theorem. S. Kobayashi [4] proved the

THEOREM 1.13. *If M is an irreducible and complete Riemannian manifold, then $A(M)$ is equal to $I(M)$, except the case M is the 1-dimensional Euclidean space.*

J. Hano [1] proved the following two theorems:

THEOREM 1.14. *Let M be a simply connected complete Riemannian manifold and $M = M_0 \times M_1 \times \ldots \times M_r$ be the de Rham decomposition of M. Then the group $A_0(M)$ is isomorphic to the direct product $A_0(M_0) \times A_0(M_1) \times \ldots \times A_0(M_r)$ and the group $I_0(M)$ is isomorphic to the direct product $I_0(M_0) \times I_0(M_1) \times \ldots \times I_0(M_r)$.*

THEOREM 1.15. *Let M be a complete Riemannian manifold. If the length of an infinitesimal affine motion $v^\varkappa$ is bounded on M, then $v^\varkappa$ is a Killing vector field.*

S. Ishihara and M. Obata [3] also obtained theorems similar to the above three theorems.

§ 2. Groups of affine motions.

An n-dimensional manifold with a linear connexion is said to have the property A (or A'), if it is possible to find an affine motion φ satisfying the following conditions:

a) φ leaves some point P of the manifold fixed.

b) The tangent space $T(P)$ at P has a base $\{X_1, X_2, \ldots, X_n\}$ such that for some real numbers ρ_i

$$\varphi X_i = \rho_i X_i \ (1 \leq i \leq n) \text{ (not summed)}$$

c) $$\rho_i \rho_j \rho_k \rho_l^{-1} \neq 1 \text{ if } j \neq k \ \ (1 \leq i, j, k, l \leq n)$$

or

c') $$\rho_i \rho_j \rho_k^{-1} \neq 1 \text{ if } i \neq j \ \ (1 \leq i, j, k, l \leq n).$$

S. Ishihara and M. Obata [1] proved the following

THEOREM 2.1. *Let M be a manifold with a linear connexion admitting a transitive group of affine motions.*
1) *If M has the property A, the curvature tensor vanishes identically.*
2) *If M has the property A', the torsion tensor vanishes identically.*
3) *If M admits a group of affine motions of the order greater than n^2, then M has the property A and A' and the group is transitive, so that M is locally Euclidean.*

In the text, we studied an A_n which is not locally Euclidean and admits a group of affine motions of the maximum order n^2. G. Vranceanu [12] studied the global properties of such spaces. D. Dumitrus [1] and S. Petrescu [1] studied such A_3 's and A_4 's in a great detail.

Y. Mutō [7] studied n-dimensional projectively Euclidean spaces D_n which admit a group G_r of affine motions of order $r = n^2 - n + 1$ and he obtained the following theorems:

THEOREM 2.2. *A necessary and sufficient condition that a projectively Euclidean space D_n $(n \geq 3)$ with asymmetric Ricci tensor admit a group G_r of affine motions of order $r = n^2 - n + 1$ is that the connexion parameters $\Gamma^{\varkappa}_{\mu\lambda}$* satisfy

$$\Gamma^{\varkappa}_{\mu\lambda} = - p_\mu A^{\varkappa}_{\lambda} - p_\lambda A^{\varkappa}_{\mu},$$

$$p_1 = ab\xi^2,\ p_2 = - ab\xi^1,\ p_3 = p_4 = \dots = p_n = 0$$

with

$$ab^2 = - 1,\ a = \pm 1$$

in a suitable coordinate system. If the space is real, we can put $a = - 1$, $b = - 1$.

THEOREM 2.3. *A necessary and sufficient condition that a projectively Euclidean space D_n $(n \geq 3)$ with symmetric Ricci tensor which is non positive admit a complete group G_r of affine motions of order $r = n^2 - n + 1$ is that the connexion parameters $\Gamma^{\varkappa}_{\mu\lambda}$ satisfy*

$$\Gamma^{\varkappa}_{\mu\lambda} = - (\nabla_\mu p) A^{\varkappa}_{\lambda} - (\nabla_\lambda p) A^{\varkappa}_{\mu}, \quad p = \tfrac{1}{2} \log (\xi^1\xi^2 - 1)$$

in a suitable coordinate system. If the space is real, we should have

$$\xi^1\xi^2 - 1 > 0.$$

THEOREM 2.4. *A necessary and sufficient condition that a projectively Euclidean D_n $(n \geq 3)$ with symmetric Ricci tensor which is non negative admit a complete group G_r of affine motions of order $r = n^2 - n + 1$ is* *that the connexion parameters* $\Gamma^{\varkappa}_{\mu\lambda}$ *satisfy*

$$\Gamma^{\varkappa}_{\mu\lambda} = -(\nabla_\mu p)A^{\varkappa}_{\lambda} - (\nabla_\lambda p)A^{\varkappa}_{\mu}, \quad p = \tfrac{1}{2}\log\{1 + (\xi^1)^2 + (\xi^2)^2\}$$

in a suitable coordinate system.

THEOREM 2.5. *A necessary and sufficient condition that a projectively Euclidean* D_n $(n \geq 3)$ *with symmetric Ricci tensor which is indefinite admit a complete group* G_r *of affine motions of order* $r = n^2 - n + 1$ *is that the connexion parameters* $\Gamma^{\varkappa}_{\mu\lambda}$ *satisfy*

$$\Gamma^{\varkappa}_{\mu\lambda} = -(\nabla_\mu p)A^{\varkappa}_{\lambda} - (\nabla_\lambda p)A^{\varkappa}_{\mu}, \quad p = \tfrac{1}{2}\log(1 - \xi^1\xi^2)$$

in a suitable coordinate system. If the space is real, we should have

$$1 - \xi^1\xi^2 > 0.$$

THEOREM 2.6. *Consider a space* A_n $(n \geq 5)$ *with a symmetric linear connexion or a projectively Euclidean* D_n $(n \geq 3)$, *a necessary and sufficient condition that the space admit a complete group* G_r *of affine motions of order* $r = n^2 - n + 1$ *is that the space be one of the spaces mentioned in Theorems* 2.2, 2.3, 2.4, 2.5.

Y. Mutō [6,9] studied also n-dimensional spaces A_n with symmetric linear connexion admitting a group G_r of affine motions of order $r > n^2 - 2n$ and obtained the following interesting theorems.

THEOREM 2.7. *If an* A_n $(n \geq 7)$ *admits a group* G_r *of affine motions of order* $r > n^2 - 2n$, *then its curvature tensor* $R_{\nu\mu\lambda}{}^{\varkappa}$ *is of the form*

$$R_{\nu\mu\lambda}{}^{\varkappa} = B_{\nu\mu\lambda}A^{\varkappa} - A^{\varkappa}_{\nu}U_{\mu\lambda} + A^{\varkappa}_{\mu}U_{\nu\lambda} + (U_{\nu\mu} - U_{\mu\nu})A^{\varkappa}_{\lambda} \tag{2.1}$$

or

$$R_{\nu\mu\lambda}{}^{\varkappa} = (V_\nu U_\mu - V_\mu U_\nu)(U_\lambda A^{\varkappa} + V_\lambda C^{\varkappa}) - A^{\varkappa}_{\nu}U_{\mu\lambda} + A^{\varkappa}_{\mu}U_{\nu\lambda} + (U_{\nu\mu} - U_{\mu\nu})A^{\varkappa}_{\lambda}. \tag{2.2}$$

The vectors $A^{\varkappa}$ *and* $C^{\varkappa}$ *and the vectors* U_λ *and* V_λ *in* (2.2) *are linearly independent respectively.*

THEOREM 2.8. *If an* A_n $(n \geq 8)$ *admits a group* G_r *of affine motions of order* $r > n^2 - 2n$, *then its curvature tensor is of the form* (2.1).

THEOREM 2.9. *If an* A_n $(n \geq 7)$ *has the curvature tensor of the form* (2.2) *then the order of the group of affine motions admitted satisfies*

$$r \leq n^2 - 3n + 8.$$

THEOREM 2.10. *A necessary and sufficient condition that an A_n ($n \geq 7$) with the curvature tensor of the form* (2.1) *admit a group G_r of affine motions of order $r > n^2 - 2n$ is that the curvature tensor be of the form*

$$R_{\nu\mu\lambda}^{\cdot\cdot\cdot\varkappa} = (V_\nu U_\mu - V_\mu U_\nu) U_\lambda A^\varkappa - A_\nu^\varkappa U_{\mu\lambda} + A_\mu^\varkappa U_{\nu\lambda} + (U_{\nu\mu} - U_{\mu\nu}) A_\lambda^\varkappa$$

or

$$R_{\nu\mu\lambda}^{\cdot\cdot\cdot\varkappa} = [(Q_\nu P_\mu - Q_\mu P_\nu) P_\lambda + (P_\nu R_\mu - P_\mu R_\nu) Q_\lambda + 2(R_\nu Q_\mu - R_\mu Q_\nu) R_\lambda - (Q_\nu P_\mu - Q_\mu P_\nu) R_\lambda] A^\varkappa - A_\nu^\varkappa P_{\mu\lambda} + A_\mu^\varkappa P_{\nu\lambda} + (P_{\nu\mu} - P_{\mu\nu}) A_\lambda^\varkappa.$$

THEOREM 2.11. *A necessary and sufficient condition that an A_n ($n \geq 7$) with non vanishing projective curvature tensor admit a group G_r of affine motions of order $r = n^2 - 2n + 5$ is that the curvature tensor be of the form*

$$R_{\nu\mu\lambda}^{\cdot\cdot\cdot\varkappa} = (V_\nu U_\mu - V_\mu U_\nu) U_\lambda A^\varkappa,$$

where $A^\varkappa$, U_λ, V_λ are covariantly constant.

Y. Mutō [10] studied also an A_n admitting a group G_r of affine motions of order $r > n^2 - pn$ and obtained the

THEOREM 2.12. *A necessary condition that an A_n ($n \geq 4$) admit a group G_r of affine motions of order $r > n^2 - p'n$ where $p' < \frac{1}{2}(n-2)$ is that the curvature tensor be of the form*

$$R_{\nu\mu\lambda}^{\cdot\cdot\cdot\varkappa} = - A_\nu^\varkappa U_{\mu\lambda} + A_\mu^\varkappa U_{\nu\lambda} + (U_{\nu\mu} - U_{\mu\nu}) A_\lambda^\varkappa + B_{\nu\mu\lambda}^{(1)} A_{(1)}^\varkappa + \dots + B_{\nu\mu\lambda}^{(p)} A_{(p)}^\varkappa \quad (p \leq p')$$

for some tensors and vectors $U_{\mu\lambda}$, $B_{\nu\mu\lambda}^{(a)}$, $A_{(a)}^\varkappa$ ($a = 1, 2, \dots, p$).

§ 3. Groups of projective motions.

We assume that an infinitesimal projective motion defined by $v^\varkappa$ leaves invariant the covariant derivative of the projective curvature tensor

$$\underset{v}{\pounds} \nabla_\omega P_{\nu\mu\lambda}^{\cdot\cdot\cdot\varkappa} = 0.$$

Then, by virtue of $\nabla_\omega P_{\nu\mu\lambda}^{\cdot\cdot\cdot\omega} = -(n-2) P_{\nu\mu\lambda}$, we have from (3.7) and

(3.9) of Chapter VI,

$$(n-2)P_{\nu\mu\lambda}^{\cdots\rho}p_\rho = 0$$

and consequently, for $n > 2$,

$$P_{\nu\mu\lambda}^{\cdots\rho}p_\rho = 0.$$

Thus on transvecting p^ω to (3.9) of Chapter VI, we find

$$-2p^\omega p_\omega P_{\nu\mu\lambda}^{\cdots\varkappa} = 0.$$

From this equation we have (K. Yano and T. Nagano [1])

THEOREM 3.1. *If a V_n $(n > 2)$ admits an infinitesimal non-affine projective motion which leaves invariant the covariant derivative of Weyl's projective curvature tensor, then the space is projectively Euclidean and therefore of constant curvature.*

THEOREM 3.2. *If a V_n $(n > 2)$ which is not of constant curvature admits an infinitesimal projective motion which leaves invariant the covariant derivative of Weyl's projective curvature tensor, then the projective motion is necessarily an affine motion.*

If the covariant derivative of Weyl's projective curvature tensor vanishes: $\nabla_\omega P_{\nu\mu\lambda}^{\cdots\varkappa} = 0$, then the condition $\underset{v}{\pounds}\nabla_\omega P_{\nu\mu\lambda}^{\cdots\varkappa} = 0$ is always satisfied. Since this is the case for a symmetric space, we have

THEOREM 3.3. *If a symmetric V_n $(n > 2)$ admits an infinitesimal non-affine projective motion, then the space is necessarily of constant curvature.*

THEOREM 3.4. *If a symmetric V_n $(n > 2)$ which is not of constant curvature admits an infinitesimal projective motion, then the projective motion is necessarily an affine motion.*

In the text, we have proved that in a compact orientable V_n, an infinitesimal affine motion is necessarily an isometry. But, here we do not have to assume the orientability of the manifold. (See, B. Kostant [1]).

Combining the above theorem and this fact, we have

THEOREM 3.5. *If a compact symmetric V_n $(n > 2)$ which is not of constant curvature admits an infinitesimal projective motion, then the projective motion is necessarily an isometry.*

We now assume that the V_n is an Einstein space with non vanishing scalar curvature and it admits an infinitesimal projective motion defined by $v^\varkappa$. We then have, from $\underset{v}{\pounds} P_{\mu\lambda} = \nabla_\mu p_\lambda$,

$$k \underset{v}{\pounds} g_{\mu\lambda} = \nabla_\mu p_\lambda$$

by virtue of

$$P_{\mu\lambda} = k g_{\lambda\mu}; \qquad k \overset{\text{def}}{=} -\frac{K}{n(n-1)} \neq 0.$$

Writing out the equation $k \underset{v}{\pounds} g_{\mu\lambda} = \nabla_\mu p_\lambda$, we find

$$k(\nabla_\mu v_\lambda + \nabla_\lambda v_\mu) = \nabla_\mu p_\lambda$$

from which

$$\nabla_\mu w_\lambda + \nabla_\lambda w_\mu = 0$$

where

$$w_\lambda = v_\lambda - \frac{1}{2k} p_\lambda.$$

Thus the vector w_λ is a Killing vector and consequently

$$p_\lambda = 2k(v_\lambda - w_\lambda)$$

defines an infinitesimal projective motion. Thus we have

THEOREM 3.6. *If an Einstein space with non vanishing scalar curvature admits an infinitesimal projective motion $v^\varkappa$, that is, if we have $\underset{v}{\pounds}\{{}^{\varkappa}_{\mu\lambda}\} = p_\mu A^\varkappa_\lambda + p_\lambda A^\varkappa_\mu$, then the vector $v^\varkappa$ is decomposed into*

$$v^\varkappa = w^\varkappa + \frac{1}{2k} p^\varkappa.$$

where $w^\varkappa$ is a Killing vector and $p^\varkappa$ is a gradient vector defining an infinitesimal projective motion.

Since $\nabla_\nu P_{\mu\lambda} = 0$ for an Einstein space, from

$$\underset{v}{\pounds} \nabla_\nu P_{\mu\lambda} = \nabla_\nu \nabla_\mu p_\lambda - 2p_\nu P_{\mu\lambda} - p_\mu P_{\nu\lambda} - p_\lambda P_{\nu\mu}, \tag{3.1}$$

we have

$$\nabla_\nu \nabla_\mu p_\lambda = k(2p_\nu g_{\mu\lambda} + p_\mu g_{\nu\lambda} + p_\lambda g_{\nu\mu}),$$

from which

$$-K_{\nu\mu\lambda}{}^{\cdot\cdot\cdot\varkappa}p_\varkappa = k(p_\nu g_{\mu\lambda} - p_\mu g_{\nu\lambda}). \tag{3.2}$$

Now transvecting $g^{\mu\lambda}$ to (3.1), we find

$$\nabla_\nu(g^{\mu\lambda}\nabla_\mu p_\lambda) = 2(n+1)kp_\nu.$$

Since p_ν is a gradient vector, putting $p_\nu = \nabla_\nu p$, we find, from the above equation,

$$\nabla_\nu[g^{\mu\lambda}\nabla_\mu\nabla_\lambda p - 2(n+1)kp] = 0,$$

from which

$$g^{\mu\lambda}\nabla_\mu\nabla_\lambda p - 2(n+1)kp = -2(n+1)kp_0$$

or

$$g^{\mu\lambda}\nabla_\mu\nabla_\lambda(p - p_0) = 2(n+1)k(p - p_0), \tag{3.3}$$

where p_0 is a constant.

On the other hand, from the theorem of Green

$$\int_{V_n} g^{\mu\lambda}\nabla_\mu\nabla_\lambda(\tfrac{1}{2}f^2)d\sigma = \int_{V_n}(fg^{\mu\lambda}\nabla_\mu\nabla_\lambda f + g^{\mu\lambda}(\nabla_\mu f)(\nabla_\lambda f)]d\sigma = 0,$$

which is valid for a function f in a compact orientable space, we see that if a function f satisfies $g^{\mu\lambda}\nabla_\mu\nabla_\lambda f = \alpha f$ where $\alpha \geq 0$, then f is a constant.

Thus if $k > 0$, that is, if $K < 0$, then from (3.3), we have

$$p - p_0 = \text{constant},$$

that is, p is a constant and consequently $p_\lambda = 0$. This means, following Theorem 5.1 of Chapter IX that the infinitesimal projective motion defined by $v^\varkappa$ is an isometry. But in a compact Einstein space with negative scalar curvature, there does not exist an isometry other than the identity. Thus we have

THEOREM 3.7. *In a compact Einstein space with negative scalar curvature, there does not exist an infinitesimal projective motion.*

Now, equation (3.2) shows that the restricted homogeneous holonomy group of an Einstein space which has non vanishing scalar curvature and which admits in infinitesimal non-affine projective motion is the special orthogonal group $SO(n)$. Since the restricted homogeneous holonomy group of a Kähler space cannot be the special orthogonal group, we have

THEOREM 3.8. *A Kähler-Einstein space with non vanishing scalar curvature cannot admit an infinitesimal non-affine projective motion.*

Thus if a Kähler-Einstein space with non-vanishing scalar curvature admits an infinitesimal projective motion, the projective motion is necessarily an affine motion. Thus we have

THEOREM 3.9. *If a compact Kähler-Einstein space with non vanishing scalar curvature admits an infinitesimal projective motion, the projective motion is necessarily an isometry.*

Since an isometry in a compact Kähler space leaves invariant the complex structure, we have

THEOREM 3.10. *In a compact Kähler-Einstein space with non vanishing scalar curvature the largest connected group of projective motions leaves invariant the complex structure.*

S. Ishihara [4] studied groups of projective motions in a space with a projective connexion and obtained the following theorems.

THEOREM 3.11. *Let M be an n-dimensional manifold with a projective connexion and G an effective and connected group of projective motions in M. Suppose moreover that $\dim G \geq n^2 + 5$ and $n \geq 3$. Then M is projectively Euclidean and* $\dim G = n^2 + 2n$, $n^2 + n$ *or* $n^2 + n - 1$ *for* $n \geq 6$; $\dim G = n^2 + 2n$ *or* $n^2 + n$ *for* $n = 5$; $\dim G = n^2 + 2n$ *for* $n = 4, 3$.

THEOREM 3.12. *Let G be an effective group of projective motions in a manifold M with a projective connexion. If the given invariant projective connexion has non trivial torsion, then* $\dim G \leq n^2$, *where* $n = \dim M$. *There exists moreover an n-dimensional manifold with a projective connexion having non-trivial torsion which admits an n^2-dimensional group of projective motions.*

THEOREM 3.13. *Let M be an n-dimensional connected manifold with a projective connexion and G a connected and effective group of projective motions of M such that* $\dim G = n^2 + 2n$. *Then G is transitive on M and M is projectively Euclidean. Furthermore, the simply connected covering manifold of M is homeomorphic to a sphere S_n of n dimensions. If moreover n is even, M is homeomorphic to S_n or to a real projective space of n dimensions.*

THEOREM 3.14. *Let G be an effective group of projective motions of an n-dimensional manifold M with an affine connexion having no torsion.*

Suppose moreover that $\dim G \geq n^2 + 5$. *Then* M *is projectively Euclidean and* $\dim G = n^2 + 2n$, $n^2 + n$ *or* $n^2 + n - 1$ *for* $n \geq 6$; $\dim G = n^2 + 2n$ *or* $n^2 + n$ *for* $n = 5$ $\dim G = n^2 + 2n$ *for* $n = 4,3$.

If φ is a transformation of a manifold M with a linear connexion which carries any torse-forming vector field along an arbitrary curve C into a torse-forming vector field along the image $\varphi(C)$ of C by φ, then φ is called a quasi-projective motion of M. If there exist two covariant vector fields p_λ and q_λ such that

$$'\Gamma^{\varkappa}_{\mu\lambda} = \Gamma^{\varkappa}_{\mu\lambda} + p_\mu A^{\varkappa}_{\lambda} + q_\lambda A^{\varkappa}_{\mu}$$

is projectively Euclidean, then $\Gamma^{\varkappa}_{\mu\lambda}$ is said to be quasi-projectively Euclidean.

THEOREM 3.15. *Let* G *be an effective group of quasi-projective motions of an n-dimensional manifold* M *with a linear connexion. Suppose moreover that* $\dim G \geq n^2 + 5$. *Then* M *is quasi-projectively Euclidean and* $\dim G = n^2 + 2n$, $n^2 + n$ or $n^2 + n - 1$ for $n = 6$; $\dim G = n^2 + 2n$ or $n^2 + n$ for $n = 5$; $\dim G = n^2 + 2n$ for $n = 4,3$.

THEOREM 3.16. *Let* G *be an effective group of quasi-projective motions of an n-dimensional manifold* M *with a linear connexion. If the torsion tensor* $S_{\mu\lambda}^{\cdot\cdot\varkappa}$ *of the linear connexion does not satisfy the equation*

$$(n-1)S_{\mu\lambda}^{\cdot\cdot\varkappa} = S_{\mu\rho}^{\cdot\cdot\rho} A^{\varkappa}_{\lambda} - S_{\lambda\rho}^{\cdot\cdot\rho} A^{\varkappa}_{\mu}$$

then $\dim G \leq n^2$.

§ 4. Groups of conformal motions.

We assume that an infinitesimal conformal motion leaves invariant the covariant derivative of the conformal curvature tensor

$$\underset{v}{\pounds} \nabla_\omega C_{\nu\mu\lambda}^{\cdot\cdot\cdot\varkappa} = 0. \tag{4.1}$$

By virtue of $\nabla_\omega C_{\nu\mu\lambda}^{\cdot\cdot\cdot\omega} = -(n-3)C_{\nu\mu\lambda}$, we have, from (3.11) and (3.13) of Chapter VII,

$$(n-3)C_{\nu\mu\lambda}^{\cdot\cdot\cdot\rho}\phi_\rho = 0$$

and consequently for $n > 3$

$$C_{\nu\mu\lambda}^{\cdot\cdot\cdot\rho}\phi_\rho = 0.$$

Thus, transvecting ϕ^ω to (3.13) of Chapter VII, we obtain

$$-2\phi^\omega \phi_\omega C_{\nu\mu\lambda}^{\cdot\cdot\cdot\varkappa} = 0. \tag{4.2}$$

From this equation, we have

THEOREM 4.1. *If a V_n $(n > 3)$ admits an infinitesimal non-homothetic conformal motion which leaves invariant the covariant derivative of Weyl's conformal curvature tensor, then the space is conformally Euclidean.*

THEOREM 4.2. *If a V_n $(n > 3)$ which is not conformally Euclidean admits an infinitesimal conformal motion which leaves invariant the covariant derivative of Weyl's conformal curvature tensor, then the conformal transformation is necessarily homothetic.*

If the covariant derivative of Weyl's conformal curvature tensor vanishes: $\nabla_\omega C_{\nu\mu\lambda}^{\cdot\cdot\cdot\varkappa} = 0$, then the condition $\underset{v}{\pounds}\nabla_\omega C_{\nu\mu\lambda}^{\cdot\cdot\cdot\varkappa} = 0$ is always satisfied. Since this is the case for a symmetric space, we have

THEOREM 4.3.[1] *If a symmetric space V_n $(n > 3)$ admits an infinitesimal non-homothetic conformal motion, then the space is conformally Euclidean.*

THEOREM 4.4. *If a symmetric V_n $(n > 3)$ which is not conformally Euclidean admits an infinitesimal conformal motion, then the conformal motion is necessarily homothetic.*

If we stand on a global point of view, the theorem corresponding to Theorem 4.3 is a corollary to the more general theorem:

THEOREM 4.5. *If a homogeneous Riemannian space V_n $(n > 3)$ admits a non-isometric conformal motion, then the space is conformally Eudlidean.*

Theorems 4.3 and 4.4 can be slightly improved in the following way. Consider the Lie derivative of the scalar $C^{\nu\mu\lambda\varkappa} C_{\nu\mu\lambda\varkappa}$ with respect to $v^\varkappa$ which defines a conformal motion, then we have

$$\underset{v}{\pounds}(C^{\nu\mu\lambda\varkappa} C_{\nu\mu\lambda\varkappa}) = -2\phi C^{\nu\mu\lambda\varkappa} C_{\nu\mu\lambda\varkappa}.$$

If the space is locally homogeneous (or symmetric), then $C^{\nu\mu\lambda\varkappa} C_{\nu\mu\lambda\varkappa}$ is a constant and consequently $\underset{v}{\pounds}(C^{\nu\mu\lambda\varkappa} C_{\nu\mu\lambda\varkappa}) = 0$. Thus we have

---

[1] T. Sumitomo [1].

THEOREM 4.6. *If a locally homogeneous (or symmetric) space* V_n ($n > 3$) *admits an infinitesimal conformal motion which is not an isometry, then the space* V_n *is conformally Euclidean.*

THEOREM 4.7. *If a locally homogeneous (or symmetric)* V_n ($n > 3$) *which is not conformally Euclidean admits an infinitesimal conformal motion, then the conformal motion is necessarily an isometry.*

An irreducible symmetric V_n is an Einstein space and if an Einstein space is conformally Euclidean, it is of constant curvature. Thus, from Theorem 4.6, we have

THEOREM 4.8. *If an irreducible symmetric* V_n ($n > 3$) *admits an infinitesimal conformal motion which is not an isometry, then the space is necessarily of constant curvature.*

We now suppose that an Einstein space V_n with non vanishing scalar curvature admits an infinitesimal conformal motion defined by $v^\varkappa$. We have, from $\dfrac{1}{n-2}\underset{v}{\pounds}L_{\mu\lambda} = \nabla_\mu \phi_\lambda$,

$$k\underset{v}{\pounds}g_{\mu\lambda} = \nabla_\mu \phi_\lambda$$

by virtue of

$$\frac{1}{n-2}L_{\mu\lambda} = kg_{\mu\lambda};\ k \overset{\text{def}}{=} -\frac{K}{2n(n-1)} \neq 0.$$

In exactly the same way as in § 3 of this Appendix we obtain,

THEOREM 4.9. *If an Einstein space with non vanishing scalar curvature admits an infinitesimal conformal motion defined by* $v^\varkappa$, *that is, if we have* $\underset{v}{\pounds}g_{\mu\lambda} = 2\phi g_{\mu\lambda}$, *then the vector* $v^\varkappa$ *is decomposed into*

$$v^\varkappa = w^\varkappa + \frac{1}{2k}\phi^\varkappa,$$

where $w^\varkappa$ *is a Killing vector and* $\phi^\varkappa$ *is a gradient vector defining an infinitesimal conformal motion.* (A. Lichnerowicz [7]).

THEOREM 4.10. *In an Einstein space with* $K > 0$, *we have*

$$L = L_1 + L_2,\ [L_1 L_1] = L_1,\ [L_1 L_2] \subset L_2,\ [L_2 L_2] \subset L_1$$

where L *is the Lie algebra of the Lie group of conformal motions.* L_1 *subalgebra defined by motions and* L_2 *vector space of the gradient of* ϕ *which appears in* $\underset{v}{\pounds}g_{\mu\lambda} = 2\phi g_{\mu\lambda}$. (A. Lichnerowicz [7]).

Since $\nabla_\nu L_{\mu\lambda} = 0$ for an Einstein space, we have, from (3.6) of § 3 of Chapter VII,

$$\nabla_\nu \nabla_\mu \phi_\lambda = 2k\phi_\nu g_{\mu\lambda},$$

from which

$$-K_{\nu\mu\lambda}^{\cdot\cdot\cdot\varkappa}\phi^\varkappa = 2k(\phi_\nu g_{\mu\lambda} - \phi_\mu g_{\nu\mu}).$$

Thus, in the same way as in § 3 of this Appendix, we obtain

THEOREM 4.11. *A Kähler-Einstein space with non vanishing scalar curvature cannot admit an infinitesimal non homothetic conformal motion.*

THEOREM 4.12. *If a compact Kähler-Einstein space with non vanishing scalar curvature admits an infinitesimal conformal motion, the conformal motion is necessarily an isometry.*

THEOREM 4.13. *In a compact Kähler-Einstein space with non vanishing scalar curvature, the largest connected group of conformal motions leaves invariant the complex structure of the space.* (A. Lichnerowicz [7]).

If we stand on a global point of view, a theorem corresponding to Theorem 4.11 is a corollary to the more general theorem:

THEOREM 4.14. *A complete Einstein space which admits a global one-parameter group of conformal motions must necessarily be a simply connected space of positive constant curvature.*

Let M denote a connected Riemannian manifold of dimension $n \geq 3$. A point P of M is called a homothetic point if the linear isotropy group of all conformal motions at P is not contained in the orthogonal group in the tangent space at P.

S. Ishihara and M. Obata [2] proved the following theorems:

THEOREM 4.15. *The conformal curvature tensor vanishes at any homothetic point.*

THEOREM 4.16. *If M contains a homothetic point and admits a transitive group of conformal motions, then M is conformally Euclidean.*

THEOREM 4.17. *Let ρ be the associated function of a conformal motion* i.e. $\varphi(g)_P = \rho(P)g_P$, *where g is the Riemannian metric tensor of M. If M is not conformally Euclidean and $\rho(P)$ satisfies the following inequality at every point of M:*

$$\rho(P) < 1 - \varepsilon \text{ or } \rho(P) > 1 + \varepsilon,$$

ε being a positive constant, then φ has no fixed point.

THEOREM 4.18. *If M is complete and is not conformally Euclidean, then the associated function of any conformal motion can take the value unity or an arbitrary value near the unity.*

THEOREM 4.19. *If M is compact orientable, then the associated function of any conformal motion takes the value unity.*

In Chapter IX of the text, we have proved the integral formulas

$$(4.3)\quad \int_{V_n} [g^{\mu\lambda}\nabla_\mu\nabla_\lambda v^\varkappa - K_\lambda{}^{\cdot\varkappa} v^\lambda)v_\varkappa + \tfrac{1}{2}(\nabla^\mu v^\lambda - \nabla^\lambda v^\mu)(\nabla_\mu v_\lambda - \nabla_\lambda v_\mu) + (\nabla_\mu v^\mu)(\nabla_\lambda v^\lambda)]d\sigma = 0,$$

$$(4.4)\quad \int_{V_n} [(g^{\mu\lambda}\nabla_\mu\nabla_\lambda v^\varkappa + K_\lambda{}^{\cdot\varkappa} v^\lambda)v_\varkappa + \tfrac{1}{2}(\nabla^\mu v^\lambda + \nabla^\lambda v^\mu)(\nabla_\mu v_\lambda + \nabla^\nu v_\mu) - (\nabla_\mu v^\mu)(\nabla_\lambda v^\lambda)]d\sigma = 0,$$

which are valid for a vector field in a compact orientable V_n.

Using exactly the same method, we can prove the integral formula:

$$(4.5)\quad \int_{V_n} \Big[\Big(g^{\mu\lambda}\nabla_\mu\nabla_\lambda v^\varkappa + K_\lambda{}^{\cdot\varkappa} v^\lambda + \frac{n-2}{n}\nabla^\varkappa\nabla_\lambda v^\lambda\Big)v_\varkappa + \tfrac{1}{2}\Big(\nabla^\mu v^\lambda + \nabla^\lambda v^\mu - \frac{2}{n}g^{\mu\lambda}\nabla_\sigma v^\sigma\Big)\Big(\nabla_\mu v_\lambda + \nabla_\lambda v_\mu - \frac{2}{n}g_{\mu\lambda}\nabla_\rho v^\rho\Big)\Big]d\sigma = 0,$$

from which we have

THEOREM 4.20. *A necessary and sufficient condition for $v^\varkappa$ in V_n to be a conformal Killing vector is that*

$$(4.6)\quad g^{\mu\lambda}\nabla_\mu\nabla_\lambda v^\varkappa + K_\lambda{}^{\cdot\varkappa} v^\lambda + \frac{n-2}{n}\nabla^\varkappa\nabla_\lambda v^\lambda = 0.$$

(A. Lichnerowicz [7], I. Sato [1]).

An infinitesimal transformation $v^\varkappa$ satisfying

$$\underset{v}{\pounds}\{{}^{\varkappa}_{\mu\lambda}\} = \phi_\mu A^\varkappa_\lambda + \phi_\lambda A^\varkappa_\mu - \phi^\varkappa g_{\mu\lambda}$$

is called a conformal collineation. From the above theorem we obtain

THEOREM 4.21. *An infinitesimal conformal collineation is a conformal motion.*

Now, for a conformal motion, we have

$$\nabla_\mu\nabla_\lambda\phi = \frac{1}{n-2}\underset{v}{\pounds}L_{\mu\lambda}$$

from which

$$g^{\mu\lambda}\nabla_\mu\nabla_\lambda\phi = -\frac{1}{2(n-1)}(\underset{v}{\pounds}K + 2K\phi).$$

Thus if K is a constant, then we have

$$g^{\mu\lambda}\nabla_\mu\nabla_\lambda\phi = -\frac{K}{n-1}\phi.$$

This equation shows that if $K < 0$, then $\phi = 0$, and the conformal motion is an isometry. If $K = 0$, then $\phi =$ constant and the conformal motion is homothetic. But in a compact space, a homothetic motion is an isometry. Thus we have

THEOREM 4.22. *If a compact V_n with $K =$ constant $\leqq 0$ admits an infinitesimal conformal motion, it is an isometry.*

§ 5. Groups of transformations in generalized spaces.

In § 8 of Chapter VIII, we have proved: In order that a general affine space of geodesics admit a group of affine motions of the maximum order $n^2 + n$, it is necessary and sufficient that the geodesics be given by the equations of the form

$$\frac{d^2\xi^\varkappa}{ds^2} + \Gamma^{\varkappa}_{\mu\lambda}(\xi)\frac{d\xi^\mu}{ds}\frac{d\xi^\lambda}{ds} = 0$$

and the space be locally an E_n.

Using the method of Y. Muto, Tanjiro Okubo [1] proved

THEOREM 5.1. *If an n-dimensional generalized space of geodesics admits a group G_r of affine motions of order r*

$$n^2 + n \geqq r > n^2, \qquad n > 2,$$

then the space is locally an ordinary E_n.

THEOREM 5.2. *If an n-dimensional generalized space of geodesics admits a group G_r of affine motions of order r*

$$n^2 \geqq r > n^2 - n + 1, \qquad n \geqq 7,$$

then the space is locally an ordinary E_n.

THEOREM 5.3. *For $n \geqq 7$, the space admitting a group G_r of affine motions of order r*

$$r = n^2 - n + 1$$

really exists; its example being furnished by the projectively Euclidean space with the connexion parameters

$$\Gamma^{\varkappa}_{\mu\lambda} = A^{\varkappa}_{\mu}\dot{\nabla}_{\lambda} p + A^{\varkappa}_{\lambda}\dot{\nabla}_{\mu} p + \xi^{\varkappa}\dot{\nabla}_{\mu}\dot{\nabla}_{\lambda} p,$$

where

$$p = (\xi^1\xi^2)^{\frac{1}{2}}.$$

§ 6. Groups of transformations in almost complex spaces.

We have defined a covariant pseudo-analytic vector field in a pseudo-Kählerian space as a vector field v^h satisfying

(6.1) $$F_j^{\cdot a}\nabla_i v_a - F_i^{\cdot a}\nabla_a v_j = 0$$

and a contravariant pseudo-analytic vector field as a vector field v^h satisfying

(6.2) $$\underset{v}{\pounds} F_i^{\cdot h} = F_a^{\cdot h}\nabla_i v^a - F_i^{\cdot a}\nabla_a v^h = 0.$$

In a compact pseudo-Kählerian space, we can prove the following integral formula (K. Yano [26])

(6.3) $$\int_{V_{2n}} [(g^{ji}\nabla_j\nabla_i v^h - K_i^{\cdot h} v^i)v_h + \tfrac{1}{2}(F^{jb}\nabla^i v_b - F^{ib}\nabla^j v_b)(F_j^{\cdot a}\nabla_i v_a - F_i^{\cdot a}{}_{;j} v_a)]d\sigma = 0,$$

(6.4) $$\int_{V_{2n}} [(g^{ji}\nabla_j\nabla_i v^h + K_i^{\cdot h} v^i)v_h + \tfrac{1}{2}(F^{jb}\nabla_b v^i - F_b^{\cdot i}\nabla^j v^b)(F_j^{\cdot a}\nabla_a v_i - F^a_{\cdot i}\nabla_j v_a)]d\sigma = 0.$$

From (6.3) we easily see that a necessary and sufficient condition for a vector field v^h in a compact pseudo-Kählerian space to be covariant pseudo-analytic is that v^h be harmonic.

From (6.4), we have

THEOREM 6.1. *A necessary and sufficient condition for a vector field v^h in a compact pseudo-Kählerian space to be contravariant pseudo-analytic is that*

(6.5) $$g^{ji}\nabla_j\nabla_i v^h + K_i^{\cdot h} v^i = 0.$$

Let a vector field v^h be given in an n-dimensional Riemannian space V_n and consider a geodesic $\xi^h = \xi^h(s)$ in V_n. The condition that the infinitesimal transformation $\xi^h \rightarrow \xi^h + v^h(\xi)dt$ transforms the geodesic $\xi^h(s)$ into a geodesic and preserves affine character of the arc lengths s

is given by

$$(6.6) \qquad (\nabla_j \nabla_i v^h + K_{kji}{}^{h} v^k) \frac{d\xi^j}{ds} \frac{d\xi^i}{ds} = 0.$$

If we take a point ξ^h and a unit vector h^h at ξ^h, the geodesic passing through ξ^h and being tangent to h^h is uniquely determined and we can consider the vector

$$(6.7) \qquad u^h = (\nabla_j \nabla_i v^h + K_{kji}{}^{h} v^k) h^j h^i$$

appearing in the left hand member of (6.6). We shall call (6.7) the geodesic deviation vector of the unit vector h^h at the point ξ^h with respect to v^h.

Now consider n mutually orthogonal unit vectors $h^h_{(a)}$ $(a = 1, 2, \ldots, \mathrm{n})$ and geodesic deviation vectors $u^h_{(a)}$ of $h^h_{(a)}$ with respect to v^h. Thus for the mean of $u^h_{(a)}$, we have

$$\frac{1}{n} \Sigma\, u^h_{(a)} = \frac{1}{n} (g^{ji} \nabla_j \nabla_i v^h + K_i{}^{h} v^i)$$

which is independent of the choise of $h^h_{(a)}$. We shal call $\frac{1}{n} \Sigma\, u^h_{(a)}$ the mean geodesic deviation vector with respect to v^h. Thus from Theorem 6.1 we have

THEOREM 6.2. *A necessary and sufficient condition for a vector field v^h in a compact pseudo-Kählerian space to be contravariant pseudo-analytic is that the mean geodesic deviation vector with respect to v^h vanish.*

Since the tensor F_{ji} in a pseudo-Kählerian space is harmonic, we have

THEOREM 6.3. *If a compact pseudo-Kählerian space admits a one-parameter group of motions, it preserves the pseudo-complex structure of the space.*

Conversely if a compact pseudo-Kählerian space admits an infinitesimal transformation $\xi^h \to \xi^h + v^h dt$ which preserves the pseudo-complex structure of the space and also the volume element, then we have

$$g^{ji} \nabla_j \nabla_i v^h + K_i{}^{h} v^i = 0, \qquad \nabla_i v^i = 0$$

and consequently the transformation is an isometry. Thus we have

THEOREM 6.4. *If an infinitesimal transformation preserves the pseudo-complex structure of a compact pseudo-Kählerian space and also the volume element, then the transformation is an isometry.*

We now consider an equation of the form

$$\Delta f \overset{\text{def}}{=} g^{ji}\nabla_j\nabla_i f = \lambda f \qquad (\lambda = \text{constant} < 0)$$

in a compact pseudo-Kählerian space, from which

$$\Delta f_h \overset{\text{def}}{=} g^{ji}\nabla_j\nabla_i f_h - K_h^{\cdot a} f_a = \lambda f_h \qquad (f_h = \nabla_h f),$$

from which

$$\Delta v_h = g^{ji}\nabla_j\nabla_i v_h - K^{a}_{\cdot h} v_a = \lambda v_h,$$

where

$$v_h = F^a_{\cdot h} f_a.$$

Substituting this equation into

$$\int_{V_{2n}} [(g^{ji}\nabla_j\nabla_i v^h + K_i^{\cdot h} v^i)v_h + \tfrac{1}{2}(\nabla^j v^i + \nabla^i v^j)(\nabla_j v_i + \nabla_i v_j) - (\nabla_j v^j)(\nabla_i v^i)]d\sigma = 0$$

and taking account of $\nabla_i v^i = 0$, we find

$$\int_{V_{2n}} [(2K_{ji} + \lambda g_{ji})v^j v^i + \tfrac{1}{2}(\nabla^j v^i + \nabla^i v^j)(\nabla_j v_i + \nabla_i v_j)]d\sigma = 0,$$

from which

THEOREM 6.5. *If, in a compact pseudo-Kählerian space, the form* $(2K_{ji} + \lambda g_{ji})v^j v^i$ *is positive definite, then the equation* $\Delta f = \lambda f$ *has no solution other than zero.*

THEOREM 6.6. *If, in a compact pseudo-Kähler-Einstein space with* $K > 0$, $\dfrac{K}{n} + \lambda > 0$, *then the equation* $\Delta f = \lambda f$ *has no solution other than zero. Consequently if the equation* $\Delta f = \lambda f$ *admits a solution other than zero, then*

$$\frac{K}{n} + \lambda \leq 0, \text{ that is, } \lambda \leq -\frac{K}{n}.$$

THEOREM 6.7. *If, in a compact pseudo-Kähler-Einstein space with* $K > 0$, *the equation* $\Delta f = -\dfrac{K}{n} f$ *admits a solution other than zero, then* $v_i = F^a_{\cdot i} f_a$ *is a Killing vector.*

Now suppose that a general compact pseudo-Kählerian space admits a Killing vector v^h, then we have

$$\nabla_j(F^a_{\cdot i} v_a) - \nabla_i(F^a_{\cdot j} v_a) = 0, \qquad \nabla_j(F^{ji} v_i) = F^{ji}\nabla_j v_i$$

by virtue of $\underset{v}{\pounds} F_{ji} = 0$, from which

THEOREM 6.8. *In a compact pseudo-Kählerian space which does not admit a parallel vector field,* $F^{ji}\nabla_j v_i \neq 0$ *for a Killing vector field* v^h.

Because if $F^{ji}\nabla_j v_i = 0$, then $F^{a}_{.i} v_a$ is harmonic and consequently so is v_i too. Thus v^h being at the same time a Killing vector and a harmonic vector, it is a parallel vector field, a fact which contradicts the hypothesis.

Now consider a compact pseudo-Kähler-Einstein space with $K > 0$ and suppose that the space admits a Killing vector field v^h, then

$$f \overset{\text{def}}{=} \frac{n}{K} F^{ji} \nabla_j v_i \neq 0.$$

On the other hand, using $\nabla_j \nabla_i v^h + K_{kji}{}^{h} v^k = 0$, we find

$$f_j \overset{\text{def}}{=} \nabla_j f = \nabla_j \left(\frac{n}{K} F^{ih} \nabla_i v_h \right) = F^{a}_{.j} v_a$$

and consequently

$$f_i = F^{a}_{.i} v_a, \quad v_i = - F^{a}_{.i} f_a,$$

from which

$$g^{ji} \nabla_j \nabla_i f = - \frac{K}{n} f. \tag{6.8}$$

Thus we have

THEOREM 6.9. *If a compact pseudo-Kähler-Einstein space with* $K > 0$ *admits a Killing vector field* v^h, *then the equation* (6.8) *admits a solution other than zero given by* $f = \frac{n}{K} F^{ji} \nabla_j v_i$ *and vice versa.*

Suppose that a compact pseudo-Kähler-Einstein space with $K > 0$ admits two Killing vectors v^h and w^h to which correspond f and g respectively, then we have

$$\begin{aligned} F^{ji} \nabla_j [v, w]_i = F^{ji} \nabla_j \underset{v}{\pounds} w_i = \underset{v}{\pounds}(F^{ji} \nabla_j w_i) &= \frac{K}{n} \underset{v}{\pounds} g \\ &= \frac{K}{n} v^i \nabla_i g = - \frac{K}{n} F^{ai} f_a \nabla_i g \\ &= - \frac{K}{n} F^{ji} (\nabla_j f)(\nabla_i g). \end{aligned}$$

Thus if we define $[f, g]$ by

$$[f, g] = -F^{ji}(\nabla_j f)(\nabla_i g),$$

we have

THEOREM 6.10. *If a compact pseudo-Kähler-Einstein space with $K > 0$ admits two Killing vectors v^h and w^h to which correspond f and g respectively, then $[v, w]^h$ and $[f, g]$ correspond to each other.*

A necessary and sufficient condition for v^h to be a contravariant pseudo-analytic vector field in a compact pseudo-Kähler-Einstein space is that

$$g^{ji}\nabla_j\nabla_i v^h + \frac{K}{2n} v^h = 0.$$

From this equation, we can easily deduce

$$g^{ji}\nabla_j\nabla_i(\nabla_a v^a) + \frac{K}{n}(\nabla_a v^a) = 0$$

and

$$g^{ji}\nabla_j\nabla_i\nabla_h(\nabla_a v^a) + \frac{K}{2n}\nabla_h(\nabla_a v^a) = 0.$$

The last equation shows that the vector $\nabla_h(\nabla_a v^a)$ is a contravariant analytic vector field.

Put

$$p^h = v^h + \frac{n}{K}\nabla^h(\nabla_a v^a), \tag{6.9}$$

then p^h is a contravariant pseudo-analytic vector field. Moreover we have

$$\nabla_h p^h = \nabla_h v^h + \frac{n}{K} g^{ji}\nabla_j\nabla_i(\nabla_a v^a) = 0$$

and consequently, p^h is a Killing vector.

Thus if we put

$$q^h = F_a^{\cdot h}\left[\frac{n}{K}\nabla^a(\nabla_i v^i)\right],$$

then g^h is also contravariant pseudo-analytic and

$$\nabla_h q^h = 0,$$

and consequently q^h is also a Killing vector.

From (6.9), we have

$$v^h = p^h + F_a^{\cdot h} q^a,$$

where p^h and q^h are both Killing vectors.

Such a decomposition of a contravariant pseudo-analytic vector is unique. Because if we have

$$v^h = {}'p^h + F_a^{\cdot h}\,{}'q^a, \qquad v^h = p^h + F_a^{\cdot h} q^a,$$

then

$$({}'p^h - p^h) + F_a^{\cdot h}({}'q^a - q^a) = 0,$$

from which

$$F^{ih}\nabla_i({}'q_h - q_h) = 0.$$

Thus ${}'q_h = q_h$ and consequently ${}'p^h = p^h$. Thus we have

THEOREM 6.11. *In a compact pseudo-Kähler-Einstein space, any contravariant pseudo-analytic vector field v^h is uniquely decomposed in the form*

$$v^h = p^h + F_a^{\cdot h} q^a,$$

where p^h and q^h are both Killing vector fields. (Y. Matsushima [1]).

A transformation φ of a pseudo-Hermitian manifold M is called a Hermitian automorphism if φ preserves both of F_{ji} and $F_i^{\cdot h}$.

S. Ishihara [2,3] proved the following theorems:

THEOREM 6.12. *Let G be a group of Hermitian automorphisms of a $2n$-dimensional pseudo-Hermitian space M. Then G is transitive on M for $n \geq 2$, if the group G is of dimension $r \geq n^2 + 2$. In case $n \geq 3$ and $n \neq 4$, there exists no group of Hermitian automorphisms of dimension r such that*

$$n^2 + 2n - 1 > r > n^2 + 2.$$

THEOREM 6.13. *Let G/H be a homogeneous pseudo-Hermitian space of $2n$-dimensions and* $\dim G = n^2 + 2n$. *Then G/H is a homogeneous pseudo-Kählerian space with constant holomorphic sectional curvature K. When $K > 0$ and G/H is simply connected, G is isomorphic locally to the unimodular unitary group in $n + 1$ complex variables and G/H is homeomorphic to $P(C, n)$. When $K < 0$, G is locally isomorphic to the identity component* in the group of all linear transformations in $n + 1$ variables $(z_1, z_2, \ldots, z_{n+1})$ *leaving invariant the form* $z_1\bar{z}_1 + \ldots + z_n\bar{z}_n - z_{n+1}\bar{z}_{n+1}$

and G/H is homeomorphic to E_{2n}. When $K = 0$, G is isomorphic to the group of all unitary motions in a unitary space of n complex dimensions and G/H is homeomorphic to E_{2n}.

THEOREM 6.14. *Let G/H be a homogeneous pseudo-Hermitian space of $2n$ dimensions and* $\dim G = n^2 + 2n - 1$ $(n > 1)$. *If $n \neq 3$, G/H is flat and homeomorphic to E_{2n} and the group G is isomorphic to the subgroup of the group of all unitary motions in a unitary space of n complex dimensions whose rotation part is the unimodular unitary group. If $n = 3$, G/H is flat or of positive constant curvature.*

In case $n = 3$ and G/H is flat, the conclusion is the same as in the general case. In case $n = 3$ and G/H is of positive constant curvature, G/H is homeomorphic to a sphere of dimension 6 and the group is isomorphic to a compact exceptional simple group of type (G).

T. Fukami and S. Ishihara [1] proved following two theorems:

THEOREM 6.15. *The almost Hermitian structure on S_6 is invariant under the group G of all automorphisms of Cayley numbers. Conversely, the group of all isometries leaving invariant the almost Hermitian structure on S_6 is isomorphic to G.*

THEOREM 6.16. *On the homogeneous almost Hermitian space $S_6 = G/H$ there exists one and only one invariant connexion Γ_{ji}^h defined by*

$$\Gamma_{ji}^h = \left\{ {h \atop ji} \right\} - \tfrac{1}{2}(\nabla_a F_j{}^h) F_i{}^a,$$

for which g_{ji} and $F_i{}^h$ are covariant constant. The covariant derivative of its torsion and curvature tensor fields are both zero, but its torsion field itself does not vanish at every point of S_6.

A. Lichnerowicz [4] proved

THEOREM 6.17. *In an irreducible pseudo-Kählerian space with $K_{ji} \neq 0$, every real infinitesimal motion is an automorphism.*

J. A. Schouten and K. Yano [4] proved

THEOREM 6.18. *In an irreducible pseudo-Kählerian space V_{2n} with n odd every real infinitesimal motion is an automorphism.*

Let M be a manifold of dimension $2m$ with the almost complex structure F. We denote by $H(P)$, $P \in M$, the homogeneous holonomy group of M with respect to a natural connexion, that is, an affine connexion with respect to which F is covariant constant. $A(M)$ denotes the group of all affine motions of M onto itself and $A_0(M)$ denotes the connected component of the identity of $A(M)$. We assume that $H(P)$ is irreducible in the real number field. Then $H(P)$ is a subgroup of the real representation $CL(m, R)$ of the complex linear group.

M. Obata [2] proved the following theorems.

THEOREM 6.19. *If m is odd or if m is even $m = 2l$ and $H(P)$ is not a subgroup of $QL(l, R)$, then $A_0(M)$ preserves the almost complex structure.*

THEOREM 6.20. *If $A_0(M)$ does not preserve the almost complex structure, then $m = 2l$ and $H(P)$ is a subgroup of $QL(l, R)$ and there exists a homomorphism of $A(M)$ into $SO(3)$.*

THEOREM 6.21. *In an irreducible pseudo-Kählerian manifold M of dimension $2m$, if m is odd or if m is even $m = 2l$ and $H(P)$ is not a subgroup of the real representation of the unitary symplectic group, then $A_0(M)$ preserves the almost complex structure.*

THEOREM 6.22. *In an irreducible pseudo-Kählerian manifold of dimension $2m$ if m is odd or if m is even $m = 2l$ and the Ricci curvature tensor does not vanish, then $A_0(M)$ preserves the almost complex structure; especially the largest connected group of isometries preserves the almost complex structure.*

THEOREM 6.23. *In an irreducible complex manifold of dimension $2m$, if m is odd or if m is even $m = 2l$ and the homogeneous holonomy group is not a subgroup of $QL(l, R)$, an infinitesimal affine transformation is always complex analytic.*

THEOREM 6.24. *In an irreducible Kählerian manifold of dimension $2m$, if m is odd or if m is even and the Ricci curvature tensor does not vanish, an infinitesimal affine transformation is always complex analytic.*

S. Kobayashi and K. Nomizu [1] studied a similar problem.

BIBLIOGRAPHY

BUSEMANN, H.

[1] Similarities and differentiability, to appear in Tôhoku Mathematical Journal.

DUMITRUS, D.

[1] Sur les espaces A_3 qui admettent une rotation. Acad. Repub. Pop. Romåne. Stud. Cerc. Mat., **4** (1953), 213—232 (Roumanian).

EGOROV, I. P.

[10] Maximally mobile Riemannian spaces V_4 of non constant curvature. Dokl. Akad. Nauk SSSR (N.S.), **103** (1955), 9—12. (Russian).

FUKAMI, T. and S. ISHIHARA

[1] Almost Hermitian structures in S^6. Tôhoku Math. Journal, **7** (1955), 151—156.

GARABEDIAN, P. R. and D. C. SPENCER

[1] A complex tensor calculus for Kähler manifolds. Acta Math., **89** (1953), 279—331.

HANO, J.

[1] On affine transformations of a Riemannian manifold. Nagoya Math. Journal, **9** (1955), 99—109.

HERMANN, R.

[2] Sur les automorphismes infinitésimaux d'une G-structure. C. R. Acad. Sci. Paris, **239** (1954), 1760—1761.

HIRAMATU, H.

[5] On Riemannian spaces admitting groups of conformal transformations, to appear in the Journal of the Mathematical Society of Japan.

ISPAS, C. I.

[1] Au sujet des dérivées de Lie et de la déformation des vecteurs contrevariants des espaces à connexion. Com. R. P. Romåne, **5** (1955), 479—488. (Russian).

ISHIHARA, S.

[2] Groups of isometries of pseudo-Hermitian spaces, I. Proc. Japan Acad. **30** (1954), 940—945.

[3] Groups of isometries of pseudo-Hermitian spaces, II. Proc. Japan Acad. **31** (1955), 418—420.

[4] Groups of projective transformations on a projectively connected manifolds. Japanese Journal of Math., **25** (1955), 37—80.

[5] Groups of projective transformations and groups of conformal transformations, to appear.

ISHIHARA, S. and T. FUKAMI

[1] Groups of affine transformations and groups of projective transformations in a space of K-spreads, to appear.

ISHIHARA, S. and M. OBATA

[1] On a homogeneous space with invariant affine connection. Proc. Japan Acad. **31** (1955), 420—425.

[2] On the group of conformal transformations of a Riemannian manifold. Proc. Japan Acad., **31** (1955), 426—429.

[3] Affine transformations in a Riemannian manifold. Tôhoku Math. Journal 2nd Series, **7** (1955), 146—150.

KALAMA, D. S.

[1] Infinitesimal deformations in a subspace V_n of a Riemannian space V_m. Acad. Roy. Belg. Bull. Cl. Sci. (5) **40** (1954), 1072—1079.

KATSURADA, Y.

[2] On the curvature of a metric space with torsion tensor admitting parallel paths. Tensor (N.S.), **5** (1955), 85—90.

KOBAYASHI, S.

[4] A theorem on the affine transformation group of a Riemannian manifold. Nagoya Math. Journal, **9** (1955), 39—41.

KOBAYASHI, S. and K. NOMIZU

[1] On automorphisms of a Kählerian structure, to appear in Nagoya Math. Journal.

KOSTANT, B.

[1] Holonomy and the Lie algebra of infinitesimal motions of a Riemannian manifold. Trans. Amer. Math. Soc., **80** (1955), 528—542.

[2] On differential geometry and homogeneous spaces, I. Proc. Nat. Acad. Sci. U.S.A., **42** (1956), 258—261.

KRAČKOVIČ, G. I.

[1] Invariant criteria of spaces V_3 with the group of motions G_4. Uspehi Mat. Nauk. (N.S.), **10,** 1 (63), (1955), 129—136, (Russian).

KUIPER, N. H. and K. YANO

[2] Two algebraic theorems with applications. Indag. Math., **18** (1956), 319—328.

LELONG-FERRAND, J.

[2] Groupes d'isométries et formes harmoniques décomposables. C. R. Acad. Sci. Paris, **240** (1955), 835—837.

LICHNEROWICZ, A.

[6] Transformations infinitésimales conformes de certaines variétés riemanniennes compactes. C. R. Acad. Sci. Paris, **241** (1955), 726—729.

[7] Some problems on transformations of Riemannian and Kählerian manifolds. Mimeographed Notes, Princeton (1956).

MATSUSHIMA, Y.

[1] Sur la structure du groupe d'homéomorphismes analytiques d'une certaine variété kaehlérienne, to appear in Nagoya Math. Journal.

MATSUSHIMA, Y. and J. HANO.

[1] Some studies on Kaehlerian homogeneous spaces. Nagoya Math. Journal, **11** (1957), 1-16.

MURAI, Y.

[1] On the group of transformations in six-dimensional space, II. Conformal group in physics. Progr. Theor. Physics, **11** (1954), 441—448.

MUTŌ, Y.

[7] On n-dimensional projectively flat spaces admitting a group of affine motions of order $r = n^2 - n - 1$. Science Reports of the Yokohama National University, Sect. I, No. 5 (1956), 1—15.

[8] On conformally curved Riemann spaces V_n, $n \geq 6$, admitting a group of motions G_r of order $r > \frac{1}{2}n(n+1) - (3n - 11)$, to appear.

[9] On some properties of a kind of affinely connected manifolds admitting a group of affine motions, II, to appear.

[10] On some properties of affinely connected manifolds admitting groups of affine motions of order $r > n^2 - pn$, to appear.

NAGAI, T.

[1] Some considerations on structures of simply transitive groups. Tensor (N.S.), **5** (1955), 91—94.

NOMIZU, K.

[7] Reduction theorem for connections and its application to the problem of isotropy and holonomy groups of a Riemannian manifold. Nagoya Math. Journal, **9** (1955), 57—66.

[8] On infinitesimal holonomy and isotropy groups, to appear in Nagoya Math. Journal.

OBATA, M.

[2] Affine transformations in an almost complex manifold with a natural connection. Journal of the Math. Soc. of Japan, **8** (1956), 345–362.

[3] Affine connections on manifolds with almost complex, quaternion or Hermitian structure, to appear.

OKUBO, T.

[1] On the order of the groups of affine collineations in the generalized spaces of paths, I, II, III, to appear in Tensor.

PETRESCU, ST.

[1] La classification des espaces à connexion affine A_2. Acad. Repub. Pop. Române. Stud. Cerc. Mat., **2** (1951), 322—363. (Roumanian).

[2] Considérations concernant les espaces à connexion projective P_2. Acad. Repub. Pop. Române. Stud. Cerc. Mat., **3** (1952), 529—558, (Roumanian).

[3] Classification des espaces P_2. Acad. Repub. Pop. Române. Stud. Cerc. Mat., **4** (1953), 453—502. (Roumanian).

[4] Classification des espaces P_2 à connexion projective. Acad. Repub. Pop. Române. Bul. Sti. Sect. Mat. Fiz., **5** (1953), 485—491. (Roumanian).

PIDEK, H.

[1] Sur un problème d'algèbre des objets géométriques de classe zéro dans l'espace X_1. Ann. Pol. Math., **9** (1954), 114—126.

[2] Sur un problème d'algèbre des objets géométriques de classe zéro dans l'espace X_m. Ann. Pol. Math., **9** (1954), 127—134.

RAŠEVSKII, P. K.

[1] Linear differential-geometric objects. Doklady Akad. Nauk. SSSR (N.S.) **97** (1954), 609—611.

SASAKI, S. and K. YANO

[1] Pseudo-analytic vectors on pseudo-Kählerian manifolds. Pacific Journal of Math., **5** (1955), 987—993.

SATO, I.

[1] On conformal Killing tensor fields. Bull. of the Yamagata Univ., **3** (1956), 175—180.

SCHOUTEN, J. A. and K. YANO

[4] On pseudo-Kählerian spaces admitting a continuous group of motions. Indag. Math., **17** (1955), 565—570.

STOJANOVITCH, R.

[1] A note on a theorem of E. Cartan on groups of stability. Tensor (N.S.) **5** (1955), 54—55.

SUMITOMO, T.

[1] On some transformations of Riemannian spaces. Tensor, **6** (1956), 136–140.

TAKENO, H.

[1] On groups of conformal transformations in spherically symmetric space-times. Tensor (N.S.), **5** (1955), 23—38.

[2] On conformal transformations in the space-time of relativistic cosmology. Tensor (N.S.), **5** (1955), 141—149.

TELEMAN, C.

[1] Les groupes transitifs de mouvements des espaces de Riemann V_5. Acad. Repub. Pop. Române. Stud. Cerc. Mat., **4** (1953), 503—526. (Roumanian).

[2] Sur les groupes maximums de mouvements des espaces de Riemann V_n. Acad. Repub. Pop. Romane. Stud. Cerc. Mat., **5** (1954), 143—171. (Roumanian).

TOMONAGA, Y.

[1] Der homogene Riemannsche Raum und seine Erweiterungen. Bull. Utsunomiya Univ., **5** (1955), 1–6.

[2] Über die symmetrischen Riemannschen Räume. Bull. Utsunomiya Univ., **6** (1956), 1–10.

VRANCEANU, G.

[7] Sur les groupes de mouvements d'un espace de Riemann. Com. Acad. R. P. Romane, **1** (1951), 137—140. (Roumanian).

[8] Sur les espaces V_4 ayant comme groupes de stabilité un G_4. Publ. Math. Univ. Debreceniensis, **3** (1953), 24—32.

[9] Les groupes transitifs de mouvements des espaces de Riemann V_5. Acad. Repub. Pop. Romane. Stud. Cerc. Math., **4** (1953), 504—526.

[10] Sur les espaces V_n à groupe simplement transitif. Rev. Math. Phys. **2** (1954), 51—58.

[11] Sur une classe d'espaces riemanniens homogènes. Acad. Repub. Pop. Romane. Stud. Cerc. Mat., **5** (1954), 173—223. (Roumanian).

[12] Propriétés différentielles globales des espaces A_n à groupe maximum G_{n^2}. Acad. Repub. Pop. Romane. Bul. Sti Sect. Sti. Mat. Fiz., **6** (1954), 49—59.

WAKAKUWA, H.

[1] On n-dimensional Riemannian spaces admitting some groups of motions of order less than $\frac{1}{2}n(n-1)$. Tôhoku Math. Journal, (2) **6** (1954), 121—134.

WATANABE, S.

[1] On special infinitesimal deformations of curves. Tensor (N.S.), **5** (1955), 95—100.

YANO, K.

[24] Gruppi di trasformazioni in spazi geometrici differenziali. Istituto Matematico, Roma, 1953—1954 (mimeographed).

[25] Groups of motions and groups of affine collineations. Convegno di Geometria Differenziale, Roma (1953).

[26] Some integral formulas and their applications, to appear in Michigan Journal of Mathematics.

YANO, K. and T. NAGANO

[1] Some theorems on projective and conformal transformations, to appear in Indagationes Mathematicae.

AUTHOR INDEX

SUBJECT INDEX

A CATALOG OF SELECTED

DOVER BOOKS

IN SCIENCE AND MATHEMATICS

Mathematics–Bestsellers

Mathematics–Logic and Problem Solving

PERPLEXING PUZZLES AND TANTALIZING TEASERS, Martin Gardner. Ninety-three riddles, mazes, illusions, tricky questions, word and picture puzzles, and other challenges offer hours of entertainment for youngsters. Filled with rib-tickling drawings. Solutions. 224pp. 5 3/8 x 8 1/2. 0-486-25637-5

MY BEST MATHEMATICAL AND LOGIC PUZZLES, Martin Gardner. The noted expert selects 70 of his favorite "short" puzzles. Includes The Returning Explorer, The Mutilated Chessboard, Scrambled Box Tops, and dozens more. Complete solutions included. 96pp. 5 3/8 x 8 1/2. 0-486-28152-3

THE LADY OR THE TIGER?: and Other Logic Puzzles, Raymond M. Smullyan. Created by a renowned puzzle master, these whimsically themed challenges involve paradoxes about probability, time, and change; metapuzzles; and self-referentiality. Nineteen chapters advance in difficulty from relatively simple to highly complex. 1982 edition. 240pp. 5 3/8 x 8 1/2. 0-486-47027-X

SATAN, CANTOR AND INFINITY: Mind-Boggling Puzzles, Raymond M. Smullyan. A renowned mathematician tells stories of knights and knaves in an entertaining look at the logical precepts behind infinity, probability, time, and change. Requires a strong background in mathematics. Complete solutions. 288pp. 5 3/8 x 8 1/2. 0-486-47036-9

THE RED BOOK OF MATHEMATICAL PROBLEMS, Kenneth S. Williams and Kenneth Hardy. Handy compilation of 100 practice problems, hints and solutions indispensable for students preparing for the William Lowell Putnam and other mathematical competitions. Preface to the First Edition. Sources. 1988 edition. 192pp. 5 3/8 x 8 1/2. 0-486-69415-1

KING ARTHUR IN SEARCH OF HIS DOG AND OTHER CURIOUS PUZZLES, Raymond M. Smullyan. This fanciful, original collection for readers of all ages features arithmetic puzzles, logic problems related to crime detection, and logic and arithmetic puzzles involving King Arthur and his Dogs of the Round Table. 160pp. 5 3/8 x 8 1/2. 0-486-47435-6

UNDECIDABLE THEORIES: Studies in Logic and the Foundation of Mathematics, Alfred Tarski in collaboration with Andrzej Mostowski and Raphael M. Robinson. This well-known book by the famed logician consists of three treatises: "A General Method in Proofs of Undecidability," "Undecidability and Essential Undecidability in Mathematics," and "Undecidability of the Elementary Theory of Groups." 1953 edition. 112pp. 5 3/8 x 8 1/2. 0-486-47703-7

LOGIC FOR MATHEMATICIANS, J. Barkley Rosser. Examination of essential topics and theorems assumes no background in logic. "Undoubtedly a major addition to the literature of mathematical logic." – *Bulletin of the American Mathematical Society.* 1978 edition. 592pp. 6 1/8 x 9 1/4. 0-486-46898-4

INTRODUCTION TO PROOF IN ABSTRACT MATHEMATICS, Andrew Wohlgemuth. This undergraduate text teaches students what constitutes an acceptable proof, and it develops their ability to do proofs of routine problems as well as those requiring creative insights. 1990 edition. 384pp. 6 1/2 x 9 1/4. 0-486-47854-8

FIRST COURSE IN MATHEMATICAL LOGIC, Patrick Suppes and Shirley Hill. Rigorous introduction is simple enough in presentation and context for wide range of students. Symbolizing sentences; logical inference; truth and validity; truth tables; terms, predicates, universal quantifiers; universal specification and laws of identity; more. 288pp. 5 3/8 x 8 1/2. 0-486-42259-3

Mathematics-Algebra and Calculus

VECTOR CALCULUS, Peter Baxandall and Hans Liebeck. This introductory text offers a rigorous, comprehensive treatment. Classical theorems of vector calculus are amply illustrated with figures, worked examples, physical applications, and exercises with hints and answers. 1986 edition. 560pp. 5 3/8 x 8 1/2. 0-486-46620-5

ADVANCED CALCULUS: An Introduction to Classical Analysis, Louis Brand. A course in analysis that focuses on the functions of a real variable, this text introduces the basic concepts in their simplest setting and illustrates its teachings with numerous examples, theorems, and proofs. 1955 edition. 592pp. 5 3/8 x 8 1/2. 0-486-44548-8

ADVANCED CALCULUS, Avner Friedman. Intended for students who have already completed a one-year course in elementary calculus, this two-part treatment advances from functions of one variable to those of several variables. Solutions. 1971 edition. 432pp. 5 3/8 x 8 1/2. 0-486-45795-8

METHODS OF MATHEMATICS APPLIED TO CALCULUS, PROBABILITY, AND STATISTICS, Richard W. Hamming. This 4-part treatment begins with algebra and analytic geometry and proceeds to an exploration of the calculus of algebraic functions and transcendental functions and applications. 1985 edition. Includes 310 figures and 18 tables. 880pp. 6 1/2 x 9 1/4. 0-486-43945-3

BASIC ALGEBRA I: Second Edition, Nathan Jacobson. A classic text and standard reference for a generation, this volume covers all undergraduate algebra topics, including groups, rings, modules, Galois theory, polynomials, linear algebra, and associative algebra. 1985 edition. 528pp. 6 1/8 x 9 1/4. 0-486-47189-6

BASIC ALGEBRA II: Second Edition, Nathan Jacobson. This classic text and standard reference comprises all subjects of a first-year graduate-level course, including in-depth coverage of groups and polynomials and extensive use of categories and functors. 1989 edition. 704pp. 6 1/8 x 9 1/4. 0-486-47187-X

CALCULUS: An Intuitive and Physical Approach (Second Edition), Morris Kline. Application-oriented introduction relates the subject as closely as possible to science with explorations of the derivative; differentiation and integration of the powers of x; theorems on differentiation, antidifferentiation; the chain rule; trigonometric functions; more. Examples. 1967 edition. 960pp. 6 1/2 x 9 1/4. 0-486-40453-6

ABSTRACT ALGEBRA AND SOLUTION BY RADICALS, John E. Maxfield and Margaret W. Maxfield. Accessible advanced undergraduate-level text starts with groups, rings, fields, and polynomials and advances to Galois theory, radicals and roots of unity, and solution by radicals. Numerous examples, illustrations, exercises, appendixes. 1971 edition. 224pp. 6 1/8 x 9 1/4. 0-486-47723-1

AN INTRODUCTION TO THE THEORY OF LINEAR SPACES, Georgi E. Shilov. Translated by Richard A. Silverman. Introductory treatment offers a clear exposition of algebra, geometry, and analysis as parts of an integrated whole rather than separate subjects. Numerous examples illustrate many different fields, and problems include hints or answers. 1961 edition. 320pp. 5 3/8 x 8 1/2. 0-486-63070-6

LINEAR ALGEBRA, Georgi E. Shilov. Covers determinants, linear spaces, systems of linear equations, linear functions of a vector argument, coordinate transformations, the canonical form of the matrix of a linear operator, bilinear and quadratic forms, and more. 387pp. 5 3/8 x 8 1/2. 0-486-63518-X

Mathematics–Probability and Statistics

BASIC PROBABILITY THEORY, Robert B. Ash. This text emphasizes the probabilistic way of thinking, rather than measure-theoretic concepts. Geared toward advanced undergraduates and graduate students, it features solutions to some of the problems. 1970 edition. 352pp. 5 3/8 x 8 1/2. 0-486-46628-0

PRINCIPLES OF STATISTICS, M. G. Bulmer. Concise description of classical statistics, from basic dice probabilities to modern regression analysis. Equal stress on theory and applications. Moderate difficulty; only basic calculus required. Includes problems with answers. 252pp. 5 5/8 x 8 1/4. 0-486-63760-3

OUTLINE OF BASIC STATISTICS: Dictionary and Formulas, John E. Freund and Frank J. Williams. Handy guide includes a 70-page outline of essential statistical formulas covering grouped and ungrouped data, finite populations, probability, and more, plus over 1,000 clear, concise definitions of statistical terms. 1966 edition. 208pp. 5 3/8 x 8 1/2. 0-486-47769-X

GOOD THINKING: The Foundations of Probability and Its Applications, Irving J. Good. This in-depth treatment of probability theory by a famous British statistician explores Keynesian principles and surveys such topics as Bayesian rationality, corroboration, hypothesis testing, and mathematical tools for induction and simplicity. 1983 edition. 352pp. 5 3/8 x 8 1/2. 0-486-47438-0

INTRODUCTION TO PROBABILITY THEORY WITH CONTEMPORARY APPLICATIONS, Lester L. Helms. Extensive discussions and clear examples, written in plain language, expose students to the rules and methods of probability. Exercises foster problem-solving skills, and all problems feature step-by-step solutions. 1997 edition. 368pp. 6 1/2 x 9 1/4. 0-486-47418-6

CHANCE, LUCK, AND STATISTICS, Horace C. Levinson. In simple, non-technical language, this volume explores the fundamentals governing chance and applies them to sports, government, and business. "Clear and lively ... remarkably accurate." – *Scientific Monthly*. 384pp. 5 3/8 x 8 1/2. 0-486-41997-5

FIFTY CHALLENGING PROBLEMS IN PROBABILITY WITH SOLUTIONS, Frederick Mosteller. Remarkable puzzlers, graded in difficulty, illustrate elementary and advanced aspects of probability. These problems were selected for originality, general interest, or because they demonstrate valuable techniques. Also includes detailed solutions. 88pp. 5 3/8 x 8 1/2. 0-486-65355-2

EXPERIMENTAL STATISTICS, Mary Gibbons Natrella. A handbook for those seeking engineering information and quantitative data for designing, developing, constructing, and testing equipment. Covers the planning of experiments, the analyzing of extreme-value data; and more. 1966 edition. Index. Includes 52 figures and 76 tables. 560pp. 8 3/8 x 11. 0-486-43937-2

STOCHASTIC MODELING: Analysis and Simulation, Barry L. Nelson. Coherent introduction to techniques also offers a guide to the mathematical, numerical, and simulation tools of systems analysis. Includes formulation of models, analysis, and interpretation of results. 1995 edition. 336pp. 6 1/8 x 9 1/4. 0-486-47770-3

INTRODUCTION TO BIOSTATISTICS: Second Edition, Robert R. Sokal and F. James Rohlf. Suitable for undergraduates with a minimal background in mathematics, this introduction ranges from descriptive statistics to fundamental distributions and the testing of hypotheses. Includes numerous worked-out problems and examples. 1987 edition. 384pp. 6 1/8 x 9 1/4. 0-486-46961-1

Mathematics-Geometry and Topology

PROBLEMS AND SOLUTIONS IN EUCLIDEAN GEOMETRY, M. N. Aref and William Wernick. Based on classical principles, this book is intended for a second course in Euclidean geometry and can be used as a refresher. More than 200 problems include hints and solutions. 1968 edition. 272pp. 5 3/8 x 8 1/2. 0-486-47720-7

TOPOLOGY OF 3-MANIFOLDS AND RELATED TOPICS, Edited by M. K. Fort, Jr. With a New Introduction by Daniel Silver. Summaries and full reports from a 1961 conference discuss decompositions and subsets of 3-space; n-manifolds; knot theory; the Poincaré conjecture; and periodic maps and isotopies. Familiarity with algebraic topology required. 1962 edition. 272pp. 6 1/8 x 9 1/4. 0-486-47753-3

POINT SET TOPOLOGY, Steven A. Gaal. Suitable for a complete course in topology, this text also functions as a self-contained treatment for independent study. Additional enrichment materials make it equally valuable as a reference. 1964 edition. 336pp. 5 3/8 x 8 1/2. 0-486-47222-1

INVITATION TO GEOMETRY, Z. A. Melzak. Intended for students of many different backgrounds with only a modest knowledge of mathematics, this text features self-contained chapters that can be adapted to several types of geometry courses. 1983 edition. 240pp. 5 3/8 x 8 1/2. 0-486-46626-4

TOPOLOGY AND GEOMETRY FOR PHYSICISTS, Charles Nash and Siddhartha Sen. Written by physicists for physics students, this text assumes no detailed background in topology or geometry. Topics include differential forms, homotopy, homology, cohomology, fiber bundles, connection and covariant derivatives, and Morse theory. 1983 edition. 320pp. 5 3/8 x 8 1/2. 0-486-47852-1

BEYOND GEOMETRY: Classic Papers from Riemann to Einstein, Edited with an Introduction and Notes by Peter Pesic. This is the only English-language collection of these 8 accessible essays. They trace seminal ideas about the foundations of geometry that led to Einstein's general theory of relativity. 224pp. 6 1/8 x 9 1/4. 0-486-45350-2

GEOMETRY FROM EUCLID TO KNOTS, Saul Stahl. This text provides a historical perspective on plane geometry and covers non-neutral Euclidean geometry, circles and regular polygons, projective geometry, symmetries, inversions, informal topology, and more. Includes 1,000 practice problems. Solutions available. 2003 edition. 480pp. 6 1/8 x 9 1/4. 0-486-47459-3

TOPOLOGICAL VECTOR SPACES, DISTRIBUTIONS AND KERNELS, François Trèves. Extending beyond the boundaries of Hilbert and Banach space theory, this text focuses on key aspects of functional analysis, particularly in regard to solving partial differential equations. 1967 edition. 592pp. 5 3/8 x 8 1/2. 0-486-45352-9

INTRODUCTION TO PROJECTIVE GEOMETRY, C. R. Wylie, Jr. This introductory volume offers strong reinforcement for its teachings, with detailed examples and numerous theorems, proofs, and exercises, plus complete answers to all odd-numbered end-of-chapter problems. 1970 edition. 576pp. 6 1/8 x 9 1/4. 0-486-46895-X

FOUNDATIONS OF GEOMETRY, C. R. Wylie, Jr. Geared toward students preparing to teach high school mathematics, this text explores the principles of Euclidean and non-Euclidean geometry and covers both generalities and specifics of the axiomatic method. 1964 edition. 352pp. 6 x 9. 0-486-47214-0

CATALOG OF DOVER BOOKS